FLUID, ELECTROLYTE, AND ACID-BASE BALANCE
A Case Study Approach

FLUID, ELECTROLYTE, AND ACID-BASE BALANCE

A Case Study Approach

Mima M. Horne, RN, MS
Lecturer, School of Nursing
University of North Carolina-Wilmington
Wilmington, North Carolina

Ursula Easterday Heitz, RN, MSN
Clinical Nurse Specialist—Care of the Adult;
Formerly at Community Hospital North
Indianapolis, Indiana

Pamela L. Swearingen, RN
Special Projects Editor

47 *illustrations*

St. Louis Baltimore Boston Chicago London Philadelphia Sydney Toronto

**Mosby
Year Book**

Dedicated to Publishing Excellence

Executive editor: Don Ladig
Developmental editor: Jeanne Rowland
Project manager: Carlotta Seely
Production editor: Amy Adams Squire
Designer: Kay Kramer
Cover design: Laura Steube
Cover art: Peter Arnold, Inc.

The cover art is a photograph of a crystal of table salt, taken through an optical microscope, using a polarized light source.

Printed in the United States of America

Mosby–Year Book, Inc.
11830 Westline Industrial Drive, St. Louis, MO 63146

Library of Congress Cataloging-in-Publication Data

Horne, Mima M.
Fluid, electrolyte, and acid-base balance: a case study approach
Mima M. Horne, Ursula Easterday Heitz, Pamela L. Swearingen.
p. cm.
Includes index.
ISBN 0-8016-5479-3
1. Water-electrolyte imbalances—Nursing. 2. Acid-base imbalances—Nursing. I. Heitz, Ursula Easterday. II. Swearingen, Pamela L. III. Title.
[DNLM: 1. Acid-Base Equilibrium—nurses' instruction. 2. Water-Electrolyte Balance—nurses' instruction. OU 105 H815f]
RC630.H66 1991
616.3'9—dc20
DNLM/DLC
for Library of Congress 90-5975
CIP

GW/D/D 9 8 7 6 5 4 3 2 1

Preface

Fluid, Electrolyte, and Acid-Base Balance: A Case Study Approach was written to help nursing students and professionals apply the scientific principles of fluid, electrolyte, and acid-base balance to the clinical setting. The book is scientifically based and focuses on nursing care for patients with a variety of pathophysiologic processes. Case studies at the end of each chapter facilitate reinforcement and application of the content, a benefit to those who find the study of fluids, electrolytes, acids, and bases one of nursing's greatest challenges.

The book uses a consistent format from chapter to chapter, discussing physiology, pathophysiology, history and risk factors, assessment, diagnostic studies, medical management, and nursing diagnoses and interventions, followed by case studies. Although the content was developed so that later chapters build on information presented in earlier chapters, important concepts are repeated, enabling each chapter to stand alone. Terms that may be new to the student are defined both within the body of the text and in the glossary. Tables and illustrations help the reader process the information presented in the narrative. The appendixes feature discussions of patient populations with specific disease processes that are particularly prone to fluid and electrolyte and acid-base disorders, the effects of age on these disorders, abbreviations, glossary, and normal values for laboratory tests that are discussed within the text.

We wish to thank the following contributors to *Pocket Guide to Fluids and Electrolytes,* sections of whose work we adapted for this book: Linda Baas, RN, MSN, CCRN, from University of Cincinnati Medical Center, Cincinnati, Ohio; Janet Hicks Keen, RN, MS, CCRN, CEN, from Georgia Baptist Medical Center in Atlanta, Georgia; Donna Kershner, RN, MS, from Cabrillo College, Santa Cruz, California; Carol E. Lang, RD, MS, from New England Deaconess Hospital, Boston, Massachusetts; Charis W. Spielman, MPH, RD, from Lifesource, Inc, Mountain View, California; Laurie Warren-Young, RN, BS, ET, from Lifesource, Inc, Mountain View, California; and Maribeth Wooldridge-King, RN, MS, from Long Island Hospital, Brooklyn, New York.

We are grateful to the following reviewers, whose suggestions helped so much during the developmental stage of the book: Barbara Brillhart, RN, PhD, University of Texas at Arlington, Arlington, Texas; Carol P. Fray, RN, MA, University of North Carolina—Charlotte, North Carolina; Mary Beth Harrington, RN, MSN, CCRN, CS, Lahey Clinic Medical Center, Burlington, Massachusetts; Edwina A. McConnell, RN,

PhD, Independent Nurse Consultant, Madison, Wisconsin; Carolyn Platis, RNC, MS, formerly with Graham Hospital School of Nursing, Canton, Illinois; Judith Sweeney, RN, MSN, Vanderbilt University, Nashville, Tennessee; and Sarah Jane Tobiason, RN, MA, Arizona State University, Tempe, Arizona.

We also thank the library staff at the Area Health Education Center, Wilmington, North Carolina, for giving generously of their time and resources; Jo Ann Galloway of the New Hanover County Public Library for her assistance with interlibrary loans; and Dixie Williamson and Evelyn Forbes of the St. Thomas Hospital Library, Nashville, Tennessee, for their time, resources, and support.

Fluid, Electrolyte, and Acid-Base Balance was written to supplement medical-surgical textbooks with the understanding that the reader has a background in pathophysiology and assessment. Our primary goals are to make information about fluid, electrolyte, and acid-base balance easy to understand and to facilitate application of this information to patient care in a variety of settings. Reviewers indicate that we have achieved these objectives; we welcome comments from our readers so that we may enhance the book's usefulness in future editions.

Mima M. Horne
Ursula Easterday Heitz
Pamela L. Swearingen

Contents

Chapter

1

Overview of Fluid and Electrolyte Balance

The cell is the fundamental functioning unit of the human body. Trillions of body cells work together to create an individual who, when healthy, is able to adapt to a constantly changing external environment, while maintaining internal balance. Since approximately 50% to 60% of the human body is water, internal balance involves the regulation of body fluids. The ability to maintain internal balance in the presence of external stressors is termed **homeostasis.**

Homeostatic balance is necessary for cells to perform their individual physiologic tasks. A steady delivery of nutrients (oxygen, glucose, fatty acids, and amino acids), the continuous removal of metabolic wastes (carbon dioxide and other end products of cellular metabolism), and the maintenance of a stable physiochemical environment are essential to homeostatic balance and thus to normal cellular function. Disruption of any of these three factors results in homeostatic imbalance with cellular dysfunction and illness.

All organs and body structures are involved in the maintenance of homeostasis. The nervous system with its three major components (sensory, integrative, and motor) enables the individual to perceive and interact with the surrounding environment and coordinates and regulates the functions of the other organ systems. Without the musculoskeletal system, the movements necessary to obtain nutrition would be impossible. The skin forms the first line of defense against a hostile external environment and prevents the loss of body fluids. The gastrointestinal tract, lungs, and liver are necessary for the acquisition, conversion, and storage of nutrients. The cardiovascular and hematologic systems deliver nutrients to the cells and carry wastes to the kidneys and lungs for excretion. In addition to the excretion of wastes, the kidneys play an especially vital role in regulating the volume and composition of body fluids. The role of the kidneys in the regulation of fluid and electrolyte balance will be discussed in greater detail in Chapter 2. The endocrine system, through the release of multiple hormones, regulates the body's metabolic functions and, like the kidneys, is involved in the regulation of the volume, concentration, and composition of body fluids.

Single-celled organisms depend on their surrounding environment for the delivery of nutrients and the removal of metabolic wastes. An amoeba living in a salty pond obtains its nutrients directly from the pond water and depends on the diluting effect of the pond for the dissipation of wastes. The amoeba, however, has little protection from changes in its environment. If the pond is flooded with rain water or dries up, the amoeba's environment is dramatically altered, and its function is affected. Multicellular organisms, on the other hand, are protected from immediate changes in their external environment by the fluids that surround their cells. It is as if multicellular organisms carry their own "pond" around with them. The volume, concentration, and composition of the "pond" can be regulated in order to maintain a stable environment for the cells with sufficient nutrients and adequate removal of wastes. Thus homeostatic balance for the human body requires continuous regulation of the fluid that surrounds the cells. In this way the fluid within the cells is protected, which ensures normal cellular function.

BODY FLUID COMPARTMENTS

It should be apparent from the previous discussion that the body's fluids are divided into two portions: the fluid contained within the cells, termed **intracellular fluid (ICF),** and the fluid outside the cells termed **extracellular fluid (ECF)** (Figure 1-1). Approximately two thirds of the body's fluid is intracellular, equaling nearly 25 L in the average (70 kg) adult male. The remaining one third is extracellular, which is equal to approximately 12 L in the average (70 kg) adult male. Body fluid in infants is divided approximately equally between intracellular and extracellular fluids. Having a larger proportion of ECF than the adult is one of several factors that place infants at increased risk for the loss of body fluids (see Appendix E for a summary of the effects of age on fluid and electrolyte balance). Since it is the ECF that interacts with the outside environment, the body is more likely to lose ECF than ICF.

Extracellular fluid is further divided into intravascular, interstitial, and transcellular. **Intravascular fluid (IVF)** is the fluid contained within the blood vessels (plasma). Blood contains both intravascular and intracellular fluid since it is a combination of plasma and cells (Figure 1-2). The average blood volume is approximately 5 to 6 L, of which 3 are plasma. The remaining 2 to 3 L consist of red blood cells (erythrocytes), which transport oxygen and act as important body **buffers; white blood cells (leukocytes), which protect the body from microorganisms and other foreign material; and platelets (thrombocytes), which assist in clot formation and help to prevent bleeding (see the accompanying box). Although the terms plasma** and **serum** will be used interchangeably in this text, it is important to understand that the two terms refer to slightly different blood components. As

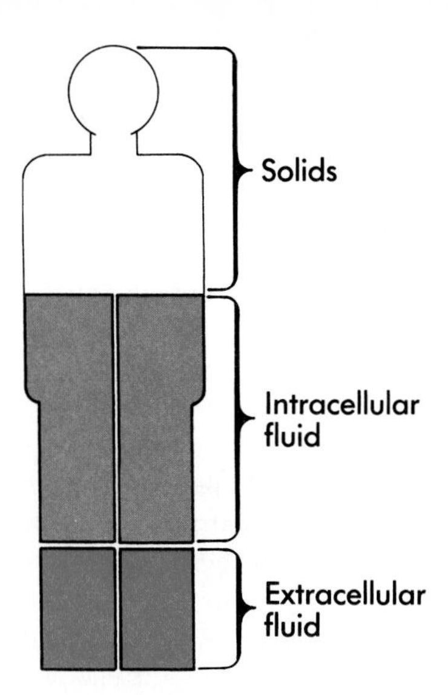

Figure 1-1 Comparison of ICF to ECF.
(From Horne M and Swearingen P: Pocket guide to fluids and electrolytes, St Louis, 1989, The CV Mosby Co.)

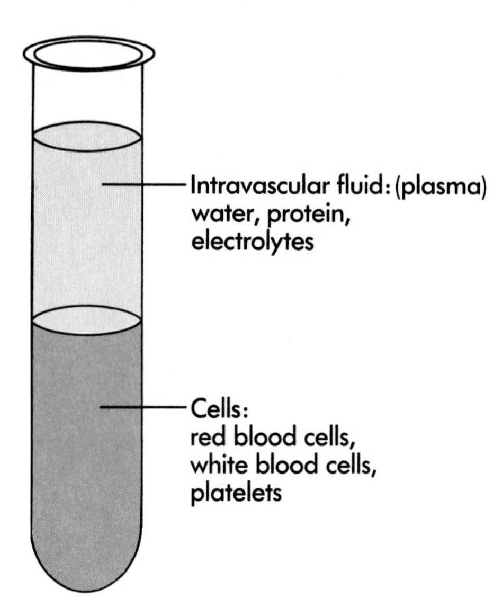

Figure 1-2 Composition of blood.

described, plasma is the fluid portion of the blood, including water, proteins, and electrolytes. A plasma sample may be obtained by centrifuging a blood specimen and removing the cellular components. In contrast, serum is the fluid that remains after a blood specimen has been allowed to form a clot. Thus serum is plasma *minus* the fibrinogen and other clotting factors.

Functions of Blood

- Delivery of nutrients (e.g., glucose and oxygen) to the tissues
- Transport of waste products to the kidneys and lungs
- Delivery of antibodies and white blood cells (WBCs) to sites of infection
- Delivery of platelets to sites of bleeding
- Transport of hormones to their sites of action
- Circulation of body heat

Interstitial fluid (ISF) is the fluid that surrounds the cells. It includes lymph fluid, and it is equal to approximately 8 L in the adult. **Transcellular fluid (TCF)** is the smallest portion of extracellular fluid. It includes cerebrospinal, pericardial, pleural, synovial, and intraocular fluids; sweat; and digestive secretions. At any given time, transcellular fluid volume is only about 1 L. However, large amounts of fluid move into and out of the transcellular space each day. For example, the gastrointestinal tract secretes and reabsorbs 6 to 8 L of fluid per day. Also, the production and loss of transcellular fluid may increase dramatically in illness. Certain cardiac and pulmonary disorders, for example, result in increased production of pericardial and pleural fluids. Increased activity and environmental temperature lead to increased production of sweat, while disorders of the gastrointestinal tract may result in the loss of fluids through vomiting and diarrhea.

COMPOSITION OF BODY FLUIDS

Water

Water is the primary constituent of the human body. All body fluids are dilute solutions of water and dissolved substances (solutes). The average 70 kg adult male is approximately 60% water by weight, while the average female is approximately 55% water by weight. The percentage of body weight that is water varies with such factors as age, gender, and body fat content. As a rule, body water decreases with increasing age. Premature infants may be as much as 80% water by weight, while the full-term infant is approximately 70% water by weight. By the age of 6 months to 1 year, the percentage of body water decreases to approximately 60, with little reduction through childhood and young adulthood.

From puberty on, females have proportionately less body water owing to proportionately greater body fat. This is due to the low water content of fat cells. While most body cells are 70% to 85% water, fat cells are composed primarily of triglycerides (Guyton, 1986). Thus an obese individual weighing 60 kg has less body water than a thin individual weighing the same amount. Body water continues to decrease in the older adult owing to the loss of lean tissue and increasing body fat. In the emaciated older adult, however, body water actually may increase (Table 1-1).

Table 1-1 Changes in total body water with age

Age	Kilogram Weight (%)
Premature infant	80
3 mo	70
6 mo	60
1-2 yr	59
11-16 yr	58
Adult	58-60
Obese adult	40-50
Emaciated adult	70-75

Groër MW: Physiology and pathophysiology of the body fluids, St Louis, 1981, The CV Mosby Co.

Solutes

In addition to water, body fluids contain two types of solutes: **electrolytes** and **nonelectrolytes**.

Electrolytes

The majority of the body's solutes are electrolytes, substances that dissociate (separate) in solution and conduct an electrical current. When electrolytes are dissolved in water, they dissociate into negatively and positively charged particles termed ions. **Cations** are ions that carry a positive charge, while **anions** are ions that carry a negative charge. Cations have given up an electron to become positively charged, while anions have gained an electron to become negatively charged. If sodium chloride is added to water, it dissociates into an equal number of positively charged sodium ions (Na^+) and negatively charged chloride ions (Cl^-) (Figure 1-3, *A*). For each cation there is a corresponding anion. Sodium and chloride are univalent ions, meaning that each possesses a single electrical charge. Other physiologically important univalent ions include potassium (K^+) and bicarbonate (HCO_3^-).

Certain electrolytes have more than one charge. Calcium, for example, is a divalent ion; it possesses two positive charges. When calcium chloride is dissolved in water, it dissociates into one calcium ion (possessing two positive charges) for every two negatively charged chloride ions (Figure 1-3, *B*). Unlike the sodium chloride solution, the number of cations and anions in the calcium chloride solution is not equal. There are twice as many anions as cations. The number of negative and positive charges, however, *is* equal. Other physiologically important divalent ions include magnesium (Mg^{2+}) and phosphate (HPO_4^{2-}).

Because electrolytes are important constituents of body fluids, changes in their concentrations may adversely affect cellular function, resulting in cellular dysfunction and illness. Many diseases, medical therapies, and nursing interventions have the

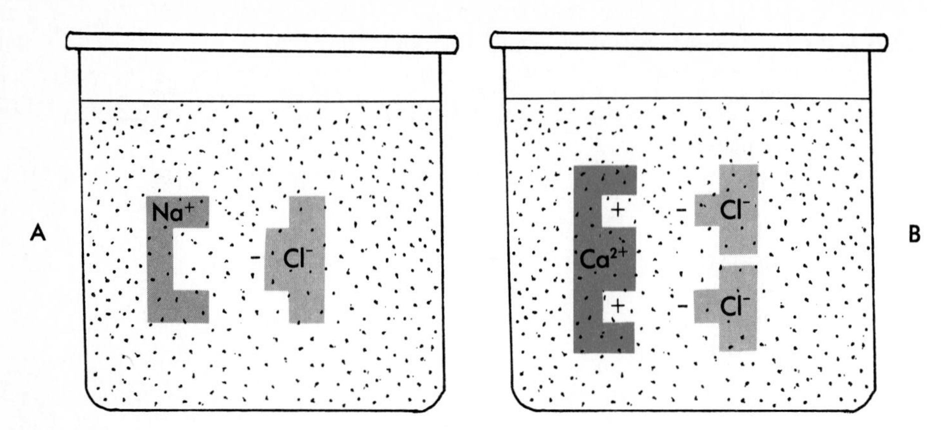

Figure 1-3 A comparison of univalent and divalent ions. **A,** Sodium (Na^+) and chloride (Cl^-) are univalent ions. **B,** Calcium (Ca^{2+}) is a divalent ion.

potential for causing an increased loss or gain of electrolytes from the body. For this reason, measurement of the electrolyte concentration of body fluids is a commonly performed laboratory test. Electrolytes may be measured in terms of their weight in a volume of solution, such as milligrams per deciliter (mg/dl) of plasma. The plasma calcium level is sometimes measured and reported this way. More commonly, electrolytes are measured and reported in milliequivalents (1 one thousandth of an equivalent). An equivalent (Eq) measures the ability of substances to combine and is equal to 1 mole of a substance (molecular weight in grams), divided by its number of charges (also called valence). Sodium, for example, has a molecular weight of 23, and thus 23 grams of sodium equals 1 mole. Because sodium is a univalent ion (i.e., it has a single charge or valence), 1 mole is equal to 1 equivalent. Or, stated another way, 23 grams of sodium is equal to 1 equivalent of sodium (see the accompanying box). Calcium is a divalent ion, meaning that 1 mole of calcium is equal to two equivalents. Or one equivalent is equal to ½ mole of calcium. Since body fluids are very dilute solutions, it is simpler to report electrolytes in milliequivalents (mEq) rather than equivalents (Eq).

Although the number of cations and anions in a solution will not necessarily be equal, the number of negative and positive charges must be equal. Since milliequivalents reflect the number of charges of a substance, the number of cations and anions expressed in milliequivalents always will be equal in body fluids (see Table 1-2). This concept has an important application in clinical medicine. While all of the major cations of the plasma are routinely measured, not all of the anions are measured. The normally unmeasured anions can be estimated by subtracting the measured anions from the sum of the cations. In this way, the increases in the normally unmeasured anions may be identified without having to measure them directly. This is termed the **anion gap** and will be discussed in detail in Chapters 5 and 17.

The electrolyte content of the two major fluid compartments is not the same (Table 1-3). The primary cation of the ECF is sodium (Na^+), while potassium (K^+) is the primary intracellular cation. This concentration difference is maintained by a pump system that is present in all cell membranes of the body, termed the **sodium-**

Table 1-2 Plasma electrolytes

Name	Symbol/Valency	mEq/L
Cations		
Sodium	Na^+	142.0
Potassium	K^+	4.5
Calcium	Ca^{2+}	5.0
Magnesium	Mg^{2+}	1.5
		153.0
Anions		
Chloride	Cl^-	104
Bicarbonate	HCO_3^-	24
Phosphate	HPO_4^{2-}	2
Sulfate	SO_4^{2-}	1
Proteinate		17
Other		5
		153

1 mole = molecular weight × grams

$$1 \text{ equivalent} = \frac{1 \text{ mole}}{\text{valence}}$$

1 mole = equivalent × valence

Molecular weight of sodium = 23; valence = 1

$$1 \text{ equivalent of sodium} = \frac{23\text{g}}{1} = 23 \text{ grams}$$

1 *mEq* sodium = 23 *milligrams* (mg)

Molecular weight of calcium = 40; valence = 2

$$1 \text{ equivalent of calcium} = \frac{40\text{g}}{2} = 20 \text{ grams}$$

1 *mEq* calcium = 20 *milligrams* (mg)

potassium pump (this concept will be discussed again under the section on active transport, p. 10). The primary extracellular anions are chloride (Cl^-) and bicarbonate (HCO_3^-), while phosphate (HPO_4^{2-}) is the primary intracellular anion (Figure 1-4). Because the electrolyte content of the plasma and interstitial fluids is essentially the same (there is a small difference due to the greater quantity of proteins found in the plasma), changes in plasma electrolytes reflect changes in the ECF. However,

Table 1-3 Primary constituents of body fluid compartments

Compartment	Na^+ (mEq/L)	K^+ (mEq/L)	Cl^- (mEq/L)	HCO_3^- (mEq/L)	HPO_4^{2-} (mEq/L)
Intravascular (plasma)	142	4.5	104	24	2.0
Interstitial	145	4.4	117	27	2.3
Intracellular (skeletal muscle cell)	12	150	4.0	12	40
Transcellular					
Gastric juice	60	7	100	0	—
Pancreatic juice	130	7	60	100	—
Sweat	45	5	58	0	—

This is a partial list. Other constituents include Ca^{2+}, Mg^{2+}, sulfates, proteinates, and organic acids. NOTE: Values given are average ones. (Modified from Rose BD: Clinical physiology of acid-base and electrolyte disorders, ed 3, New York, 1989, McGraw-Hill Co.)

changes in plasma electrolytes do not necessarily indicate similar changes in the composition of the ICF. Understanding the differences between the ICF and ECF is important in anticipating the types of imbalances that can occur with certain disorders such as rapid tissue breakdown or acid-base imbalance. In these situations, electrolytes may be released from, or move into or out of the cells, significantly altering the content of the ECF. In acidosis, for example, there is a movement of potassium out of the cell and into the ECF. In this case, an increase in plasma potassium does not indicate a corresponding increase in intracellular potassium, nor does it reflect an overall gain in potassium by the body. The effects of acidosis on potassium balance are described in detail in Chapters 9 and 17.

Nonelectrolytes

In addition to electrolytes, body fluids contain dissolved substances that do not dissociate into ions. These nonionizing substances include urea, glucose, creatinine, and billirubin. Since they do not separate into particles that carry a charge, they are not measured in milliequivalents. Instead, they are measured in weight per solution (mg/dl).

FACTORS THAT AFFECT MOVEMENT OF WATER AND SOLUTES

Membranes

Each of the fluid compartments is separated by a *selectively permeable* membrane that permits the movement of water and some solutes. Although small molecules such as urea and water move freely between all compartments, certain substances, such as protein, do not cross membranes as readily. Selective permeability of membranes helps to maintain the unique composition of each compartment, while allowing for the movement of nutrients from the plasma to the cells and the movement of waste

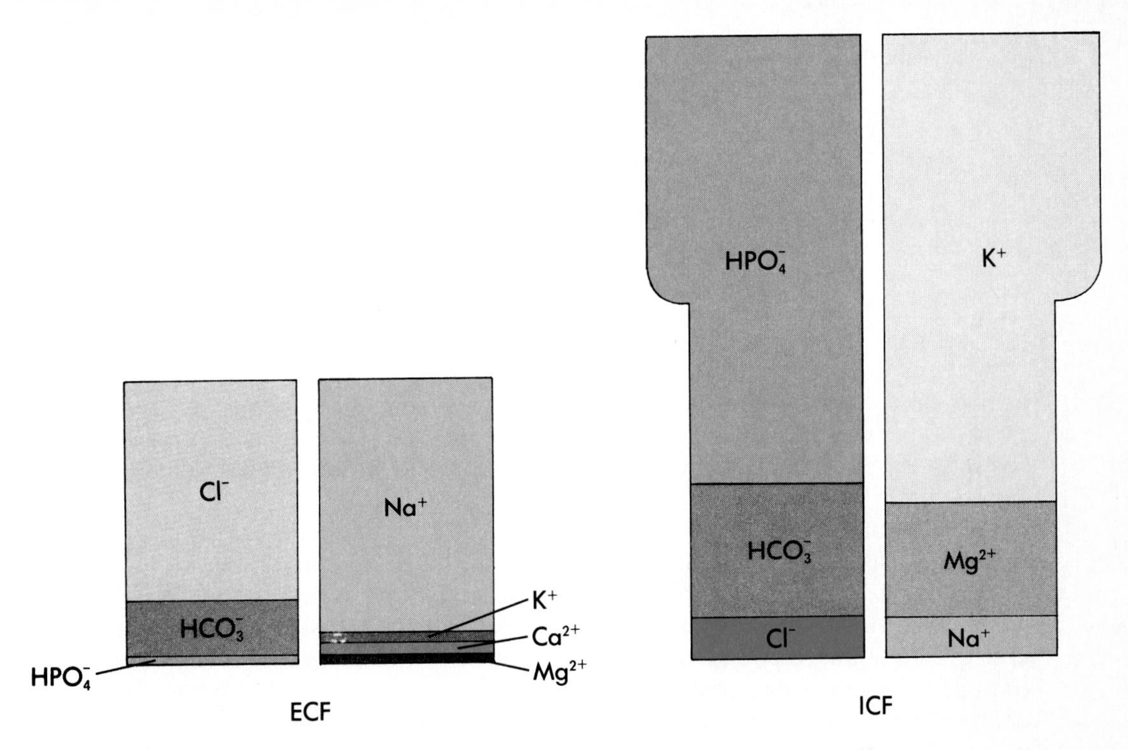

Figure 1-4 Relative electrolyte composition of body fluid.
(Modified from Horne M and Swearingen P: Pocket guide to fluids and electrolytes, St. Louis, 1989, The CV Mosby Co.)

products out of the cells and into the plasma. The body's semipermeable membranes include cell membranes, which separate intracellular fluid from interstitial fluid and are composed of lipids and protein; capillary membranes, which separate intravascular fluid from interstitial fluid; and epithelial cell membranes, which separate the transcellular fluid from the other fluid compartments. Examples of epithelial membranes include the mucosal epithelium of the stomach and intestines, the synovial membrane, and the renal tubules.

Transport Processes

In addition to membrane selectivity, the movement of water and solutes is determined by several transport processes.

Diffusion

Diffusion is the random movement of particles in all directions through a solution or gas. Particles move from an area of high concentration to an area of low concentration along a **concentration gradient.** The energy for diffusion is produced

Factors that Increase Diffusion

- Increased temperature
- Increased concentration of the particle
- Decreased size or molecular weight of the particle
- Increased surface area available for diffusion
- Decreased distance across which the particle mass must diffuse

NOTE: Opposite factors will act to reduce diffusion.

by thermal energy. An example of diffusion is the movement of oxygen from the alveoli of the lung to the blood of the pulmonary capillaries. See the accompanying box for a list of factors that increase diffusion (opposite factors will act to reduce diffusion).

Cell walls are composed of sheets of lipids with many minute protein pores. Substances may diffuse across the cell wall under the following conditions: if they are small enough (e.g., water and urea) to pass through the protein pores (termed *simple diffusion*); if they are lipid soluble (e.g., oxygen and carbon dioxide), another example of simple diffusion; or by means of a carrier substance, termed *facilitated diffusion*. Large lipid-insoluble substances such as amino acids and glucose must diffuse into the cell via a carrier substance. Glucose, for example, combines with a carrier protein on the outside of the cell to become lipid soluble. Once inside the cell, glucose breaks away from the carrier and the carrier is then free to facilitate diffusion of additional glucose. (Note that the glucose carrier protein is *not* insulin. Insulin is also necessary for glucose utilization by certain cells.)

As with simple diffusion, facilitated diffusion requires the presence of a concentration gradient that favors diffusion. The rate of facilitated diffusion, however, depends on the availability of the carrier substance. If there is a large concentration gradient (i.e., the difference between the areas of high and low concentration is great), the carrier can become saturated (used up) and diffusion will decrease despite the presence of a favorable concentration gradient. Significant amounts of glucose will move into the cell, for example, only if there is a favorable concentration gradient and available carrier proteins.

Active transport

In the absence of a favorable electrochemical or concentration gradient, movement of solutes across cell membranes will require energy. This is termed **active transport,** and like facilitated diffusion, it depends on the availability of carrier substances. The energy required for active transport is provided by adenosine triphosphate (ATP). In turn, the formation of ATP is dependent on the cell receiving adequate oxygen and nutrients (glucose, fatty acids, or amino acids). Active transport is vital for maintaining the unique composition of both the ECF and ICF. Many important solutes, including sodium, potassium, calcium, hydrogen, glucose, and amino acids, are transported actively across cell membranes. As discussed, a special sodium-potassium pump exists in cell membranes that *actively* transports sodium out of the cell and potassium in, maintaining the concentration difference for these ions. Because it is capable of splitting ATP molecules, the carrier substance for this mechanism is termed *sodium-potassium ATPase.* Intracellular magnesium ATP provides the necessary energy for the transport process. The effects of magnesium deficiency on the function of the sodium-potassium pump will be discussed in Chapter 12.

The renal tubules depend on active transport to save several important solutes (e.g., sodium, glucose) that would otherwise be lost in the urine. Normally, all the glucose filtered by the renal glomeruli is reabsorbed (saved) by means of active transport, thus enabling the excretion of glucose-free urine. As with facilitated diffusion, though, this mechanism can become overwhelmed or saturated. Saturation begins when the blood sugar exceeds *approximately* 180 to 200 mg/dl, causing glucose to appear in the urine. The kidneys' use of active transport will be discussed again in Chapter 2.

Filtration

Filtration is the movement of *water and* **solutes** from an area of high fluid **hydraulic pressure** to an area of low fluid hydraulic pressure. Hydraulic pressure is a combination of the pressure created by the force of gravity (weight) acting on the fluid **(hydrostatic pressure)** and the pressure created by the pump action of the heart. The nature of the membrane will determine which solutes are filtered. Whether a given solute will cross a membrane depends on its size, lipid solubility, and electrical charge. Filtration is important in directing fluid out of the arterial end of the capillaries. It is also the force that enables the kidneys to filter 180 L of plasma per day.

Osmosis

Osmosis is the movement of *water* across a semipermeable membrane from an area of lower solute concentration to an area of higher solute concentration. Osmosis can occur across any membrane when solute concentrations on either side of the membrane change. An important example of osmosis is the formation of a concentrated urine by the renal tubule, a process that will be discussed in Chapter 3. Although the term osmosis specifically refers to the movement of water only, to a much smaller degree, osmosis also affects the movement of solutes. The forces of friction cause some solutes to be carried along with the water. This is termed *solvent drag.*

The following are terms that are associated with osmosis: **osmotic pressure,** which is the amount of hydrostatic pressure required to stop the osmotic flow of water; and **oncotic pressure,** which is the osmotic pressure exerted by colloids (proteins). Albumin, for example, exerts oncotic pressure within the blood vessels and helps hold the water content of the blood in the intravascular space. A third term, **osmotic diuresis,** is increased urine output caused by substances such as mannitol, glucose, or contrast media, which are excreted in the urine and reduce renal water reabsorption. An osmotic diuresis occurs in uncontrolled diabetes mellitus, for example, due to the presence of excess glucose in the renal tubule. When the blood sugar is within normal range, all the glucose that is filtered by the kidney is reabsorbed (saved) via active transport. In hyperglycemia (blood sugar >180 to 200 mg/dl), however, the kidneys' ability to reabsorb glucose is overwhelmed (recall that the carrier substance becomes saturated). The glucose that is not reabsorbed remains in the tubule and acts osmotically to hold water that otherwise would be reabsorbed. The net result is glucosuria and polyuria.

Concentration of Body Fluids

Osmolality

As discussed above, changes in the concentration of body fluids will affect the movement of water between fluid compartments due to the process of osmosis. **Osmolality** is a measurement of the concentration of body fluids (the ratio of water to solutes). It is reported in milliosmoles (1 one thousandth of an osmole) per kilogram of water (mOsm/kg). Just as equivalent reflects the number of charges in a mole of a substance, an osmole is equal to the number of particles in a solution when a mole of a substance has been added to water. One osmole is equal to the number of particles in 1 mole (molecular weight in grams) of a nonionizing substance (e.g., glucose, sodium, calcium, or urea). If 1 mole of urea is added to water, it will produce a solution containing 1 osmole (Figure 1-5, *A*). In contrast, if 1 mole of a substance that ionizes (separates into more than 1 particle) is added to water, it will create a solution containing more than 1 osmole. When 1 mole of sodium chloride is added to water, for example, it dissociates into an equal number of sodium and chloride ions, and, assuming that complete dissociation occurs, it produces a solution that contains 2 osmoles (Figure 1-5, *B*).

Osmolarity, another term used to describe the concentration of solutions, reflects the number of particles in a *liter of solution* (remember that osmolality is a measurement of the number of particles in a *kilogram of water*). The osmolarity of body fluids is reported in milliosmoles per liter (mOsm/L) of solution. Since body fluids are relatively dilute, the difference between their osmolality and osmolarity is relatively small. Osmolality is the measurement most often used in clinical medicine.

Changes in extracellular osmolality lead to changes in intracellular volume, owing to osmotic shifts of water. Decreased extracellular osmolality causes a movement of water from the ECF to the ICF, while increased extracellular osmolality causes a movement from the ICF to the ECF. Water will continue to move until the two compartments reach equilibrium. The intravenous administration of so-

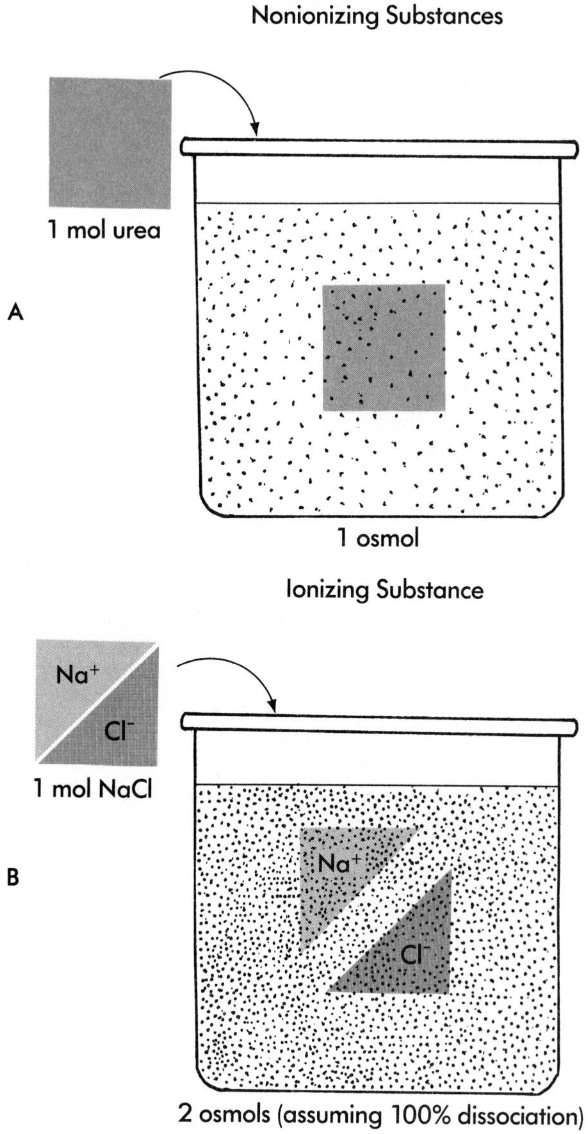

Nonionizing Substances

1 mol urea

A

1 osmol

Ionizing Substance

Na^+

Cl^-

1 mol NaCl

B

Na^+

Cl^-

2 osmols (assuming 100% dissociation)

Figure 1-5 A comparison of the osmolality of solutions composed of nonionizing vs. ionizing substances. **A,** Solution containing nonionizing substance (urea). **B,** Solution containing ionizing substance (NaCl).

dium bicarbonate (used to treat acidosis), for example, will increase the osmolality of the ECF, causing a shift of water out of the ICF until the two compartments are again at equilibrium (Figure 1-6, *A-C*). Thus intravenous administration of sodium bicarbonate can result in potentially dangerous intravascular fluid volume excess.

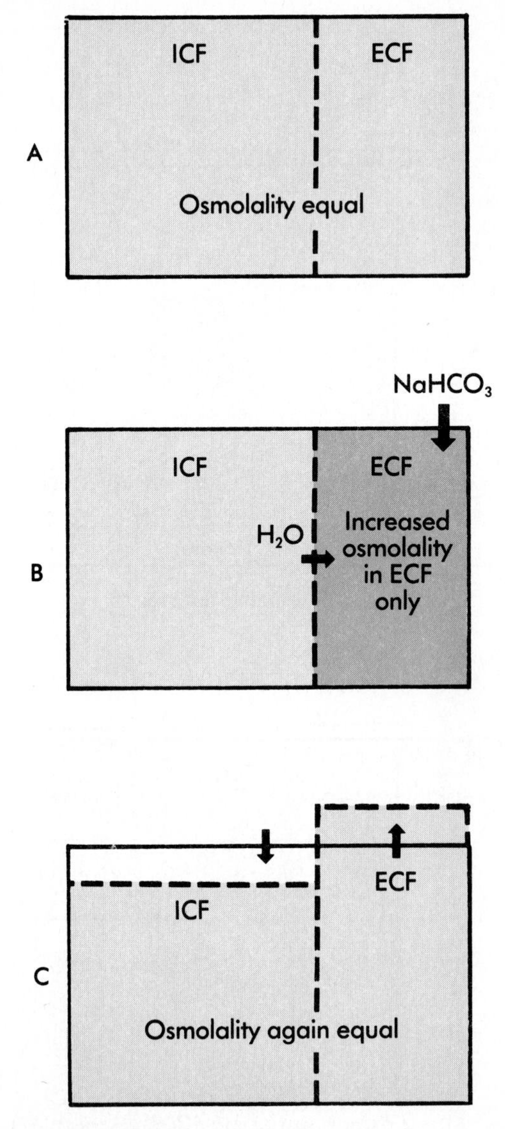

Figure 1-6, A-C Changes in ECF osmolality and volume after administration of intravenous sodium bicarbonate ($NaHCO_3$).

Osmolality of the ECF may be determined by measuring serum osmolality. Sodium is the primary determinant of ECF osmolality. Because it is limited primarily to the ECF, sodium acts to hold water in that compartment. Potassium helps to maintain the volume of intracellular fluid, and the plasma proteins help to maintain the volume of the intravascular space.

Tonicity

Small molecules, such as urea, that readily cross all membranes, quickly equilibrate between compartments and ultimately have little effect on the movement of water. These small molecules are termed *ineffective osmoles* (Geheb, 1987). In contrast, sodium and glucose are examples of effective osmoles. They do not cross the cell membrane quickly and will, therefore, affect the movement of water. The **effective osmolality** (i.e., the number of solutes present in a solution that will act osmotically to cause the movement of water) is dependent not only on the number of solutes, but also on the permeability of the membrane to these solutes. **Tonicity** is another term for effective osmolality.

Solutions are either isotonic, hypotonic, or hypertonic. **Isotonic solutions** are those that have an effective osmolality similar to that of body fluids (approximately 280 to 300 mOsm/kg). An example is normal saline, which is a 0.9% NaCl solution. *Hypotonic solutions* have an effective osmolality less than body fluids. An example is 0.45% NaCl solution. *Hypertonic solutions* have an effective osmolality greater than body fluids. An example is 3% NaCl solution. The concepts of osmolality and tonicity will be discussed in greater detail in Chapter 3.

Case Study 1-1

Mannitol is a unique medication used in a variety of clinical settings. An alcohol commercially prepared from dextrose, mannitol acts as an osmotic agent. It is not absorbed through the gastrointestinal tract and must be administered parenterally. It distributes throughout the ECF with very little entering the cells. Only a small percentage of intravenous mannitol is metabolized; the majority is excreted intact by the kidneys.

1. **Question:** If a substance such as mannitol is added to the ECF, what will happen to the osmolality (concentration) of the ECF?

   Answer: It will increase (Figure 1-7, *A*). Remember that osmolality measures the number of particles in a solution.

2. **Question:** If very little of the substance enters the ICF, what will happen initially to the osmolality (concentration) of the ICF?

   Answer: Initially there will be very little change. Thus the osmolality of the ECF will be greater than the osmolality of the ICF.

3. **Question:** There is now a concentration difference (gradient) between the two compartments. Remember that the two compartments are separated by a semipermeable membrane that allows the movement of water. In which direction will the water move?

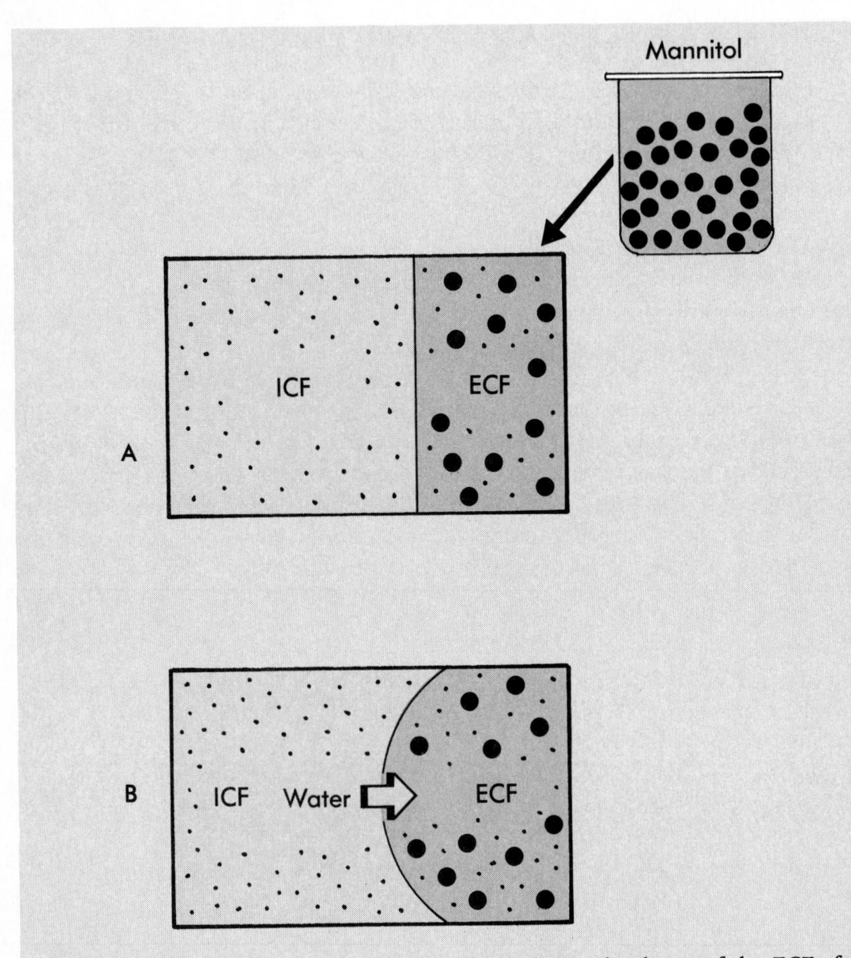

Figure 1-7, A and B Increase in the osmolality and volume of the ECF after administration of intravenous mannitol.

Answer: The increased osmolality of the ECF will cause a movement of water from the intracellular space into the extracellular space (Figure 1-7, *B*). This is the rationale for using mannitol in the treatment of cerebral edema. Water moves out of the brain cells and into the ECF in response to the osmotic effect of mannitol. Mannitol also works by the same mechanism to reduce intraocular pressure. An increase in the volume of the ECF may be harmful for certain patients, such as individuals with congestive heart failure, due to an increase in blood pressure and cardiac workload. Mannitol must be given cautiously in these individuals.

4. **Question:** Mannitol is classified as an osmotic diuretic. What will happen to a patient's urine output after the administration of mannitol?

   Answer: It will increase. Mannitol is freely filtered by the kidney and is not reabsorbed (saved). The high concentration within the renal tubule acts osmotically to inhibit the reabsorption of some water and sodium. Other therapeutic uses of mannitol include the treatment of acute renal failure and prevention of sudden changes in plasma osmolality during hemodialysis, known as dialysis disequilibrium syndrome.

References

Arieff AI and De Fronzo RA: Fluid, electrolyte, and acid-base disorders, New York, 1985, Churchill Livingstone, Inc.

Geheb MA: Clinical approach to the hyperosmolar patient, Crit Care Clin 5:797-815, 1987.

Goodman LS and Gillman A: The pharmacological basis of therapeutics, ed 7, New York, 1985, Macmillan.

Groer MW: Physiology and pathophysiology of the body fluids, St Louis, 1981, The CV Mosby Co.

Guyton AC: Textbook of medical physiology, ed 7, Philadelphia, 1986, WB Saunders Co.

Horne M and Swearingen PL: Pocket guide to fluids and electrolytes, St Louis, 1989, The CV Mosby Co.

Keyes JL: Fluid, electrolyte, and acid-base regulation, Monterey, Calif, 1985, Wadsworth Co.

Kokko JP and others: Fluids and electrolytes, Philadelphia, 1986, WB Saunders Co.

Luckmann J and Sorensen KC: Medical-surgical nursing: a pathophysiologic approach, ed 3, Philadelphia, 1987, WB Saunders Co.

Maxwell MH and others: Clinical disorders of fluid and electrolyte metabolism, ed 4, New York, 1987, McGraw-Hill Co.

Metheny NM: Fluid and electrolyte balance: nursing considerations, Philadelphia, 1987, JB Lippincott Co.

Nissenson AR and Kleeman CR: Mannitol, West J Med 131:277-284, 1979.

Rose BD: Clinical physiology of acid-base and electrolyte disorders, ed 3, New York, 1989, McGraw-Hill Co.

Smith K: Fluids and electrolytes: a conceptual approach, New York, 1980, Churchill Livingstone, Inc.

Vander AJ: Renal physiology, ed 3, New York, 1985, McGraw-Hill Co.

Warren SE and Blantz RL: Mannitol, Arch Intern Med 141:493-497, 1981.

Chapter

2

Renal Physiology and Pathophysiology

FUNCTIONS OF THE KIDNEYS

Regulation of Water and Electrolytes

Although all organs and body systems are involved in the maintenance of homeostasis, the kidneys have a central role in the regulation of body fluids. The body gains water via the gastrointestinal (GI) tract or through therapeutic parenteral administration. Some additional water is produced metabolically. Normally, water is lost from the body via the skin, lungs, GI tract, and kidneys. Only thirst and urinary output are physiologically regulated in order to maintain a balance between intake and loss of water. Thirst is important in the body's response to water deficits. However, water ingestion is also strongly influenced by social and emotional factors. Thus the kidneys act as the primary regulators of water balance through the regulation of urine output. Similarly, the kidneys are pivotal to the regulation of electrolytes. The ions sodium, potassium, chloride, magnesium, and phosphorus are regulated primarily by the kidneys. As with water excretion, urinary excretion of these electrolytes is adjusted so that a balance is maintained between overall intake and output.

Regulation of Acid-base Balance

The kidneys work with the lungs to maintain the concentration of hydrogen ions in body fluids and thus promote acid-base balance. They do this by excreting a portion of the daily load of metabolically produced hydrogen ions and by generating bicarbonate ions. The excretion of hydrogen ions and generation of bicarbonate is increased any time there is an abnormal production of acids (as in diabetic ketoacidosis) or increased retention of carbon dioxide (as in respiratory failure). See Chapter 13 for additional information concerning the role of the kidneys in the regulation of acid-base balance.

Excretion of Metabolic Wastes

The kidneys have several other functions in addition to the regulation of water and electrolyte balance. Each of these functions assists in the maintenance of homeostasis. The function most often associated with the kidneys is the excretion of metabolic wastes. Metabolic wastes are substances produced within the cell that serve no known biologic function. The most common metabolic wastes are urea (from the breakdown of proteins), creatinine (from muscle creatine), and uric acid (from nucleic acids). Other substances excreted by the kidneys include bilirubin and hormone metabolites. Some of the substances excreted by the kidneys are quite harmless, such as urea. Others are toxic to tissues and their retention causes many of the symptoms observed in renal failure. Certain of these toxins have yet to be identified. Plasma urea nitrogen and creatinine levels are useful as indicators of renal function (see Chapter 5 for a discussion of these two tests). These tests commonly are performed to identify and monitor renal failure since increases in their levels suggest a reduction in renal function.

Activation of Vitamin D

Although not the primary regulator, the kidneys participate in the regulation of calcium ions via their role in vitamin D metabolism. Vitamin D is not a single compound, but rather a group of closely related compounds. Vitamin D_3 (cholecalciferol), the form of vitamin D synthesized by the skin and contained in food, is not the metabolically active form of vitamin D. Rather, vitamin D_3 must be altered by the liver and then by the kidneys before it can act on the GI tract to increase the absorption of calcium. Activation of vitamin D by the kidneys is regulated by parathyroid hormone (PTH). PTH production is, in turn, regulated by changes in the plasma calcium levels. A decreased plasma calcium level, for example, stimulates the parathyroid gland to secrete PTH. Increased PTH levels increase the activation of vitamin D by the kidneys, increasing plasma calcium levels. PTH also decreases the excretion of calcium by the kidneys. Several additional factors are involved in the regulation of plasma calcium levels and these will be discussed in greater detail in Chapter 10.

Additional Functions

The kidneys are also responsible for the excretion of foreign chemicals, such as medications, food additives, and pesticides; regulation of blood pressure through the release of renin (this will be discussed in greater detail in Chapter 3); and regulation of erythrocyte production through the secretion of erythropoietin. The importance of the kidneys' role in these functions is not obvious in the healthy individual. When the kidneys fail, however, abnormalities in these functions become readily apparent. For example, the presence of renal failure necessitates the adjustment of certain medication dosages to avoid medication toxicity, which may occur because of decreased urinary elimination and altered absorption, metabolism, and protein binding. Dosage modification may be achieved by increasing the interval between doses. For instance, the medication may be given every 12 or 24 hours instead of every 6 hours, or the amount of individual doses may be decreased. (If the usual dose is 50 mg once a day, the dose may be reduced to 25 mg per day.) See the accompanying box for a list of commonly prescribed medications that require dosage modification in renal failure. Renal failure often results in the development of hypertension, which may be caused by excess fluid retention or increased production of renin (Wright, 1988).

Erythropoietin is produced by the kidneys and released in response to decreased oxygen delivery. It acts directly on precursor cells in the bone marrow to increase the production of erythrocytes, thereby increasing the oxygen-carrying capacity of the blood. Although the etiology of anemia in renal failure is multifactorial, decreased production of erythropoietin is considered the primary cause (Richards, 1986). Recombinant human erythropoietin (Epogen), recently approved by the FDA, is now in use for replacement therapy in the treatment of the anemia of chronic renal failure. The primary complication of erythropoietin replacement therapy is the development of hypertension (Watson, 1989).

Drugs that Require Dosage Modification in Renal Failure

Antimicrobials
- amikacin
- gentamicin
- kanamycin
- tobramycin
- amphotericin B
- vancomycin
- lincomycin
- sulfonamides
- ethambutol
- penicillins

Cardiovascular Agents
- digoxin
- digitoxin
- procainamide
- guanethidine

Analgesics
- meperidine
- methadone

Sedatives
- phenobarbital
- meprobamate

Miscellaneous
- insulin
- cimetidine
- clofibrate
- neostigmine

Drugs to Avoid
- tetracycline
- nitrofurantoin
- spironolactone
- amiloride
- aspirin
- lithium carbonate
- cisplatin
- phenylbutazone
- nonsteroidal antiinflammatory agents
- magnesium-containing medications

STRUCTURE OF THE KIDNEYS AND URINARY SYSTEM

Location

The kidneys are located outside of the peritoneal cavity, along the posterior abdominal wall, one on either side of the vertebral column. Although the kidneys usually are paired organs, rare individuals are born with a single functioning kidney. The right kidney is 1 to 2 cm lower than the left, owing to the position of the liver. Because of its lower position, the right kidney is easier to palpate on physical assessment. The upper portions of the kidneys are protected by the last two ribs and the lower portions by the heavy back muscles. The costovertebral angle (CVA), the angle created by the twelfth rib and the vertebral column, provides a convenient reference for identifying the location of the kidneys (Figure 2-1).

Gross Structures

The kidneys are covered in a thin, fibrous capsule and are surrounded by a layer of fat. These and other structures help to maintain the kidneys' position. An adrenal gland sits on top of each kidney, and although it functions independently of the kidney, hormones secreted by the adrenal gland affect renal function. The interaction between adrenal hormones and renal function will be discussed in greater detail in Chapter 3.

The concave (indented) portion of the kidney is termed the *hilum* and is located

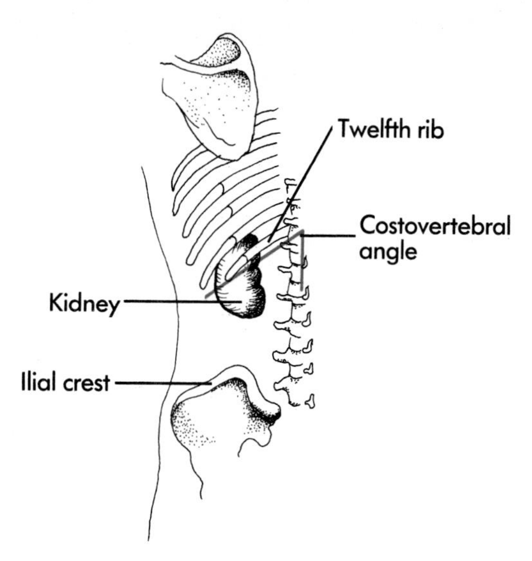

Figure 2-1 Costovertebral angle: Location of the kidney.

(Modified from Bates B: A guide to physical examination and history taking, ed 4, Philadelphia, 1987, JB Lippincott.)

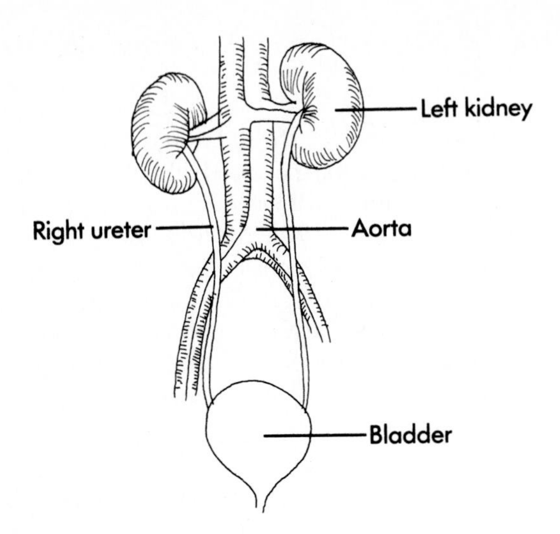

Figure 2-2 Structure of the urinary system.

where the renal artery, renal vein, nerves, lymphatics, and ureter enter and exit the kidneys. Each kidney has its own ureter, which carries urine to the bladder. The ureters are narrowed extentions of the calyces and renal pelvis (Figure 2-2). They transport urine to the bladder via continuous peristaltic action. Urine is stored in the bladder until voluntary micturition or catheterization occurs. The desire to urinate is experienced when the bladder has collected approximately 300 ml of urine. The urethra transports urine from the bladder to the outside of the body.

MACROSCOPIC ANATOMY OF THE KIDNEYS

Each kidney is composed of two portions: the cortex or outer portion and the medulla or inner portion (Figure 2-3). The cortex is approximately 1 cm wide, while the medulla is approximately 5 cm wide and is divided into approximately 8 to 12 triangular portions termed *pyramids.* The tips of the pyramids form the papillae. Urine leaves the papillae and is collected in the minor calyces. There are usually no significant changes in the composition or the volume of urine after it leaves the papillae. The minor calyces merge to form the major calyces, which in turn merge to form the renal pelvis. The renal pelvis holds approximately 3 to 5 ml of urine at a time. It narrows to form the ureter. Because the pelvis, ureters, and bladder are somewhat permeable to water and urea, small amounts of urea may diffuse out of the urine, while small amounts of water may diffuse into the urine.

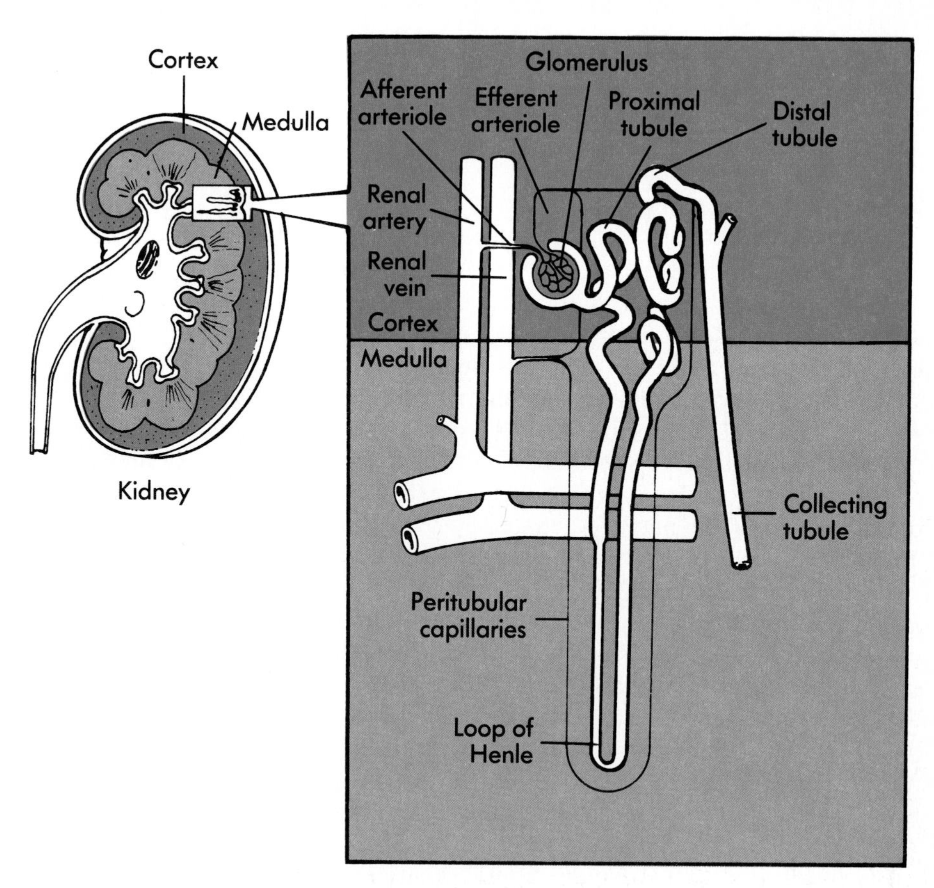

Figure 2-3 Macroscopic and microscopic structure of the kidney.
(From Horne M and Swearingen P: Pocket guide to fluids and electrolytes, St Louis, 1989, The CV Mosby Co.)

MICROSCOPIC ANATOMY OF THE KIDNEYS

Nephrons

The nephron is the functioning unit of the kidney. Each kidney contains more than a million nephrons, many more than are necessary for survival. Kidney transplant recipients and living related transplant donors, for example, are able to maintain essentially normal renal function with a single kidney. For the purposes of this discussion, the physiology of the kidney will be reviewed as if the kidney were composed of a single nephron. Keep in mind, however, that the kidney is composed not only of many nephrons, but also of more than one type of nephron. The majority of the nephrons are cortical, which means that they are located primarily in the cortex. Others are juxtamedullary nephrons, which begin just inside the cortex and extend deep into the medulla. Although all nephrons are involved in the formation of urine, the juxtamedullary nephrons, because of their location within the medulla, have a greater role in the formation of a concentrated urine.

Nephron Components

Each of the nephrons is composed of a filtering portion and a tubular portion. Because urine is formed by the filtration of blood, urine production depends on blood flow to the nephron. Blood enters the kidney via the renal artery, which branches into successively smaller arteries as it extends into the cortex. In the cortex, the interlobular arteries give rise to a series of special arterioles termed *afferent arterioles.* Each afferent arteriole delivers blood to the filtering portion of the nephron. This filtering portion is a compact group of interconnected capillaries called the glomerulus. The glomerulus is surrounded by a hollow capsule (Bowman's capsule), which begins the tubular portion of the nephron. The glomerulus allows the filtration of *essentially* protein-free plasma. (Recall from Chapter 1 that plasma is the portion of the blood that includes water, protein, and electrolytes but excludes the cells.) Approximately 20% of the plasma delivered to the glomerulus by the afferent arterioles is filtered into the tubular portion of the nephron. The remaining plasma and other blood components leave via the *efferent arteriole.* The filtered plasma, termed *filtrate*, moves from Bowman's capsule into the proximal tubule, then into the loop of Henle and the distal tubule, and finally through the collecting ducts (see Figure 2-3). As the filtrate moves through each of these portions of the tubule, it undergoes many changes. The final urine that enters the calyces is markedly different in volume, composition, and concentration from the initial filtrate.

The blood leaving the efferent arteriole enters the peritubular capillaries (the capillaries that surround the tubule). Water and solutes move between the peritubular capillary blood and the tubule. Thus substances that are not filtered at the glomerulus may still enter the tubule and end up in the final urine. This is how certain medications, including furosemide (Lasix), are excreted. Likewise, substances in the plasma, such as glucose, that are filtered into the tubule may return to the peritubular capillary blood and not be lost in the urine. These concepts will be discussed in greater detail in the next section. From the peritubular capillaries, the blood enters the venous system and leaves the kidney via the renal vein.

FORMATION OF URINE

The kidneys receive approximately 20% to 25% of the cardiac output or about 1,200 ml/min in the average adult. Assuming a hematocrit of 48%, approximately 625 ml of plasma move through all the glomeruli each minute. (Remember that if the hematocrit is 48%, then the plasma volume is 52% of the total blood volume. Fifty-two percent of 1,200 ml is equal to 625 ml.) Of the approximate 625 ml of plasma, 20%, or 125 ml, is filtered into the tubules per minute. This is termed the **glomerular filtration rate (GFR)** and varies with renal function. If the glomeruli are damaged, as in renal disease, the GFR will drop, and the kidneys will be less able to excrete metabolic wastes and alter the composition of body fluids. The GFR may be estimated by means of a creatinine clearance test (see the case study at the end of this chapter and Chapter 5 for a discussion of creatinine clearance). As already described, the filtrate is radically altered as it moves through the tubule. It is reduced in volume by better than 99%, so that only about 1 ml of urine is produced each minute. Thus the average urine output in the adult is said to be about 60 ml/hr (*average range* 40 to 80 ml/hr). It is vital to understand, though, that urine output is affected by many factors, such as fluid intake, other fluid losses, medications, and renal blood flow. An individual who is severely fluid restricted would not be expected to produce as much as 60 ml of urine per hour, nor would the individual receiving excess volumes of fluid be expected to excrete as little as 60 ml/hr. *The appropriateness of a given urine output must be assessed in relation to the individual's overall physiologic status.*

Filtration

As discussed previously, the transition from blood flow to urine formation begins with glomerular filtration. What and how much is filtered at the glomerulus is determined by the pressures across the glomerular capillary membrane and the anatomy of that membrane. The glomerular capillary membrane is a selectively permeable membrane composed of two thin layers: capillary endothelium and basement membrane. (The epithelial cells of Bowman's capsule act as a third thin barrier to filtration.) Water, electrolytes, urea, and other small substances readily cross the membrane, while larger substances, such as proteins, protein-bound substances, and blood cells are restricted due to the size of the pores in the endothelium and epithelium. In addition, the negatively charged basement membrane repels large, negatively charged substances, such as proteins, and allows the filtration of neutral and positively charged substances. Filtrate is *essentially protein-free* plasma. Filtered solutes are found in the same concentration in the filtrate as in the plasma. Damage to the glomerular membrane, as occurs in such diseases as glomerulonephritis, diabetic nephropathy, and lipoid nephrosis, allows the movement of normally nonfiltered substances into the urine. Hence the presence of protein or blood cells in the urine suggests renal disease involving the glomerulus (Seldin and Giebisch, 1985).

The structure of the glomerular membrane determines which substances are filtered. Membrane permeability, combined with surface area and pressures across the membrane, determines the amount that is filtered. The pressure that determines the movement of fluid across the glomerular capillary membrane and into Bowman's capsule is the net difference between the hydraulic and oncotic pressures across the membrane (Figure 2-4).

Recall from Chapter 1 that hydraulic pressure is the pressure created by the weight of fluid and the pressure created by the pump action of the heart, while oncotic

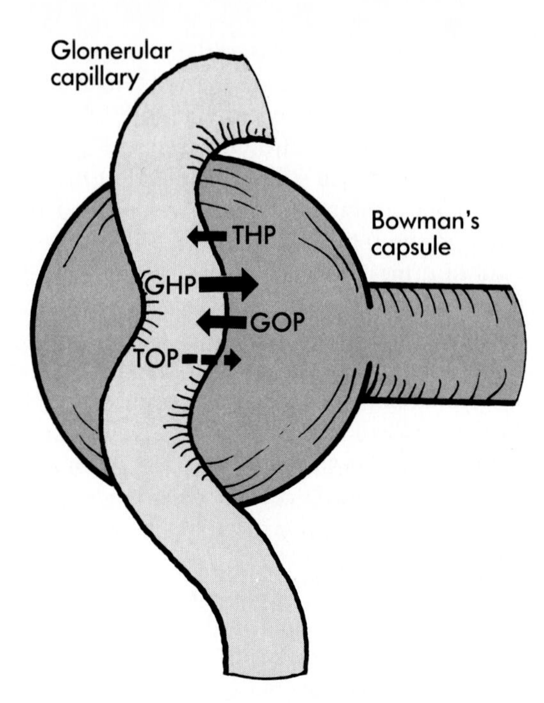

Figure 2-4 Net glomerular filtration pressure: (glomerular hydraulic pressure [GHP] + tubular oncotic pressure [TOP]) – (tubular hydraulic pressure [THP] + glomerular oncotic pressure [GOP]).

pressure is the osmotic pressure created by the plasma proteins. Also remember that osmotic pressure is defined as the amount of fluid pressure necessary to stop the osmotic flow of water. The hydraulic pressure within the glomerular capillary and the oncotic pressure within Bowman's capsule will induce filtration across the glomerular membrane. Since little protein is filtered into Bowman's capsule, the effect of oncotic pressure within Bowman's capsule is minimal. The forces that oppose filtration are the oncotic pressure within the glomerular capillary blood and the hydraulic pressure within Bowman's capsule. Only in the case of urinary system blockage causing a backup of fluid within the tubule does the hydraulic pressure within Bowman's capsule significantly alter GFR. The single most important factor affecting filtration (and the factor that has the most physiologic regulation) is the hydraulic pressure (i.e., blood pressure) within the glomerular capillary (Vander, 1985). Increased glomerular capillary hydraulic pressure causes an increase in glomerular filtration. Likewise, a decrease in glomerular capillary hydraulic pressure will cause a decrease in glomerular filtration. Thus changes in systemic blood pressure may result in similar changes in glomerular filtration and, ultimately, urine formation. Net glomerular filtration pressure (NGFP) equals pressures that favor filtration minus pressures that limit filtration (see Figure 2-4).

The kidneys are able to minimize the effects of changes in systemic pressure by regulating their own blood flow. This is known as autoregulation, and it is believed to be accomplished via changes in afferent arteriolar resistance. Unfortunately, large fluctuations in blood pressure, stimulation of the sympathetic nervous system, and administration of certain medications limit or override the kidneys' ability to maintain a constant glomerular pressure. When an individual's mean arterial blood pressure (the average pressure throughout the cardiac cycle) drops below 70 mm Hg, autoregulation is virtually absent and glomerular filtration will drop as the blood pressure drops. Hypotension also causes an increase in sympathetic nervous system discharge with the release of potent vasoconstrictors. Constriction of both the afferent and efferent arterioles somewhat minimizes the hypotension-induced drop in GFR, but more importantly, it shunts blood flow away from the kidney to help maintain vital cerebral and coronary perfusion. Thus the individual in shock is unable to maintain normal renal function, and if hypotension is prolonged, acute renal failure may develop. The nurse must also bear in mind that medications that alter renal hemodynamics (e.g., nonsteroidal antiinflammatory drugs, catecholamines) may limit the kidneys' ability to function optimally.

Glomerular filtration is only the first step in the formation of urine. Assuming that the kidneys are healthy and filter approximately 20% of the plasma they receive each minute, they will produce 160 to 180 L of filtrate over a 24-hour period (Fernandez and Cox, 1984). This is equal to approximately 30 times the entire plasma volume. From this 180 L of plasma-like filtrate, the kidneys produce 1½ to 2 L of urine. The contribution of the kidneys to the regulation of body fluids is evidenced by the fact that the composition of urine is markedly different from that of plasma. In addition to a high concentration of metabolic wastes, water and electrolytes are present in quantities quite different from that of plasma. The composition of the urine is varied to maintain a balance between overall intake and output of water and electrolytes.

Reabsorption

There are two processes involved in the transformation of filtrate to urine. These were mentioned earlier in this chapter, although not identified by name. The first is tubular reabsorption, or the movement of water and solutes from the filtrate to the peritubular capillary circulation (Figure 2-5). If the kidneys were not capable of reabsorption, theoretically the entire plasma volume would be excreted within 30 minutes! Tubular reabsorption is important not only in adjusting the volume of the filtrate, but also in adjusting the composition. Substances may be reabsorbed actively or passively. As discussed in Chapter 1, active transport requires energy expenditure. The majority of the kidneys' energy requirements go toward active reabsorption (i.e., active transport) of sodium. Active reabsorption of positively charged sodium ions, negatively charged bicarbonate ions, glucose, and amino acids creates electrical, osmotic, and concentration gradients that then allow for the passive reabsorption of certain other substances such as water, negatively charged chloride ions, and urea. The very small amount of protein that is filtered by the glomerulus is also actively reabsorbed.

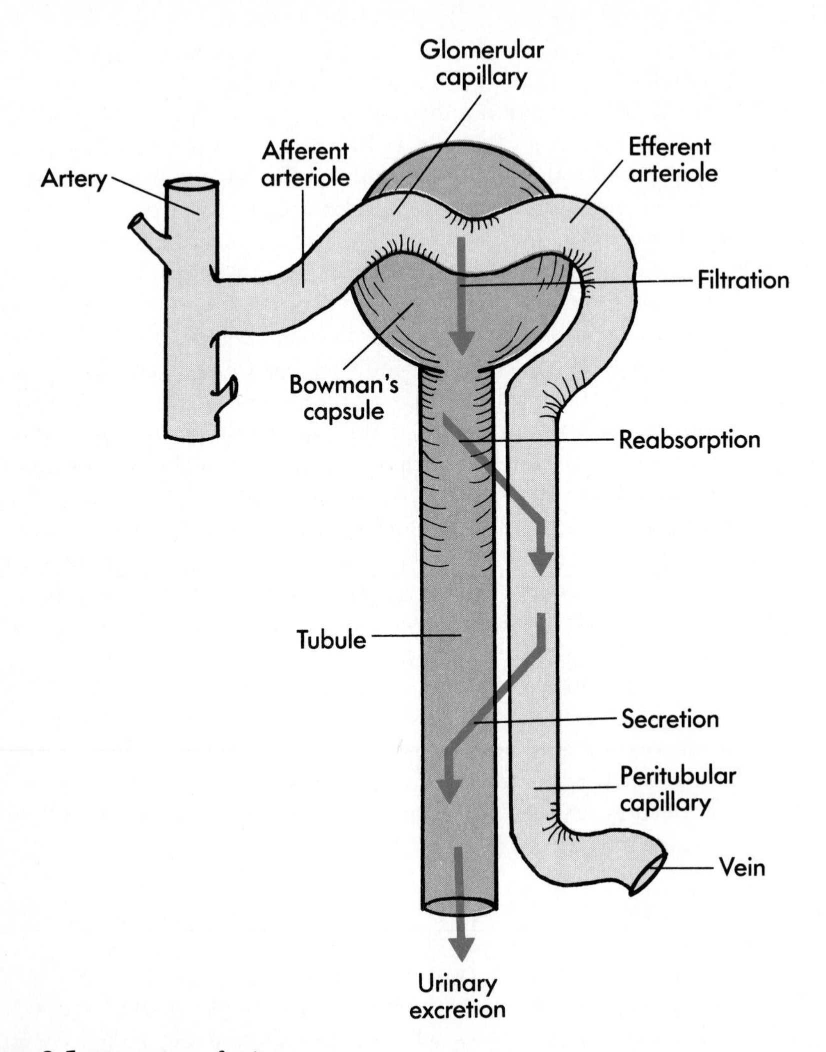

Figure 2-5 Formation of urine.
(Modified from Vander AJ: Renal physiology, ed 3, New York, 1985, McGraw-Hill.)

The kidneys have an unlimited ability to reabsorb sodium. This is not true for all substances that are actively reabsorbed. In Chapter 1 the example of glucose was used to demonstrate that active transport may be limited by the availability of carrier substances. Normally, urine is glucose free. There is a sufficient amount of the carrier substance available to permit active transport of all filtered glucose. If, however, the blood sugar is elevated past 180 to 200 mg/dl, glucose begins to appear in the urine. At a blood sugar of approximately twice this value, the carrier substance is fully saturated; further increases in plasma glucose cause proportionate increases in glucose excretion. The point at which glucose first appears in the urine is termed the *renal threshold for glucose.* For many years, individuals with diabetes mellitus determined their daily insulin dosages based in part on the amount of glucose "spilled" into their urine. With the advent of home glucose monitoring, this method is no longer encouraged, because urine glucose testing can be unreliable. The renal threshold for glucose varies somewhat from one diabetic individual to the next and may be altered dramatically with reduced renal function.

Secretion

The second renal process necessary for transforming filtrate to final urine is termed secretion, the movement of substances from the peritubular capillary circulation into the tubule (see Figure 2-5). Potassium and hydrogen ions are two important solutes whose urinary excretion is regulated by means of secretion. All the potassium and hydrogen ions filtered at the glomerulus are reabsorbed and returned to the circulation. Urinary excretion of these two substances does *not* depend on the amount filtered at the glomerulus, but on the amount secreted into the lumen of the tubule. Also many medications are handled via secretion. A medication that cannot be filtered at the glomerulus because of size or protein binding still may be excreted in the urine by means of tubular secretion.

Transformation of Filtrate to Urine

As filtrate moves through the tubule, varying amounts of water and solutes are reabsorbed. In the proximal tubule, 60% to 70% of the filtered sodium (Na^+), potassium (K^+), chloride (Cl^-), and water are reabsorbed. This is also where 90% of the filtered bicarbonate is reabsorbed and hydrogen ions are secreted. Reabsorption of glucose, amino acids, organic acids, and divalent ions also occurs in the proximal tubule. The fluid leaving the proximal tubule and entering the loop of Henle has been dramatically reduced in volume but remains isotonic in concentration. As the fluid moves down the descending portion of the loop, the tubular contents become increasingly hypertonic due to the movement of water out of the tubule. Reabsorption of 20% to 30% of the filtered sodium, chloride, and potassium takes place in the ascending portion of the loop. This is the site of action of the loop diuretics furosemide (Lasix) and ethacrynic acid (Edecrin). The overall affect of the loop is that more solute is reabsorbed than water. Thus the fluid leaving the loop and entering the distal tubule is hypotonic.

Solute reabsorption continues in the distal tubule. Approximately 5% of the filtered load of sodium is reabsorbed in the early portion of the distal tubule. The thiazide and thiazide-like diuretics work here. Based on their site of action, it is easy to understand why they exert a weaker diuretic action than the loop diuretics, which affect 20% to 30% of the filtered sodium load (see Chapter 7 for a discussion of diuretic therapy). The early distal tubule is relatively impermeable to water; thus the fluid within the tubule becomes increasingly hypotonic as it moves toward the late distal tubule. The late distal tubule (referred to by some physiologists as the initial collecting duct) is the site of net potassium and hydrogen ion secretion and aldosterone-mediated sodium regulation. Under the influence of aldosterone, sodium is reabsorbed and potassium and hydrogen ions are secreted. Additional potassium is secreted, depending on potassium concentrations within the renal tubular cell. The late distal tubule and collecting ducts are the sites of antidiuretic hormone (ADH)-regulated water reabsorption, which ultimately determines urinary concentration. The hormones aldosterone and ADH will be discussed in detail in Chapter 3. See specific electrolyte chapters (Chapters 8-12) for a discussion of their handling by the kidneys. From the collecting ducts, final urine passes through the tips of the papillae into the minor calyces.

Case Study 2-1

H.R. is a 24-year-old male computer programmer who arrived at a local emergency clinic with a 2-week history of nausea, fatigue, malaise, decreasing urine output ("less than one cup in the morning" for the last 3 days), and increasingly severe headaches. Despite his complaints of nausea, H.R. stated that he had maintained a "fairly normal" intake of food and fluids. His blood pressure was 210/120 mm Hg, and he had 2+ pretibial edema. Laboratory tests showed evidence of renal failure. He was admitted to the hospital for evaluation and treatment.

1. **Question:** Given H.R.'s diagnosis of renal failure, what would you expect his BUN (blood urea nitrogen), creatinine, and uric acid levels to be like: increased, decreased, or normal?

   Answer: They would be increased since these are all examples of waste products produced by the body and normally excreted by the kidneys. H.R.'s levels were:

H.R.'s Levels (mg/dl)	Normal Levels (mg/dl)
BUN 137	6-20
Creatinine 8.9	0.6-1.5
Uric acid 15	Male = 2.1-7.5

2. **Question:** Which altered renal function would explain H.R.'s elevated blood pressure and edema?

   Answer: The regulation of sodium and water balance. Remember that the kidneys maintain a balance between intake and output of sodium and water by regulating the concentration and volume of urine output and that the average urine output is approximately 1,500 ml/day. H.R. was producing less than one cup of urine per day, despite a near-normal food and fluid intake. Thus he was retaining fluid normally excreted by the kidneys. This fluid expanded both the intravascular and interstitial fluid volumes, causing hypertension and edema. Excess production of renin also may have contributed to the development of H.R.'s hypertension.

Renal failure may be either acute or chronic. Acute renal failure is a sudden loss of renal function that is usually reversible. The causes of acute renal failure may be classified as prerenal (any condition that results in decreased renal perfusion), intrarenal (actual parenchymal damage), and postrenal (any condition that causes obstruction to urine flow) (Table 2-1). Chronic renal insufficiency (CRI) is a progressive, irreversible loss of renal function that may develop over days to years. There are many causes of CRI, but the most common include glomerulonephritis, diabetic nephropathy, hypertension, and polycystic kidney disease. CRI usually progresses to end-stage renal disease (ESRD), requiring dialysis or a kidney transplant to sustain life. Prior to ESRD the individual with CRI may lead a relatively normal life managed by diet and medications.

One of the first tests ordered by the nephrologist (a physician who specializes in the treatment of kidney disorders) was a routine urinalysis (UA), since it may provide information concerning the cause, severity, and chronicity of renal failure. A UA includes measurement of urine specific gravity (a test that evaluates the kidneys' ability to concentrate urine—see Chapter 5 for additional information concerning laboratory tests); a semiquantitative estimate of pH, glucose, red blood cells, white blood cells, ketone bodies, and bilirubin; and microscopic examination of the urinary sediment.

3. Question: The presence of a large amount of protein in the urine would suggest damage to which portion of the nephron?

Answer: The glomerulus. Normally only a tiny amount of protein is filtered at the glomerulus—usually less than 100 mg of protein are excreted per day. Damage to the glomerulus will result in increased filtering and excretion of protein. While proteinuria also may occur with damage to other portions of the nephron or with conditions such as multiple myeloma that result in an abnormal production of proteins, it usually indicates altered glomerular function. An individual with renal disease complicated by severe proteinuria may excrete as much as 10 g of protein in 24 hours.

After a complete review of H.R.'s history, physical findings, x-rays, and laboratory results, no specific cause of renal failure could be identified. A percutaneous renal biopsy was then performed in an attempt to identify a treatable cause. The results of the biopsy would provide information concerning the cause and extent of the disease and might help direct treatment. The results of H.R.'s renal biopsy showed extensive, irreversible renal damage consistent

Table 2-1 Causes of acute renal failure

Prerenal (Decreased renal perfusion)	Intrarenal (Parenchymal damage; acute tubular necrosis)	Postrenal (Obstruction)
Hypovolemia ■ GI losses ■ Hemorrhage ■ Third space (interstitial) losses (burns, peritonitis) ■ Dehydration from diuretic use **Hepatorenal Syndrome** **Edema-forming Conditions** ■ Congestive heart failure ■ Cirrhosis ■ Nephrotic syndrome **Renal Vascular Disorders** ■ Renal artery stenosis ■ Renal artery thrombosis ■ Renal vein thrombosis	**Nephrotoxic Agents** ■ Antibiotics (aminoglycosides, sulfonamides, methicillin) ■ Diuretics (e.g., furosemide) ■ Nonsteroidal anti-inflammatory drugs (e.g., ibuprofen) ■ Contrast media ■ Heavy metals (lead, gold, mercury) ■ Organic solvents (carbon tetrachloride, ethylene glycol) **Infection (gram-negative sepsis), pancreatitis, peritonitis** **Transfusion Reaction (hemolysis)** **Rhabdomyolysis with Myoglobinuria (severe muscle injury)** ■ Trauma ■ Exertion ■ Seizures ■ Drug-related: heroin, barbiturates, IV amphetamines, succinylcholine **Glomerular Diseases** ■ Poststreptococcal glomerulonephritis ■ IgA nephropathy (e.g., Berger's disease) ■ Lupus glomerulonephritis ■ Serum sickness **Ischemic Injury (prolonged prerenal)**	**Calculi** **Tumor** **Benign Prostatic Hypertrophy** **Necrotizing Papillitis** **Urethral Strictures** **Blood Clots** **Retroperitoneal Fibrosis**

From Horne M and Swearingen PL: Pocket guide to fluids and electrolytes, St Louis, 1989, The CV Mosby Co.

with chronic renal failure. A creatinine clearance test also was performed to estimate the severity of his disease and determine the need for permanent renal replacement therapy, such as dialysis or transplantation.

4. **Question:** A creatinine clearance study estimates which aspect of urine formation?

   Answer: The glomerular filtration rate (GFR).

5. **Question:** Why would it be helpful to have an estimate of H.R.'s GFR?

   Answer: The GFR represents the amount of plasma filtered by the kidney and varies directly with renal function. Remember that the GFR is the sum of the filtration from all of the functioning nephrons. If the GFR is reduced to one third of normal, the creatinine clearance also will be reduced to approximately one third the normal value. The creatinine clearance test is actually a measure of the volume of blood completely cleared of creatinine over a specific period of time, usually 24 hours. It is expressed in ml/min and factored to consider body surface area. The value is determined by comparing the plasma creatinine level drawn sometime during the test period with the amount of creatinine excreted in the urine during the full test period, using the following formula:

$$\frac{\textbf{Urine creatinine} \times \textbf{Urine volume}}{\textbf{Plasma creatinine}}$$

The normal range for the adult male is 107 to 141 ml/min, while 87 to 132 ml/min is the normal range for adult females. An average value for creatinine clearance and GFR is said to be 125 ml/min. Because renal function decreases with advancing age, creatinine clearance drops approximately 10% with each decade past the age of 50 (Corbet, 1987).

A renal clearance study may be performed on any substance that is excreted by the kidney. But to obtain an accurate estimate of glomerular filtration, a clearance study should be performed using a substance that is freely filtered at the glomerulus and neither secreted nor reabsorbed by the tubule. Net secretion or reabsorption would affect the overall amount of the substance found in the urine, and thus urine concentrations would not accurately represent filtration. Creatinine is freely filtered at the glomerulus, not reabsorbed, and only slightly secreted by the tubule; therefore it provides a reasonable estimate of glomerular filtration. In contrast, approximately 40% to 50% of the filtered urea is reabsorbed by the tubule. The percent of reabsorption varies with sodium and water reabsorption, so that less urea appears in the urine in volume depleted states. Urea clearance, therefore, provides a less reliable estimate of GFR. See Chapter 5 for additional information concerning factors that affect urea excretion.

H.R.'s creatinine clearance is 8 ml/min (normal range for a male = 107-141 ml/min), a level consistent with ESRD. He is to have surgery for the placement of a permanent vascular access for hemodialysis.

6. **Question:** Based on the functions of the kidney discussed in this chapter, what additional pathophysiologic changes might H.R. experience?

Answer:

a. Electrolyte imbalances, such as hyperkalemia (increased plasma potassium), hyperphosphatemia (increased plasma phosphorous), and hypermagnesemia (increased plasma magnesium) due to the kidneys' decreased ability to excrete excess amounts of these ions, and hypocalcemia (decreased plasma calcium), due to decreased activation of vitamin D. (Hyperphosphatemia will also contribute to the development of hypocalcemia. This will be discussed in Chapters 10 and 11). Although the kidneys retain sodium, typically hypernatremia is not seen. The serum sodium level is usually normal or may even be low because of concomitant water retention.
b. Acid-base imbalance, specifically metabolic acidosis caused by the kidneys' decreased ability to excrete a portion of the daily load of hydrogen.
c. Anemia due to the kidneys' decreased production of erythropoietin.
d. Symptomatic retention of metabolic wastes, termed *uremia.* Uremia adversely affects all body systems. Symptoms of uremia include nausea, anorexia, increased susceptibility to infection, increased bruising due to platelet dysfunction and increased capillary fragility, yellow skin, pallor, pruritis, malaise, decreased ability to concentrate, confusion, irritability, tremors, peripheral neuropathy, stomatitis, uremic halitosis, and bone disease.
e. Altered renal handling of medications.

References

Beeuwkes R and Rosen S: The structure and function of the human kidney. In Flamenbaum W and Hamburger RJ: Nephrology: an approach to the patient with renal disease, Philadelphia, 1982, JB Lippincott Co.

Beyer M: Diabetic nephropathy, Pediatr Clin North Am 31(3):635-650, 1984.

Corbett JV: Laboratory tests and diagnostic procedures with nursing diagnoses, ed 2, Norwalk, Conn and Los Altos, Calif, 1987, Appleton & Lange.

Fernandez P and Cox M: Basic concepts of renal physiology, Int Anesthesiol Clin 22(1):1-34, 1984.

Horne M and Swearingen P: Pocket guide to fluids and electrolytes, St Louis, 1989, The CV Mosby Co.

Klahr S: The kidney and body fluids in health and disease, New York, 1983, Plenum Publishing Corp.

Lancaster LE: Renal and endocrine regulation of water and electrolyte balance, Nurs Clin North Am 22(4):761-772, 1987.

Leaf A and Cotran R: Renal pathophysiology, ed 3, New York, 1985, Oxford University Press.

Richards C: Comprehensive nephrology nursing, Boston, 1986, Little, Brown & Co.

Rose BD: Clinical physiology of acid-base and electrolyte disorders, ed 3, New York, 1989, McGraw-Hill, Inc.

Schoengrund L and Balzer P: Renal problems in critical care, New York, 1985, John Wiley & Sons, Inc.

Seldin DW and Giebisch G: The kidney: physiology and pathophysiology, vol 2, New York, 1985, Raven Press.

Valtin H: Renal function: mechanisms preserving fluid and solute balance in health, ed 2, Boston, 1983, Little, Brown & Co.

Vander AJ: Renal physiology, ed 3, New York, 1985, McGraw-Hill, Inc.

Watson AJ: Adverse effects of therapy for the correction of anemia in hemodialysis patients, Seminars Nephrol 9(1):30-34, 1989.

Wright M: Physiology for general practitioners, 2: cardiovascular and renal systems, Fam Pract 5(2):145-153, 1988.

Chapter

3

Regulation of Volume and Osmolality

To provide an optimum environment for the body's cells, the composition, concentration, and volume of the extracellular fluid (ECF) are regulated by a combination of renal, metabolic, and neurologic functions. The ECF is continuously altered and then modified as the body reacts with its surrounding environment. Since the primary constituents of the ECF are water and sodium (and sodium's accompanying anions), their regulation is crucial for maintaining volume and concentration (i.e., osmolality) of the ECF. Regulation of the composition of the ECF depends on the regulation of the individual electrolytes and hydrogen ions (see Chapters 8-13).

REGULATION OF VOLUME

As a rule, large fluctuations can occur in the volume of the interstitial portion of the ECF without markedly affecting body functions. This is especially true if the changes occur slowly. Individuals with cirrhosis, for example, often are able to tolerate significant amounts of ascitic fluid. One exception to this generalization, however, is the expansion of the pulmonary interstitium. Significant increases in the volume of the pulmonary interstitium result in movement of fluid into the alveoli of the lung, causing decreased oxygenation of the blood.

The vascular portion of the ECF is less tolerant of change than the interstitial portion and must be maintained carefully to ensure that the tissues receive an adequate supply of nutrients and continuous removal of metabolic wastes without compromising the cardiovascular system. The kidneys are the primary regulators of volume and are responsible for the daily regulation of vascular volume. Increased intake of fluid causes an increase in cardiac output and blood pressure. This results in an increased excretion of sodium and water by the kidneys. Decreases in blood volume cause a reduction in cardiac output and arterial pressure, leading to the retention of sodium and water by the kidneys. This simple feedback loop is facilitated and modified by the actions of the sympathetic nervous system and several hormones.

Effective Circulating Volume

The portion of the vascular volume that actually perfuses tissue is termed **effective circulating volume (ECV).** ECV usually varies with ECF volume. Sudden loss of blood due to hemorrhage, for example, results in reduced ECF volume and decreased perfusion of the tissues. However, the two volumes can change independently of each other. For example, opposite changes occur in the volumes of the ECF and ECV in congestive heart failure. As the heart fails and is unable to move arterial blood forward, the volume of blood that actually reaches and perfuses the tissues (ECV) falls as blood pools in the venous circuit. Although the overall vascular volume has not changed, the volume of blood delivered to the tissues has decreased. The kidneys respond to this decreased perfusion by conserving sodium and water. Since the heart is unable to pump this added volume effectively, the pressure within the venous circuit increases. With increasing venous pressure, there is a movement of fluid into the interstitial space. Thus both the interstitial and intravascular portions of the ECF have increased, and yet the ECV remains decreased. The physiologic mechanisms involved in these changes will be discussed in subsequent sections.

Since tissue perfusion and cellular function depend on adequate ECV, it is not surprising that multiple physiologic mechanisms monitor and regulate the ECV. Changes in the ECV are sensed by specialized receptors located in the carotid sinuses, aortic arch, and renal vessels. These receptors do not measure total volume, but rather, respond to changes in blood pressure via changes in stretch in the arterial walls. Increases in ECV cause an increase in blood pressure and thus stretch at these receptors. In contrast, a decrease in ECV causes a decrease in pressure and stretch. Changes in pressure sensed by these receptors lead to changes in cardiac output, vascular resistance, thirst, and renal handling of sodium and water. These changes are mediated by a combination of interrelated neurologic and hormonal functions described in subsequent sections.

Sympathetic Nervous System

The sympathetic nervous system provides the initial compensatory response to rapid or short-term changes in the effective circulating volume (i.e., changes in blood pressure). Alterations in ECV lead to changes in the responsiveness of the sympathetic nervous system (sympathetic tone). Decreased ECV, for example, results in stimulation of the sympathetic nervous system (increased sympathetic tone). Increased sympathetic tone in turn leads to changes in cardiac and vascular function that result in increased blood pressure.

Blood pressure is the product of **cardiac output (CO)** multiplied by **systemic vascular resistance (SVR).** Cardiac output is determined by multiplying heart rate (HR) by stroke volume (SV), which is the volume of blood moved with each contraction of the left ventricle.

$$\textbf{Mean arterial pressure (MAP)} = \textbf{CO (HR} \times \textbf{SV)} \times \textbf{SVR}$$

Thus changes in stroke volume, heart rate, or vascular resistance affect blood pressure. Stimulation of the sympathetic nervous system raises blood pressure by causing a direct increase in heart rate, stroke volume, and systemic vascular resistance. In addition to direct action on the heart and blood vessels, stimulation of the sympathetic nervous system acts on the kidneys to increase the release of renin. Increased renin levels also will cause a rise in blood pressure due to the actions of angiotensin and aldosterone. Increases in ECV result in both a decrease in sympathetic tone and an increase in the release of atrial natriuretic factor.

Renin-Angiotensin-Aldosterone

Renin is a proteolytic enzyme produced and released by a specialized group of cells located within the afferent arterioles of the kidney. Renin is released in response to decreased renal perfusion (secondary to a reduction in effective circulating volume) and increased sympathetic nervous system stimulation. It acts on angiotensinogen produced in the liver to create **angiotensin I.** In turn, angiotensin I is converted to **angiotensin II** by an enzyme located in the lungs and other tissues. Angiotensin II is a potent vasoconstrictor and increases in its concentration result in increased blood pressure. Increased angiotensin II levels are in part responsible for the increases in blood pressure that occur in certain disorders associated with elevated renin levels. In renal artery stenosis, for example, decreased perfusion to the affected kidney results in an increased release of renin and the production of angiotensin II, which contribute to the marked hypertension so often observed in this condition. Certain antihypertensive medications (e.g., captopril) work in part by preventing the conversion of angiotensin I to angiotensin II.

Renin also raises blood pressure through the actions of **aldosterone** (Figure 3-1). Aldosterone is a mineralocorticoid hormone released by the adrenal cortex that acts on the distal portion of the renal tubule to increase the reabsorption (saving) of sodium and the secretion and excretion of potassium and hydrogen. Since sodium retention leads to osmotic water retention, aldosterone acts as a volume regulator (Figure 3-2). The release of aldosterone is regulated by several factors, but the most important stimuli to its release are increased renin, angiotensin II, and serum potassium levels. This is logical since aldosterone acts to

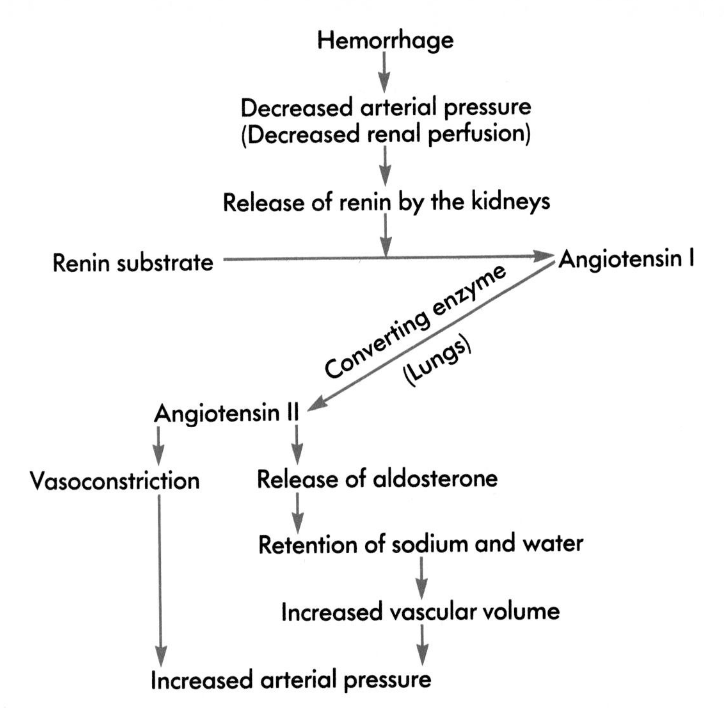

Figure 3-1 Action of the renin-angiotensin-aldosterone system: a clinical example. (From Horne M and Swearingen P: Pocket guide to fluids and electrolytes, St Louis, 1989, The CV Mosby Co.)

restore volume and excrete potassium. Two additional factors that increase the release of aldosterone are decreased plasma sodium levels and increased adrenocorticotropic hormone (ACTH) (Vander, 1985). The actions of the renin-angiotensin-aldosterone system are summarized in Figure 3-1. The diuretic spironolactone (Aldactone) works by blocking the action of aldosterone on the renal tubule; thus it results in the increased excretion of sodium and water and decreased excretion of potassium.

Atrial Natriuretic Factor

Atrial natriuretic factor (ANF), also referred to as *atrial natriuretic peptide,* is a recently identified hormone that is released by the cardiac atria. In contrast to the renin-angiotensin-aldosterone system, ANF acts to reduce blood pressure and vascular volume. Its actions include the following: (1) increased excretion of sodium and water by the kidney secondary to increased filtration, (2) decreased synthesis of renin and decreased release of aldosterone, and (3) direct vasodilatation (Sagnella and others, 1987).

ANF is released in response to any condition that causes an increase in atrial stretch

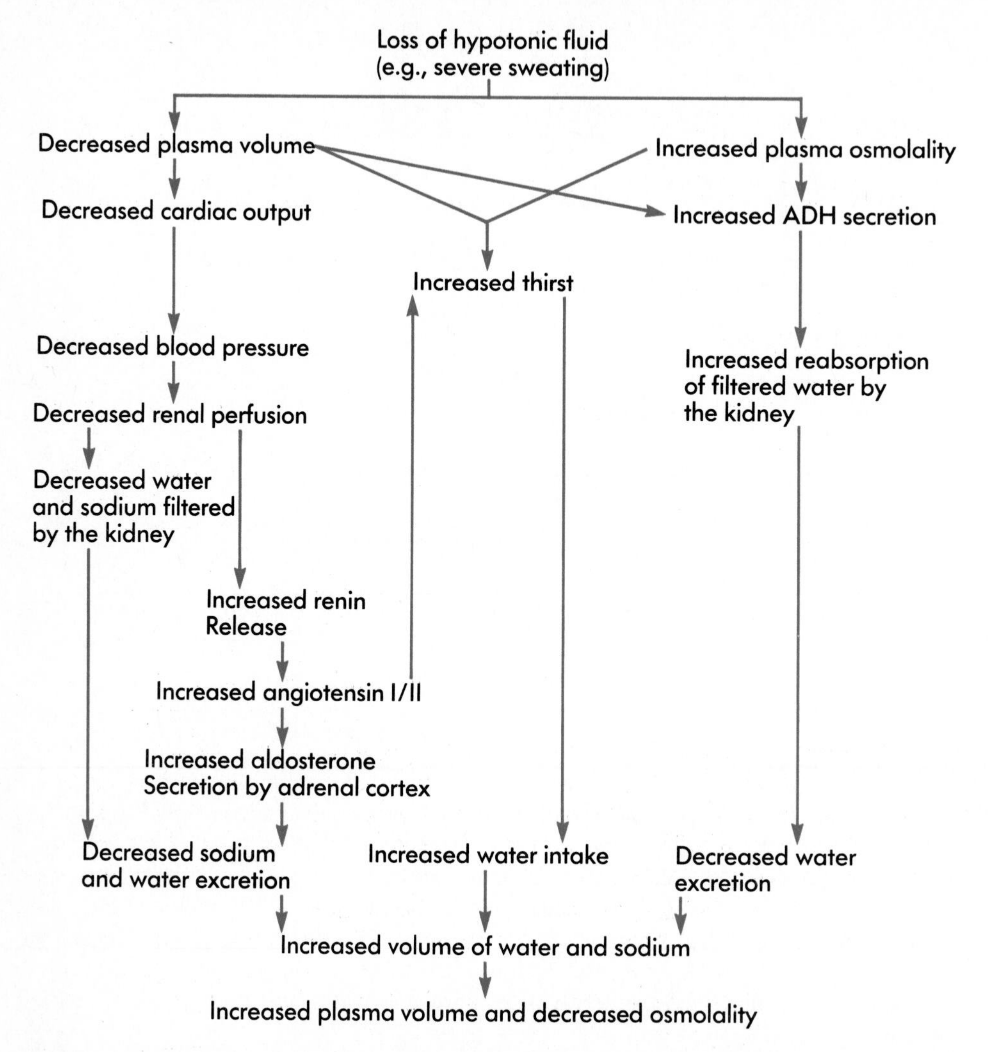

Figure 3-2 Regulation of fluid volume and osmolality: a clinical example.
(From Horne M and Swearingen P: Pocket guide to fluids and electrolytes, St Louis, 1989, The CV Mosby Co.)

or pressure, including volume expansion, vasoconstrictor agents, and atrial tachycardia (Needleman and Greenwald, 1986). If **analogs** (substances with similar structure and function) of ANF can be developed, potentially they may be useful in the management of hypertension, renal failure, and other volume overload states. See Figure 3-3 for a depiction of how ANF acts to reduce vascular volume and blood pressure. Although increased release of ANF has been well documented in acute volume overload, its role in the day-to-day regulation of fluid volume has yet to be identified.

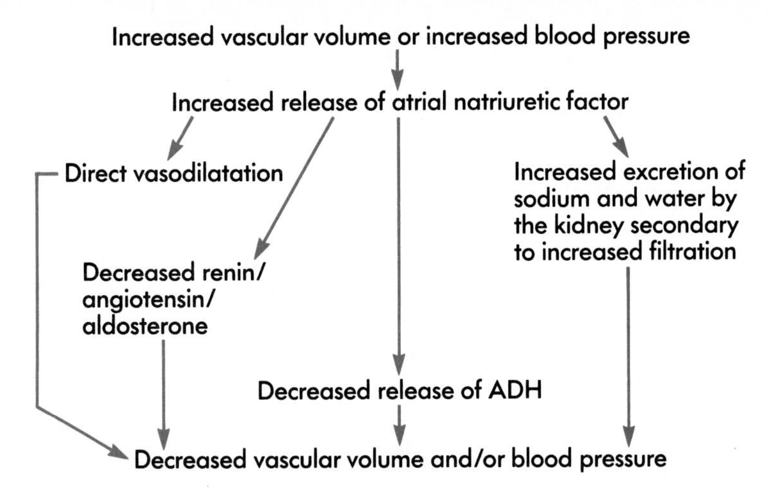

Figure 3-3 Action of the ANF in decreasing vascular volume and blood pressure.
(From Horne M and Swearingen P: Pocket guide to fluids and electrolytes, St Louis, 1989, The CV Mosby Co.)

Renal Regulation of Sodium Excretion

In addition to the hormones discussed above, factors occurring within the kidney will affect urinary excretion of sodium, and thus, vascular volume. Increases in the amount of fluid filtered at the glomerulus will tend to increase the excretion of sodium. Changes in intrarenal hemodynamics and the distribution of blood flow within the kidneys also will affect the reabsorption or excretion of sodium (Rose, 1989). It should be clear at this point that the regulation of volume is closely tied to the excretion of sodium. Factors that increase the reabsorption (saving) of sodium increase the ECF volume, while factors that increase the excretion of sodium decrease the ECF volume.

Antidiuretic Hormone (ADH) and Thirst

Both ADH and thirst assist in the regulation of volume, primarily through the regulation of water. For a discussion of their actions, see the next section.

REGULATION OF ECF OSMOLALITY

As discussed in Chapter 1, osmolality measures the concentration of body fluids (i.e., the ratio of water to solutes). *Changes* in ECF osmolality will affect the movement of water into or out of the cells. Increased ECF osmolality, for example, will cause cells to shrivel, while decreased ECF osmolality will cause cells to swell. Although the total amount of solute and the types of solute found in each body fluid compartment will

$$\textbf{A} \quad \text{ECF osmolality} = 2 \times \text{plasma sodium}$$

$$\textbf{B} \quad \text{ECF osmolality} = 2\,(\text{plasma Na}^+) + \frac{\text{Plasma glucose}}{18} + \frac{\text{BUN}}{2.8}$$

Figure 3-4 A, Rough estimate of ECF osmolality. B, More accurate estimate.

vary greatly, in the steady state the osmolalities of the intracellular fluid (ICF) and the ECF are equal. This occurs because water rapidly crosses all membranes, maintaining a constant water-to-solute ratio between compartments (see Figure 1-4). Thus it is the *changes* in ECF osmolality that affect the movement of water between compartments.

Osmolality is determined by the number of solutes in a solution, not by their relative weight or size. Therefore the primary determinants of the osmolality of a given compartment are the solutes found in the greatest quantity in that compartment. Since sodium and its accompanying anions are the primary solutes of the ECF, they are also the primary determinants of ECF osmolality. The ECF osmolality may be estimated by doubling the serum sodium (remember that plasma is similar in composition to the ECF as a whole and that doubling the sodium takes into account both sodium and its accompanying anions) (Figure 3-4, *A*).

The nonelectrolytes glucose and urea also contribute to ECF osmolality and should be considered for a more accurate estimate of ECF osmolality. Since glucose and urea are measured in weight (mg/dl), their values must be converted to concentration (number of particles) by dividing their weight per liter of solution by their molecular weight. Hence glucose is divided by a factor of 18 and the blood urea nitrogen (BUN) is divided by a factor of 2.8 (Figure 3-4, *B*). This expanded equation gives a close approximation of the serum osmolality as measured in the laboratory. Remember, though, that urea is a small molecule and, like water, it will cross all cell membranes readily. The urea concentration within the ECF and ICF will be equal and urea will have no effect on the movement of water. Thus urea is considered an *ineffective osmole.* Other substances that may raise the plasma osmolality yet not affect the movement of water (and therefore are *ineffective osmoles*) include ethanol, methanol, and ethylene glycol.

As discussed in Chapter 1, *tonicity* is another term for *effective osmolality* (i.e., changes in concentration that will cause water to move from one compartment to another). It is critical that ECF tonicity be maintained within a narrow range to avoid sudden shifts of fluid that might adversely affect cellular function. When red blood cells are added to a markedly hypertonic solution, for example, they shrivel (become crenated). In contrast, when red blood cells are added to a markedly hypotonic fluid, such as pure water, the cells burst (become hemolyzed). Red blood cells retain their normal shape and function when added to isotonic solutions (Figure 3-5). For this reason intravenous fluids have a limited range of tonicity.

Altered tonicity is usually evidenced by changes in central nervous system cell function. Two control systems work together to maintain normal tonicity through the regulation of the solute-to-water ratio: antidiuretic hormone (ADH) and thirst.

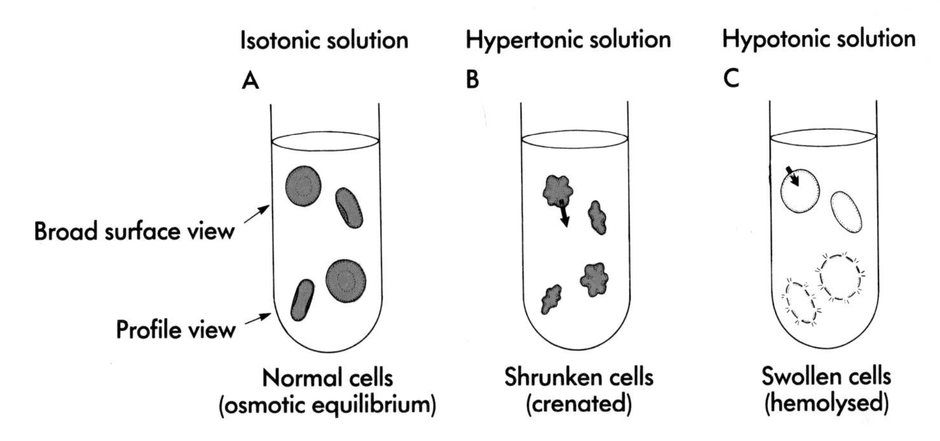

Figure 3-5 Effect of osmotic pressure on the cells.
(From Horne M and Swearingen P: Pocket guide to fluids and electrolytes, St Louis, 1989, The CV Mosby Co.)

Antidiuretic Hormone (ADH)

Antidiuretic hormone (ADH) is a hormone produced by the hypothalamus and secreted into the general circulation by the posterior pituitary gland. It acts on the distal tubule and collecting duct in the kidney to increase the reabsorption (saving) of water and enable the excretion of a concentrated urine. ADH also is an arterial vasoconstrictor that acts to raise blood pressure by increasing vascular resistance; hence its other name, vasopressin. ADH is regulated primarily by changes in plasma osmolality and ECV. Increases in plasma osmolality, as sensed by osmoreceptors located in the hypothalamus, and decreases in ECV both cause an increase in the release of ADH. This is logical since retention of water will cause a drop in plasma osmolality and an increase in volume. Additional factors that increase the release of ADH include hypotension, pain, stress, surgery, and certain medications, such as amitriptyline (Elavil) and morphine (see the accompanying boxes at the bottom of this page and the top of the next).

In addition to medications that affect the release of ADH, there also are medications that suppress or enhance the action of ADH on the renal tubule (see the accompanying

Factors that Increase the Release of ADH

- Increased plasma osmolality
- Decreased effective circulating volume
- Decreased blood pressure
- Stress and pain
- Medications including morphine, amitriptyline (Elavil), and haloperidol (Haldol)
- Surgery and certain anesthetics
- Positive pressure ventilators

Factors that Decrease the Release of ADH

- Decreased plasma osmolality
- Increased effective circulating volume
- Increased blood pressure
- Medications including phenytoin and ethyl alcohol (ETOH)

Medications that Alter the Action of ADH

Suppress	Enhance
lithium	chlorpropamide
demeclocycline	indomethacin
methoxyflurane	

box). ADH also may be produced ectopically by certain types of tumors. This occurs most commonly in oat cell carcinoma of the lung (Rose, 1989). Conditions in which there is an increased hypothalamic release, enhanced action, or ectopic production of ADH result in inappropriate water retention and are termed **syndrome of inappropriate antidiuretic hormone (SIADH).** SIADH is typified by the finding of a decreased serum sodium combined with an inappropriately concentrated urine. A deficiency in the synthesis or release of ADH or a decrease in the kidneys' responsiveness to ADH will result in decreased water reabsorption by the kidneys. This condition is termed diabetes insipidus and is typified by the production of an abnormally large quantity of very dilute urine.

Thirst

Along with ADH, thirst acts to regulate the ECF concentration and is stimulated essentially by the same factors that increase the release of ADH: increased plasma osmolality, volume depletion, and hypotension. Increased angiotensin II levels and dry mucous membranes (the sensation of a dry mouth) also stimulate thirst. Thirst is not as carefully regulated as ADH since it is affected strongly by social and emotional factors. However, thirst does provide the primary protection against hyperosmolality (an elevated osmolality). Symptomatic hyperosmolality only occurs in individuals who do not have a normal thirst mechanism or who do not have adequate access to water. Thus hyperosmolality typically occurs in infants or comatose patients who are unable to ask for water. Alert individuals with diabetes insipidus, for example, who excrete a large amount of abnormally dilute urine due to altered ADH function will maintain a relatively normal osmolality and volume as long as they are able to drink and satisfy their thirst. See Figure 3-2 for a clinical example depicting the actions of ADH and thirst in the regulation of plasma osmolality.

ABNORMALITIES OF ECF OSMOLALITY AND TONICITY

Changes in ECF osmolality may occur with or without changes in ECF tonicity. Increases or decreases in *effective osmoles* (e.g., sodium, glucose) will cause changes in both osmolality and tonicity. Increases or decreases in *ineffective osmoles* (e.g., urea, ethanol) will cause changes in the osmolality without changing tonicity. Thus individuals with elevated serum osmolalities due to retention of urea or ingestion of ethanol will not exhibit the neurologic symptoms of hypertonicity, although they may exhibit neurologic symptoms related to the toxicity caused by the retention of urea or ingestion of ethanol. In contrast, individuals with elevated serum osmolality secondary to increased serum sodium levels or hyperglycemia (elevated blood glucose) may exhibit the neurologic symptoms of hypertonicity. The severity of the signs and symptoms of hypotonicity or hypertonicity will depend on how much and how quickly the serum osmolality has increased or decreased. Symptoms may range from mild confusion to frank coma and seizures. Clinical hypotonicity occurs when there is an abnormal gain in water or loss of sodium-rich fluids with replacement by water only. Clinical hypertonicity may develop because of the loss of water (e.g., diabetes insipidus), loss of hypotonic body fluids (e.g., sweating, diarrhea), or gain of effective solutes (e.g., hyperglycemia or administration of hypertonic NaCl, $NaHCO_3$, or mannitol). See Chapter 8 for additional information and a complete discussion of the signs and symptoms of hyponatremia (decreased serum sodium) and hypernatremia (elevated serum sodium), the most common causes of altered ECF tonicity.

Case Study 3-1

M.B. is a 20-year-old comatose woman admitted to the intensive care unit with a medical diagnosis of ketoacidosis secondary to Type I (insulin-dependent) diabetes mellitus. She has been a known diabetic for approximately 6 months, and her family informs you that she has had a great deal of difficulty coping with the diagnosis and necessary treatment. Three days ago she broke up with her steady boyfriend of 2 years. M.B.'s mother states that she is relieved that the relationship has broken up since she feels that marriage and childbearing no longer are appropriate for M.B. but admits that M.B. has been very depressed since the breakup. She also states that despite constant thirst and hunger, M.B. seems to be losing weight.

Ketoacidosis may develop in the individual with Type I (insulin-dependent) diabetes secondary to an absolute lack of insulin (e.g., insulin is not administered) or to a relative lack of insulin (e.g., physical or emotional stress has caused an increased need for insulin). Infection is the most common precipitating factor.

M.B.'s admission laboratory work shows the following, with normal values in parentheses:

Serum glucose 643 mg/dl (65-110 mg/dl)
Serum acetone 4+ (negative)
Serum sodium 142 mEq/L (137-147 mEq/L)
Serum potassium 5.6 mEq/L (3.5-5.0 mEq/L)
Serum chloride 99 mEq/L (95-108 mEq/L)
BUN 30 mg/dl (6-20 mg/dl)
Serum osmolality 340 mEq/L (280-300 mEq/L)
WBCs 12,000 μl (4,500-11,000 μl)
M.B.'s admission vital signs:
BP: 90/54 mm Hg
Heart rate: 124 bpm
Respirations: 28 breaths/min
Rectal temperature: 36.8° C (98.4° F)

1. Question: At the time of admission, the exact cause of M.B.'s diabetic ketoacidosis is unknown. Based on her history, what are some potential causes?

Answer:

a. Emotional stress secondary to breakup with boyfriend, leading to an increased need for insulin.
b. Noncompliance with diet and insulin.
c. Infection, as suggested by elevated WBCs.

2. Question: Identify at least two nursing diagnoses that may relate to these potential causes.

Answer:

a. **Knowledge deficit:** Importance of diet, insulin treatment, and minimizing emotional stress
b. **Noncompliance** with diet and insulin treatment
c. **Ineffective individual coping** related to severity of diagnosis, daily insulin therapy, and breakup with boyfriend
d. **Potential for infection** related to increased risk secondary to altered glucose metabolism
e. **Altered role performance** related to perceived inability to marry and have a family
f. **Ineffective family coping** related to serious illness of family member

Although M.B. exhibits no obvious signs of infection (e.g., she is afebrile), she does have an elevated WBC count; therefore routine blood, urine, and sputum cultures are obtained. Infection may be masked in ketoacidosis because of the absence of fever. Insulin deficiency in the Type I diabetic leads to hyperglycemia because of decreased glucose utilization by the cells and increased hepatic glucose production, as well as to ketonemia (increased ketones in the blood) due to altered fat metabolism with increased production of ketones (acetone, acetoacetic acid, and β-hydroxybuturic acid). Hyperglycemia and ketonemia in turn lead to a combination of potentially life-threatening fluid and electrolyte imbalances.

3. **Question:** What are the primary determinants of serum osmolality?

   Answer:

a. Sodium and its accompanying anions
b. Glucose
c. Urea

4. **Question:** If the serum glucose increases, what will happen to serum osmolality?

   Answer: It will increase.

5. **Question:** If serum osmolality increases, what will happen to the ICF?

   Answer: There will be a shift of fluid out of the cells. If this continues long enough, significant intracellular dehydration will develop.

6. **Question:** As fluid shifts out of the ICF and into the IVF, what will happen to the serum sodium?

   Answer: It will drop. Note that there is no change in total body sodium, but rather a change in sodium-to-water ratio in the IVF. This is termed *factitious or pseudohyponatremia* (hyponatremia is low sodium) and for every 100 mg/dl increase in serum glucose above normal there may be a 10% drop in serum sodium. For additional information on causes of pseudohyponatremia, see Chapter 8. Since M.B. is not eating, she will also suffer an actual loss of sodium secondary to increased urinary losses.

Although there is a shift of fluid into the IVF, IVF overload does not develop since the presence of ketones and excess glucose in the urine causes an osmotic diuresis with the loss of large quantities of water and electrolytes in the urine. The DKA patient's symptoms usually reflect a severe fluid volume deficit.

7. **Question:** Identify the two physiologic factors that caused M.B. to be thirsty.

   Answer:

a. Increased serum osmolality secondary to increased glucose and loss of water in the urine.
b. Decreased vascular volume.

   NOTE: It is important that all diabetic individuals be taught to recognize increased thirst as an early symptom of hyperglycemia.

Although large quantities of water and electrolytes are lost with an osmotic diuresis, water is lost in excess of electrolytes; that is, the patient experiences a hypotonic fluid loss. As dehydration progresses, the serum sodium begins to increase. The serum sodium level reflects the ratio of sodium to water in the serum, not necessarily total body stores. Thus any condition that results in a loss of water in excess of sodium will result in hypernatremia (elevated serum sodium). This hypernatremia develops even though there may be a loss of sodium from the body.

8. Question: If water is lost in excess of electrolytes, what will happen to serum osmolality?

Answer: It will increase. Thus a combination of hyperglycemia *and* increased water loss may cause hyperosmolality to develop in the individual with DKA. Changes in serum osmolality, rather than in blood glucose levels, correlates best with the neurologic changes that occur with DKA. In other words, the individual with a blood glucose of 400 mg/dl and a fairly normal serum osmolality is less likely to exhibit neurologic symptoms than the individual with the same blood glucose and a significantly elevated serum osmolality. Acidosis secondary to the retention of ketoacids also will contribute to impaired consciousness.

9. Question: How does M.B.'s estimated (calculated) osmolality compare with her measured serum osmolality?

Answer: Her calculated osmolality is 330.4, approximately 10 mOsm less than her measured osmolality. Remember that the serum osmolality may be estimated using the following formula:

$$\textbf{Calculated osmolality equals} = 2(Na^+) + \frac{\text{glucose}}{18} + \frac{\text{BUN}}{2.8}$$

$$\begin{aligned}\textbf{M.B.'s calculated osmolality} &= 2(142) + 643/18 + 30/2.8 \\ &= 284 + 35.7 + 10.7 \\ &= 330.4\end{aligned}$$

The difference between the calculated osmolality and the actual measured osmolality is usually approximately 10 mOsm and reflects the other solutes in the plasma (e.g., cations other than sodium).

M.B. shows signs and symptoms of severe dehydration. Although low-dose IV insulin will be started at once, the immediate goal of treatment is to restore intravascular volume and prevent hypovolemic shock. Shock leads to an increased release of the hormones that raise blood sugar (e.g., catecholamines) and contributes to the development of acidosis secondary to the production of lactic acid. Prolonged hypotension also may lead to the development of acute renal failure.

10. Question: M.B.'s severe dehydration will result in a reduction in ECV. What is the body's first compensatory response to a sudden reduction in ECV?

Answer: Increased sympathetic nervous system discharge. M.B.'s rapid heart rate indicates increased sympathetic tone.

11. Question: What additional hormone systems will act to increase ECV, helping M.B. compensate for her loss of volume?

Answer:

a. The renin-angiotensin-aldosterone system. Recall that renin is released in response to both sympathetic nervous system stimulation and a reduction in effective circulating volume (decreased renal perfusion). It acts on a substrate to increase the level of angiotensin, a potent vasoconstrictor that in turn increases the release of aldosterone by the adrenal gland. Aldosterone increases the conservation of sodium and water by the kidneys, thus increasing volume.

b. ADH, which is released in response to an increase in blood osmolality and a decrease in effective circulating volume. ADH acts on the renal tubule to decrease urinary excretion of water.

12. Question: Since M.B. has experienced a hypotonic fluid loss, what type of fluid therapy would you expect her to receive?

Answer: Hypotonic fluid therapy.

Although the fluid loss is hypotonic, isotonic sodium chloride (*normal saline*) may be used initially to replace intravascular volume as quickly as possible and prevent too rapid a fall in serum osmolality. Remember that the serum osmolality has increased due to hyperglycemia and water loss. As insulin is administered, the blood sugar will decrease. If water is replaced rapidly (via hypotonic fluids) simultaneous with the insulin administration, there can be a dramatic drop in serum osmolality. This can lead to a sudden shift of fluid back into the cells with resultant cerebral edema. After initial replacement with normal saline, M.B. will be given 0.45 NaCl, which better approximates the concentration of her deficit. When her blood sugar approaches 250 to 300 mg/dl, she will be given dextrose-containing solutions to prevent development of hypoglycemia.

Since M.B. has clinical signs of hypertonicity and severe fluid volume deficit, the physician has prescribed 2 L of normal saline to be administered at a rate of 300 ml/hr, followed by 0.45 NaCl at 300 ml/hr until her heart rate decreases, BP increases, and urine output approaches normal. An important nursing function will be to monitor M.B. for signs of fluid volume excess secondary to excessive or too rapid fluid replacement.

M.B. is at risk for developing several electrolyte and acid-base imbalances. Discussion of her case will be continued in Chapter 11.

Case Study 3-2

L.V. is a 27-year-old woman admitted to the hospital because of persistant severe headaches and progressive lethargy. She has a tentative diagnosis of encephalitis (inflammation of the brain) with an unknown cause. A brain biopsy is performed in an attempt to identify a treatable cause of her encephalitis. Despite supportive therapy, her condition rapidly deteriorates, she develops the signs and symptoms of increased intracranial pressure, and she suffers a cardiopulmonary arrest. L.V. is successfully resuscitated, but her neurologic status has deteriorated, and she is transferred to the medical intensive care unit for close monitoring and ventilatory support. Twenty-four hours after transfer, L.V.'s urine output suddenly increases from an average of approximately 80 ml/hr to over 500 ml/hr. Her urine is no longer straw colored, but now looks like pure water. A stat serum electrolyte panel reveals a serum sodium of 142 mEq/L (normal is 137-147 mEq/L). This is an increase from the value obtained earlier that day (134 mEq/L). A diagnosis of diabetes insipidus is now added to her problem list.

1. **Question:** Diabetes insipidus occurs as a result of a deficiency of which hormone?

   Answer: ADH—antidiuretc hormone.

2. **Question:** What are the actions of ADH?

   Answer: ADH increases the reabsorption of water by the renal tubule and permits the excretion of a concentrated urine. A lack of ADH results in the excretion of an inappropriately dilute (watery) urine.

3. **Question:** From where is ADH produced and released?

   Answer: It is produced in the hypothalamus and released from the posterior pituitary gland. Damage to either of these structures or the tract that connects them may result in a decrease in available ADH. L.V.'s diabetes insipidus is most likely the result of anoxic brain damage suffered during her cardiopulmonary arrest.

4. **Question:** What will happen to the plasma sodium and plasma osmolality if an inappropriately large amount of water is lost in the urine?

   Answer: Their values will increase. The total body sodium content has not changed. However, the volume of water that it is dissolved in has decreased. If L.V. were not treated for her diabetes insipidus, her large urine output would continue and her serum sodium would increase further. L.V.'s medical orders were changed to include vasopressin (Pitressin), 5 units IM every 8 hours. Vasopressin is an aqueous solution of the hormone ADH. Its duration of action is only 2 to 8 hours and it is given IM or SC. Vasopressin tannate (Pitressin Tannate in oil) is a longer acting preparation that may be given IM. In addition to its renal actions, vasopressin causes vasoconstriction with decreased blood flow to splanchnic (internal organs), coronary, and gastrointestinal beds; increases motility of the bowel; and in high doses may cause uterine contractions. Nursing care for L.V. should include careful monitoring of I&O, vital signs, and electrocardiogram.

5. Question: What are some of the potential nursing diagnoses for L.V.?

Answer:

a. **Fluid volume deficit** related to increased urinary losses secondary to diabetes insipidus
b. **Potential fluid volume excess** related to fluid retention secondary to vasopressin therapy
c. **Altered tissue perfusion:** Cardiopulmonary, renal, gastrointestinal, and peripheral, related to vasoconstriction secondary to vasopressin therapy
d. **Potential for decreased cardiac output** related to decreased vascular volume secondary to increased urinary losses

References

Arieff AI and De Fronzo RA: Fluid, electrolyte, and acid-base disorders, New York, 1985, Churchill Livingstone, Inc.

Cogan MG: Atrial natriuretic factor, West J Med 144:591-595, May 1986.

Cowley AW and others: Osmoregulation during high salt intake: relative importance of drinking and vasopressin secretion, Am J Physiol 25:878-886, 1986.

Geheb MA: Clinical approach to the hyperosmolar patient, Crit Care Clin 5:797-815, 1987.

Goldberger E: A primer of water, electrolyte, and acid-base syndromes, ed 7, Philadelphia, 1986, Lea & Febiger.

Groer MW: Physiology and pathophysiology of the body fluids, St Louis, 1981, The CV Mosby Co.

Guyton AC: Textbook of medical physiology, ed 7, Philadelphia, 1986, WB Saunders Co.

Hartshorn J and Hartshorn E: Vasopressin in the treatment of diabetes insipidus, J Neurosci Nurs 20:58-59, 1988.

Horne M and Swearingen P: Pocket guide to fluids and electrolytes, St Louis, 1989, The CV Mosby Co.

Keyes JL: Fluid, electrolyte, and acid-base regulation, Monterey, Calif, 1985, Wadsworth, Inc.

Kershner D: Endocrinologic dysfunctions. In Swearingen PL and others: Manual of critical care: applying nursing diagnoses to adult critical illness, St Louis, 1988, The CV Mosby Co.

Kruse AP: Atrial natriuretic factors, Prog Cardiovasc Nurs 3:39-44, 1988.

Maxwell MH and others: Clinical disorders of fluid and electrolyte metabolism, ed 4, New York, 1987, McGraw-Hill, Inc.

Moses AM and others: Acid-base and electrolyte disorders associated with endocrine disease: pituitary and thyroid. In Arieff AI and De Fronzo RA: Fluid, electrolyte, and acid-base disorders, New York, 1985, Churchill Livingstone, Inc.

Needleman P and Greenwald JE: Atriopeptin: a cardiac hormone intimately involved in fluid, electrolyte, and blood pressure homeostasis, N Engl J Med 314:828-834, 1986.

Rose BD: Clinical physiology of acid-base and electrolyte disorders, ed 3, New York, 1989, McGraw-Hill, Inc.

Sagnella GA and others: Plasma atrial natriuretic peptide: its relationship to changes in sodium intake, plasma renin activity, and aldosterone in man, Clin Sci 72:25-30, 1987.

Smith K: Fluids and electrolytes: a conceptual approach, New York, 1980, Churchill Livingstone, Inc.

Vander AJ: Renal physiology, ed 3, New York, 1985, McGraw-Hill, Inc.

Vokes TJ and Robertson GL: Disorders of antidiuretic hormone, Endocrinol Metab Clin North Am 17(2):281-299, 1988.

Williams GH and Dlughy RG: Hypertensive states: associated fluid and electrolyte imbalances. In Maxwell MH and others: Clinical disorders of fluid and electrolyte metabolism, ed 4, New York, 1987, McGraw-Hill, Inc.

Wright M: Physiology for general practitioners, II: Cardiovascular and renal systems, Family Pract 5(2):145-153, 1988.

Chapter

4

Fluid Gains and Losses

In health there is a steady state or balance between the fluids gained and lost by the body. As discussed in Chapters 2 and 3, the volume, concentration, and composition of body fluids are regulated so that output matches intake and balance is maintained. Loss of hypotonic fluids, for example, leads to decreased water excretion and increased thirst. Fluid imbalance occurs when the body is unable to compensate for normal or abnormal fluid losses or gains. The bedside nurse is in a unique position to identify individuals at risk for developing fluid imbalance through careful documentation of all intake and output. This chapter will review the means of fluid gain and loss, while Chapter 5 will discuss specific nursing procedures related to documenting intake and output.

FLUID GAINS

Oxidative Metabolism

In the adult approximately 300 ml of water are produced daily by the oxidation of carbohydrates, proteins, and fat; that is, oxygen combines with some of the hydrogen in these substances to produce water. Endogenous production of water may increase to as much as 800 ml/day in the individual who is hypercatabolic, e.g., the person with fever. In the infant and child, endogenous water production is estimated at 20 ml for every kcal that is metabolized (Siegel and Lattanzi, 1985). The amount of water produced by oxidative metabolism and the water released with the breakdown of cells is insufficient to compensate for the body's obligatory fluid losses; thus some additional oral, parenteral, or enteral intake is necessary to maintain body fluid. Under the best of conditions individuals may survive weeks without food intake but only days without water intake.

Oral Fluids

Approximately 1,100 to 1,400 ml of fluid are consumed orally per day. Fluid intake varies greatly since thirst is not accurately regulated in humans and is affected by social and emotional, as well as physiologic, factors (see Chapter 3).

Solid Food

Fluid is gained through the consumption of solid food, which provides approximately 800 to 1,000 ml of water each day. Meat, for example, is approximately 70% water and fruits and vegetables are over 90% water by weight.

Fluid Therapy

Fluid also may be gained through parenteral or enteral routes and by means of irrigants that are retained. If, for example, a nasogastric (NG) tube is irrigated and an equal amount is not withdrawn and discarded, the extra irrigant must be considered a fluid gain. Fluids also may be gained through medications or enemas. The practice of administering repeated tap water enemas for bowel cleansing, for example, can result in significant water gain. Newborn infants in critical care settings may gain small but clinically significant amounts of fluid from bronchial lavage and flushing of arterial catheters. Likewise, adults in critical care settings may gain fluid through pressure monitoring systems that provide a small but continuous delivery of fluid. Frequent flushing of these devices provides additional fluid that is difficult to quantify. These volumes may be minimized, though, by limiting the duration and frequency of flushes. Determining the volume of a given fluid therapy also requires consideration of all sources of fluid gained, e.g., intravenous fluids, tube feedings, medications, flushes, and irrigants. See Chapter 7 for a discussion of fluid therapy.

FLUID LOSSES

Kidneys

The kidneys are the primary regulators of fluid and electrolyte balance. Approximately 180 liters of plasma are filtered daily by the adult kidneys. From this volume, approximately 1,500 ml of urine are produced and excreted each day. Hourly urine output has an *average range* of 40 to 80 ml in the adult and 0.5 to 2 ml/kg/hr in the child (Hazinski, 1988). The volume, composition, and concentration of urine varies greatly and will depend on intake and other fluid losses. Urine values (volume and concentration) always should be evaluated in relation to the body's need to conserve or excrete fluid. The dehydrated individual who needs to conserve fluid, for example, would be expected to excrete less urine than the individual who is adequately hydrated.

The concentration of urine may range from 50 to 1,400 mOsm/kg, although the average is approximately 500 to 800 mOsm/kg in the adult. In contrast, infants and small children are only capable of maximally concentrating their urine to approximately 400 to 600 mOsm/L. Renal concentrating ability also decreases with advancing age. Limited urinary concentrating ability in infants and the elderly means that they are less capable of conserving water in response to water deprivation and thus are at increased risk for the development of fluid volume deficit. While sodium is the primary determinant of ECF osmolality or concentration, metabolic wastes are the primary determinant of urinary osmolality or concentration. Therefore in severe hypovolemia or hypotension, for example, the kidneys are able to excrete a concentrated, yet relatively sodium-free, urine.

Oliguria

At maximal urinary concentration, approximately 400 ml of urine must be produced to excrete the daily load of metabolic waste. A urinary output of less than 400 ml/24 hr in the adult is termed **oliguria,** and it signals the retention of metabolic wastes. Infants, the elderly, and individuals with reduced renal function who cannot maximally concentrate their urine will have greater obligatory water losses—that is, they will need to produce a proportionately larger volume of urine to excrete their daily load of metabolic wastes. The excretion of less than 30 ml of urine an hour for 2 consecutive hours suggests the development of oliguria.

While oliguria does indicate renal dysfunction (either due to actual renal damage, decreased renal perfusion, or obstruction to urinary flow), a normal volume of urine does not rule it out. Under most conditions urine is not concentrated to the physiologic maximum, and a urine output of greater than 400 ml does not necessarily indicate adequate excretion of metabolic wastes. Some forms of renal failure, for example, are accompanied by normal or even increased volumes of urine. Remember that normally 180 L of plasma are filtered to produce only 1½ L of urine. Severely diseased kidneys may filter only a few liters of plasma and still produce this amount.

Polyuria and anuria

Polyuria is an abnormally large urine output that may occur in renal disease, after the release of a urinary tract obstruction, with the administration of diuretics, or with urinary excretion of substances such as glucose or contrast media that induce an osmotic diuresis. **Anuria** is the production of less than 100 ml of urine in 24 hours. While it may occur with intrarenal failure, it may also indicate urinary tract obstruction. Polyruia alternating with oliguria or anuria also suggests possible urinary tract obstruction.

Skin

An average of 500 to 600 ml of sensible and insensible fluid are lost via the skin each day. Intact skin provides an important barrier to the loss of ECF. Altered skin integrity can result in dramatic loss of fluid when large areas of skin are damaged, as in major burns.

Insensible fluid

Insensible fluid loss is evaporative and occurs without the individual's awareness. It should be considered pure water loss, since it is nearly electrolyte free. Insensible water is lost at a rate of approximately 6 ml/kg/24 hr in the average adult, but it can increase significantly with fever or burns. Use of phototherapy or radiant warmers for infants also will increase insensible water loss secondary to increased body heat.

Sensible fluid

Sensible fluid (i.e., sweat) is important in dissipating body heat, and, like insensible fluid, it is hypotonic. Sensible fluid, however, does contain a significant amount of electrolytes (see Table 1-5). The rate of sensible fluid loss varies greatly with the individual's activity level and the ambient temperature. In extreme cases sensible fluid loss may be as great as 2 L in an hour.

Lungs

Approximately 400 ml of insensible fluid are lost through the lungs each day. This amount may increase somewhat with increased respiratory depth or dry climate. Water loss from the lungs may be reduced in individuals breathing humidified air. The individual breathing humidified air via mechanical ventilation may have no insensible fluid loss from the lungs. Abnormal production of respiratory secretions, as occurs with respiratory tract infections, is another possible source of abnormal fluid loss.

The total volume of insensible fluid loss from both the skin and lungs may be estimated at 10 ml/kg/24 hr in the adult. Note that the average amount of fluid lost from the skin and lungs is equal to approximately the same amount as that gained in solid food (800 to 1,000 ml). Cox (1987) suggests that insensible fluid loss need not be considered when determining the fluid requirements for the individual on a normal diet. However, insensible fluid loss should be considered in determining the fluid requirements of persons receiving enteral or parenteral fluids.

Gastrointestinal (GI) Tract

The GI tract plays an important role in maintaining fluid and electrolyte balance since in health it is the primary site of fluid and electrolyte gain. As described, approximately 2 L of fluid are gained each day through the consumption of fluids and solid food. Additionally, 6 to 8 L of fluid are secreted into and reabsorbed out of the GI tract daily, equaling approximately half of the extracellular fluid (ECF) volume. Despite the large volume of fluid that moves through it daily, in health the GI tract contributes minimally to normal daily fluid loss (approximately 100 to 200 ml in the adult and older child, 50 to 75 ml in the preschooler, and 5 ml/kg/day in the infant). In disease, however, the GI tract becomes the most common site of abnormal fluid and electrolyte loss, which may result in profound fluid and electrolyte imbalance.

The composition of GI fluid will vary, depending on its location within the GI tract (Table 4-1). Like the nephron, the GI tract can be divided into a series of functional sections that secrete and reabsorb water and electrolytes in varying amounts. Thus the nature of the fluid and electrolyte imbalance that occurs with abnormal GI loss will depend on the site of fluid loss. For simplicity, GI disorders causing fluid and electrolyte loss may be differentiated into loss of upper GI fluids and loss of lower GI fluids. Above the pylorus, the losses are isotonic and rich in sodium, potassium, chloride, and hydrogen. Below the pylorus, losses are isotonic and rich in sodium, potassium, and bicarbonate. Losses from the large intestine are hypotonic. See Appendix B for additional information.

Upper GI fluids

Upper GI fluids include saliva and gastric juices. Approximately 1 liter of saliva is produced each day by the sublingual, submandibular, and parotid glands. Saliva participates in the digestive process by initiating the breakdown of starches. Abnormal loss of saliva may occur in individuals who are unable to swallow their oral secretions (e.g., those who are comatose). Food and saliva move through the esophagus to the stomach by means of peristalsis. Once in the stomach, the food and saliva are exposed to acidic gastric fluid. Approximately 1.5 to 2.5 L of fluid are produced daily by the stomach. Gastric juices contain hydrochloric acid (HCl), which aids digestion by breaking down food. Because of the acidic content of gastric fluid, abnormal losses may result in the development of **alkalosis** (see Chapter 18).

Only alcohol and a limited amount of water are absorbed from the stomach. Most of the stomach contents move by peristalsis into the small intestine. Although the fluid and electrolyte content of food is variable, gastric contents mix and become similar in concentration to the ECF, owing to osmotic shifts of water into and out of the stomach. Loss of upper GI fluids occurs primarily from problems such as vomiting and procedures such as nasogastric suction. See accompanying box for potential causes of vomiting.

Gastric suction

Gastric suction, whether via a nasogastric or orogastric tube, is a common medical procedure used to decompress the stomach. Removal of gastric contents can lead to multiple fluid and electrolyte disturbances and requires adequate parenteral replacement. This may be accomplished by administering an IV solution developed specifically for gastric replacement, such as Isolyte G, manufactured by McGaw, or

Table 4-1 Volume and composition of gastrointestinal secretions

Secretion	$L/24^0$	Na^+ mEq/L	K^+ mEq/L	Cl^- mEq/L	HCO_3^- mEq/L
Saliva	1	40	15	30	0
Gastric juice	1.5-2.5	40	7	100	0
Pancreatic juice	1-2	130	7	60	100
Bile	0.5	150	7	80	30
Intestinal secretions	2-3	140	5	variable	variable

Values are approximate.

Potential Causes of Vomiting

- Gastrointestinal infection
- Gastritis
- Inner ear infection or disorder
- Certain medications (e.g., chemotherapy)
- Pregnancy
- Small bowel obstruction
- Pyloric stenosis
- Uremia
- Binge-purge syndrome
- Pancreatitis
- Hepatitis
- Diabetic ketoacidosis

customizing a solution to meet the individual's needs based on serum electrolyte levels. See accompanying box for nursing considerations that are important for minimizing electrolyte imbalance with gastric suction.

Lower GI fluids

Lower GI fluids include bile, pancreatic juice, and intestinal secretions. Pancreatic juice is high in bicarbonate, which neutralizes the acidic gastric contents as they enter the duodenum. Pancreatic juice also contains enzymes that aid in the digestion of protein (trypsin), starches (amylase), and fats (lipase). Approximately 1 L of bile is released by the gallbladder into the duodenum each day. Bile provides a means of excreting bilirubin (a breakdown product of hemoglobin) and aids in the digestion of fats via emulsification by bile salts. Abnormal loss of bile occurs with drainage of the common bile duct after gallbladder surgery.

In addition to bile and pancreatic juice, the intestines contain secretions produced by the intestinal glands. These glands secrete mucus, which helps protect the intestinal

Nursing Considerations for Suctioning Gastric Contents

1. Keep the individual NPO.
2. Avoid giving ice by mouth or irrigating the catheter with plain water since these actions will increase the loss of electrolytes due to *wash out.* If patients are allowed ice chips (by physician prescription), give small amounts (less than 1 oz) hourly. The physician may prescribe ice chips made from specific electrolyte solutions.
3. Provide frequent oral care and lip balm to minimize thirst and maximize comfort.
4. Irrigate gastric tube with normal saline (isotonic NaCl solution) only. If the catheter is irrigated and an equal amount is not withdrawn and discarded, the extra irrigant should be added to the intake record to avoid overestimation of net fluid loss.
5. Measure and record all sources of intake and output separately. An accurate record is essential for determining parenteral fluid replacement and avoiding fluid and electrolyte imbalance.
6. Obtain daily weight measurements on all individuals with unusual fluid losses or gains. Changes in daily weights may provide an early clue to fluid imbalance.

mucosa, hormones (e.g., secretin), electrolytes, and digestive enzymes. Approximately 2 to 3 L of intestinal secretions are produced each day. The secretion of isotonic intestinal fluids may increase dramatically in certain diseases such as cholera and other intestinal infections, after administration of certain medications such as laxatives, and with bowel obstruction (see Appendix B).

The intestines are the primary site of nutrient and water absorption. The majority of the GI contents are absorbed in the small intestines. Approximately 8 L of fluid enter the small intestines daily, of which 75% is reabsorbed (and returned to the ECF). The colon receives only 1 to 2 L of fluid from the ileum; of this, normally all but 100 to 200 ml of water and a small quantity of electrolytes are absorbed, creating formed stool.

Abnormal losses from the small intestines of as much as 2,000 ml/day may occur with the creation of a new ileostomy (Smith and Johnson, 1986), or with short bowel syndrome. Over time, ileostomy losses normally decrease to only 300 to 500 ml/day. However, ileostomy losses of water and electrolytes remain greater than with normal stool. Fistulas (abnormal passage between the bowel and skin) also may result in the loss of several liters of intestinal fluid. Diarrhea, however, is the most common cause of lower GI fluid loss, especially in children. In developed countries diarrhea accounts for a significant percentage of pediatric hospital admissions and clinic visits. In underdeveloped nations, diarrhea is a major cause of infant deaths (see accompanying box for potential causes of diarrhea and Appendix B for further discussion of diarrhea).

Vomiting can contribute to lower GI fluid loss since both gastric and duodenal fluid may be lost. Losses from both the upper and lower GI tract also can occur with bowel obstruction. Acid-base balance is usually maintained when both gastric and duodenal contents are lost, since a loss of both hydrogen (acid) and **bicarbonate** (alkali) occurs.

Potential Causes of Diarrhea

Osmotic Diarrhea

- Certain medications (e.g., lactulose, sorbitol)
- Malabsorption or maldigestion syndromes

Secretory Diarrhea

- GI infection
- Inflammatory bowel disease
- Emotional stress
- Pancreatic insufficiency
- Intestinal obstruction
- Abuse of laxatives (e.g., bisacodyl, castor oil)
- Carcinoma

Although the absorption of certain electrolytes (e.g., calcium, phosphorus, and magnesium) is in part hormonally regulated, the amount of fluid, electrolytes, and other nutrients absorbed is largely dependent on the quantity presented to the GI tract. The greater the intake of water and electrolytes, the greater the absorption and retention.

Additional Losses

Significant amounts of fluid may be lost through draining wounds and fistulas, increased evaporative loss from large open wounds, or external bleeding.

Third-space Losses

The loss of ECF into a normally nonequilibrating space is termed **third-space fluid shift.** Although this fluid has not been lost from the body, it is temporarily unavailable to the ICF or ECF for its use. Third-space fluid losses must be considered when evaluating the adequacy of fluid therapy. Some examples of third-space fluid shifts include internal bleeding, accumulation of fluid in the bowel after acute bowel obstruction, pleural effusion, pericardial effusion, accumulation of fluid in the peritoneal cavity with acute peritonitis, and plasma-to-interstitial fluid shift during the first 2 to 3 days following a burn injury. Third-space fluid shifts also may occur after surgical or traumatic tissue injury. Tissue injury causes a temporary increase in capillary permeability, with loss of fluid and proteins into the interstitial fluid space. There is usually a return to normal capillary permeability at about 48 to 72 hours and remobilization of fluid. See Table 4-2 for more information.

Table 4-2 Common disorders associated with shifting of fluid into the third space

Disorder	Pathophysiologic Process
Peritonitis	Trapping of fluid and electrolytes in the peritoneal cavity owing to damage to or inflammation of the peritoneum. As many as 6 L of fluid can accumulate, depending on degree of acuity.
Bowel obstruction	Loss of lower GI fluid due to sequestering of same in the distended bowel. Several liters may accumulate in the intestinal lumen, leading to a dramatic increase in lumen pressure with eventual damage to intestinal mucosa.
Burns	Temporary sequestering of fluid in the interstitial space owing to increased capillary permeability or decreased vascular colloid osmotic pressure.
Ascites	Accumulation of several liters of fluid in the peritoneal cavity, occurring in severe hepatic cirrhosis. Ascites occurs in cirrhosis due to hepatic venous obstruction and retention of sodium and water.
Fractured hip	Loss of intravascular volume owing to extensive bleeding into the joint.
Carcinoma	Trapping of fluid in the interstitial space owing to lymphatic or venous obstruction.
Major surgery involving extensive tissue trauma	Abnormal sequestration of fluid at the surgical site, owing to extensive tissue involvement (e.g., with major abdominal surgery). It also can occur with the loss of ECF into the wall and lumen of the bowel during bowel surgery.

There is no third space per se, but rather it is a concept describing fluid that is temporarily unavailable either to the ICF or ECF. Because third-space fluids are unavailable to the body for its use, the patient will exhibit clinical indicators associated with fluid volume deficit, with the exception of weight loss.

Normally, third-space fluid is unmeasurable, although it may be estimated by careful comparison of overall intake and output and changes in body weight. An early clue to the development of third-space fluid shift may be a decreasing urine output despite seemingly adequate fluid therapy (Young & Flynn, 1988). If there has been a shift of fluid out of the intravascular space, the kidneys receive less blood flow and will attempt to compensate by decreasing urine output. Additional clues to a third-space fluid shift include the other symptoms of intravascular fluid volume deficit (e.g., increased heart rate, decreased blood pressure, decreased central venous pressure) and swelling or edema. Occasionally, certain types of third-space fluid are removed by needle aspiration (e.g., thoracentesis for removal of pleural fluid, pericardicentesis for the removal of pericardial fluid, and paracentesis for removal of fluid from the peritoneal cavity). See Table 4-3 for normal fluid gains and losses.

Table 4-3 Average daily fluid gains and losses in the adult

Fluid Gains		Fluid Losses	
Oxidative metabolism	300 ml	Kidneys	1,200-1,500 ml
Oral fluids	1,100-1,400 ml	Skin	500-600 ml
Solid foods	800-1,000 ml	Lungs	400 ml
TOTAL	2,200-2,700 ml	GI	100-200 ml
		TOTAL	2,200-2,700 ml

Case Study 4-1

S.H. is a 13-month-old toddler brought to the pediatrician's office by his mother. The office nurse obtained the following information: the child had had a *runny nose* and *wet cough* for the last 5 days, had been febrile (rectal temperatures 101° to 103.4° F) with a poor appetite for the last 2 days, and was wheezing since the previous evening. His mother stated that she had treated S.H.'s fever with acetaminophen every 4 to 6 hours and noted *sweaty* episodes after several of the doses. The mother also noted that S.H.'s diapers were drier than usual and that his most recent stool appeared *constipated.* Physical assessment revealed a child with a moderately severe respiratory tract infection as evidenced by slightly labored breathing and a productive cough.

1. **Question:** What unusual fluid gains or losses, if any, might S.H. be experiencing?

Answer:

a. Increased loss due to abnormal production of respiratory secretions.
b. Possible increased insensible loss from the lungs due to increased respiratory depth.
c. Increased loss via the skin secondary to fever and increased sweating.
d. Decreased oral intake.
e. Decreased urinary and gastrointestinal losses as probable compensation for increased losses and decreased intake.

S.H. was started on amoxicillin, a broad-spectrum penicillin antibiotic, 3 times a day. Over the next 4 days S.H.'s respiratory symptoms dramatically improved and his temperature returned to normal. However, his appetite remained poor, and he is now vomiting and has profuse diarrhea. Because of the frequent loose stools, S.H.'s parents are unsure when he last urinated.

2. **Question:** What are some potential nursing diagnoses for S.H.?

Answer:

a. **Fluid volume deficit** related to normal and abnormal fluid losses and decreased intake
b. **Alteration in nutrition:** Less than body requirements related to poor appetite, vomiting, and diarrhea
c. **Diarrhea**
d. **Alteration in pattern of urinary elimination:** Oliguria

S.H. does show signs of moderate fluid volume deficit: dry skin and mucous membranes, sunken eyeballs, decreased skin turgor, soft anterior fontanelle, crying without tears, and irritability. His physical assessment and current weight are consistent with a fluid loss of approximately 9% to 10%. Based on S.H.'s condition and his parents' level of fatigue, the decision is made to admit S.H. to the hospital for parenteral rehydration therapy.

3. Question: What additional nursing diagnoses are applicable for this family?

Answer:

a. **Sleep pattern disturbance** related to need to attend to sick child
b. **Fatigue** related to altered sleep pattern
c. **Anxiety** related to uncertainty of outcome for their child
d. **Powerlessness** related to lack of control over child's illness
e. **Potential parental role conflict** related to separation from child due to child's illness and perceived need to comply with agency policies regarding care and decision making for child

Case Study 4-2

C.H. is a 68-year-old man admitted to the hospital for treatment of acute respiratory failure. Within an hour of hospitalization he is intubated and placed on a respirator. Two weeks later a tracheostomy is performed because of his continued need for ventilatory support. His hospital stay is complicated by the development of cardiac failure, several episodes of pneumonia, and psychosis. C.H. is treated with broad-spectrum antibiotics for his repeated bouts of pneumonia. He is supported nutritionally with continuous nasogastric tube feedings administered via a flexible small-bore feeding tube. Repeated attempts to feed C.H. orally are unsuccessful, owing to problems with aspiration. His hospitalization is further complicated by the development of persistent diarrhea.

1. Question: What are some potential nursing diagnoses for the patient with diarrhea?

Answer:

a. **Diarrhea**
b. **Fluid volume deficit** related to increased loss of water in the stool
c. **Potential for impaired skin integrity** related to risk of excoriation of the anal area
d. **Alteration in nutrition:** Less than body requirements related to poor absorption of nutrients
e. **Sensory/perceptual alterations** related to motor and mentation disturbances secondary to abnormal electrolyte loss

2. **Question:** Which electrolyte imbalances may develop as a result of the diarrhea?

Answer:

a. Hyponatremia (low serum sodium). See Chapter 8.
b. Hypokalemia (low serum potassium). See Chapter 9.
c. In addition, hypocalcemia (low serum calcium—see Chapter 10) and hypomagnesemia (low serum magnesium—see Chapter 12) may develop.

3. **Question:** Which acid-base imbalance may develop secondary to diarrhea and why?

Answer: Metabolic acidosis due to the loss of fluids rich in bicarbonate. Remember that the lower GI tract secretes bicarbonate in order to buffer the acid contents of the upper GI tract. See Chapter 17 for detail. Two potential causes of C.H.'s diarrhea include too rapid an administration of tube feedings and overgrowth infection of intestinal pathogens secondary to the use of broad-spectrum antibiotics.

4. **Question:** What are some nursing considerations for the individual with diarrhea?

Answer:

a. Assess and document the following: duration of the diarrhea, associated symptoms (e.g., nausea, cramping, fever) and frequency, character, and consistency of the stools.
b. Restrict oral or enteral intake during the acute phase. For individuals on tube feedings the following should be implemented: control the infusion rate to provide frequent small feedings or a slow continuous drip; administer feedings at room temperature (feedings that are too hot or too cold may cause nausea or discomfort); dilute the feedings to the appropriate strength (concentrations should be either isoosmolar or hypoosmolar); hang no more than a 6-hour supply at a time and refrigerate any opened, unused feeding solution to prevent spoilage (bacterial growth); rinse infusion set with each use and replace every 24 hours to minimize contamination and bacterial growth; monitor for retained feedings by assessing for abdominal distention and aspirating stomach contents prior to each feeding or at least every 4 hours for the individual receiving continuous feedings.

c. Avoid cold liquids or foods that stimulate the GI tract. Offer foods such as tea, toast, pudding, and custard.
d. Administer prescribed antidiarrheal medications such as diphenoxylate hydrochloride (Lomotil), loperamide hydrochloride (Imodium), or kaolin with pectin (Kaopectate).
e. Obtain prescribed stool cultures.
f. Monitor for indicators of fluid volume deficit and electrolyte imbalance. These will be discussed in detail in subsequent chapters.

C.H.'s tube feedings were changed from a full-strength formula to a half-strength dilution (half water and half formula), and the rate was decreased from 75 ml to 50 ml/hr in an attempt to control his diarrhea. Despite these measures, C.H. continues to pass liquid stools nearly hourly. C.H. is started on Imodium and a stool culture is obtained. The culture is positive for yeast *(Candida albicans)*. Gastrointestinal candidiasis may develop when broad-spectrum antibiotics inhibit the growth of normal intestinal flora. C.H. is to receive nystatin, an antifungal agent, via NG tube.

References

Barkin, RM: Acute infectious diarrheal disease in children, J Emer Med, vol 3, pp 1-9, 1985.

Binder HJ: The pathophysiology of diarrhea, Hosp Practice 19(10):107-113, 116-118, Oct 1984.

Corbett JV: Laboratory tests and diagnostic procedures with nursing implications, ed 2, Norwalk, Conn and San Mateo, Calif, 1987, Appleton & Lange.

Cox P: Insensible water loss and its assessment in adult patients: a review, Acta Anaesthesiol Scand 31:771-776, 1987.

Ellis K: One piece of paper we can't do without, RN 51(8):110, 1988.

Goldberger E: A primer of water, electrolyte, and acid-base syndromes, ed 7, Philadelphia, 1986, Lea & Febiger.

Guyton AC: Textbook of medical physiology, ed 7, Philadelphia, 1986, WB Saunders Co.

Hazinski MF: Understanding fluid balance in the seriously ill child, Pediatr Nurs 14(3):231-236, 1988.

Horne M and Swearingen PL: Pocket guide to fluids and electrolytes, St Louis, 1989, The CV Mosby Co.

Lattanzi WE and Siegel NJ: A practical guide to fluid and electrolyte therapy, Curr Probl Pediatr 16(1):1-43, 1986.

Metheny NM: Twenty ways to prevent tube feeding complications, Nursing 1985 Jan pp 47-50.

Noble-Jamieson CM and others: Hidden sources of fluid and sodium intake in ill newborns, Arch Dis Child 61(7):695-696, 1986.

Pritchard V: Geriatric infections: the gastrointestinal tract, RN 51(4):58-60, April 1988.

Rose BD: Clinical physiology of acid-base balance and electrolyte disorders, ed 3, New York, 1989, McGraw-Hill, Inc.

Rowe JW: Aging and renal function. In Arief AL and De Fronzo RA: Fluid electrolyte and acid base disorders, New York, 1985, Churchill Livingstone Inc.

Siegel NJ and Lattanzi WE: Fluid and electrolyte therapy in children. In Arief AL and De Fronzo RA: Fluid electrolyte and acid base disorders, New York, 1985, Churchill Livingstone, Inc.

Smith DB and Johnson DE: Ostomy care and the cancer patient: surgical and clinical considerations, 1986, Grune and Stratton.

Swearingen PL: Addison-Wesley photo-atlas of nursing procedures, ed 2, Redwood City, Calif, 1991, Addison-Wesley Publishing Co, Inc.

Thompson J and others: Clinical nursing, St Louis, 1986, The CV Mosby Co.

Vander AJ: Renal physiology, ed 3, New York, 1985, McGraw-Hill, Inc.

Van Ruden KT: The effect of hemodynamic monitoring system flush solution on the fluid status of cardiac surgery patients, Focus Crit Care 12(4):20-23, 1985.

Young ME and Flynn KT: Third-spacing: when the body conceals fluid loss, RN 51(8):46-48, 1988.

Chapter

5

Nursing Assessment of the Individual at Risk for Developing Fluid and Electrolyte Disturbances

Fluid and electrolyte homeostasis is essential for health and well-being. Unfortunately, fluid and electrolyte disturbances are potential complications of almost all disease states and medical therapies. Nurses in all areas of practice must be diligent in their assessment of persons at risk for developing fluid and electrolyte disturbances. The specifics of assessment will vary with the individual fluid or electrolyte imbalance. An elevated plasma potassium level, for example, may be best identified by EKG changes, while the first clues to an elevated plasma calcium level may be personality changes. Not only is it important to recognize the physical signs and symptoms of an imbalance, but also one must identify the individual at risk so that preventive nursing measures may be taken. Often potential imbalances may be avoided or minimized by the appropriate nursing interventions. Identification of individuals at risk begins with a thorough nursing history.

NURSING HISTORY

Each of the following dimensions of the health history should be considered.

Physiologic

Does the individual have any diseases or disorders, such as ulcerative colitis or diabetes mellitus, that may cause a disturbance in fluid and electrolyte homeostasis? Is the individual receiving any medications or therapy, such as diuretics or NG suction, that may affect fluid and electrolyte balance? What is his or her past medical history? Refer to Appendix A for a discussion of fluid and electrolyte imbalances associated with specific medical diagnoses.

Developmental

Is the individual at increased risk due to his or her age or social situation—an elderly adult who lives alone, for instance? Fluid volume deficit is more common in infants, small children, and older adults than it is in young or middle-aged adults. Both the very young and very old have a decreased ability to concentrate their urine. This means that they are less able to compensate for fluid loss by reducing their urine output. Additionally, the older adult often has a diminished thirst mechanism and is less likely to develop thirst and seek fluids in response to a volume deficit. Immobility and confusion also may contribute to poor oral intake. Likewise, infants are dependent on others to provide them with adequate fluids and have a limited ability to communicate their thirst. Nurses should consider that the elderly may purposely reduce fluid intake to minimize episodes of incontinence or nocturia or decrease the need to seek assistance using the toilet. A study evaluating factors affecting fluid intake in 67 institutionalized elderly individuals identified advancing age as a risk factor for inadequate water intake (Gaspar, 1988). See Appendix E for the effects of age on fluid and electrolyte balance.

Psychologic

Are there behavioral or emotional problems that may increase the risk of fluid and electrolyte disturbances, such as denial and noncompliance with a medical regimen in a diabetic teenager? Chronic alcohol abuse, especially during acute withdrawal, is associated with several fluid and electrolyte disturbances, including decreased plasma magnesium and decreased plasma phosphorus.

Spiritual

Does the individual have any beliefs, values, or practices that may affect his or her ability to comply with medical interventions, for instance, the Jehovah's Witness with GI bleeding who refuses human blood products?

Sociocultural

Are there any social, cultural, financial, or educational factors that place the individual at increased risk or affect his or her ability to comply with medical therapy (e.g., the patient on a fixed income who, in an attempt to save money, fills only the digoxin and diuretic prescription, but not the potassium supplement prescription)?

CLINICAL ASSESSMENT

After obtaining a complete nursing history to identify an individual at risk, the next step is to review the clinical indicators. Two of the most important tools for the clinical assessment of fluid balance problems are simple nursing procedures that may be initiated without a physician's order. They are daily weights and intake and output. Hemodynamic monitoring is an invasive means of evaluating fluid balance disorders.

Daily Weight

Acute weight changes usually are indicative of acute fluid changes. Each kg of weight lost or gained suggests 1 L of fluid lost or gained. Thus a 2-kg acute weight loss indicates a 2-L fluid loss. An exception to this is individuals who are not eating or are not being maintained on enteral or parenteral nutrition. They may lose 0.3 to 0.5 kg a day due to the loss of body tissue. This weight loss should not be mistaken for fluid loss.

Weight gains do not necessarily indicate an increase in vascular volume, but rather, an increase in total body volume that may be located in any of the fluid compartments. Individuals with hepatic cirrhosis, for example, may retain fluid and gain weight because of the development of ascites (fluid in the peritoneal cavity), but they still may not be overloaded in their intravascular space. For accuracy and consistency, weight should be measured at the same time of day, preferably before breakfast. The scale should be balanced before each use, and the individual should be weighed wearing approximately the same clothing. The type of scale (e.g., standing, bed, chair) should be noted so that whenever possible the same scale can be used.

When obtaining a health history, it is important to gather information concerning recent weight patterns. Even if the individual or family member is unsure of actual weight changes, information such as increasing tightness or looseness of clothing or rings may suggest weight changes. Variations in weight are helpful both in identifying and measuring fluid imbalance. This is especially true when fluid losses have gone unrecorded or are difficult to estimate or measure. Acute weight changes combined with physical assessment, for example, may be used in the outpatient setting to determine the volume of fluid necessary to rehydrate a child with a history of diarrhea.

The significance of a given weight loss or gain will vary with age and body size. A weight loss of 50 g or greater per day is considered significant in the infant, as compared with 200 g or greater per day in the child or 500 g or greater per day in the adolescent (Hazinski, 1988). An acute weight loss of 2% to 5% of body weight reflects a mild volume deficit in the adult, while a loss of 5% to 10% is moderate, 10% to 15% is severe, and 15% to 20% is fatal.

Intake and Output (I&O)

Intake and output (I&O) should be recorded for any individual at risk for fluid and electrolyte imbalance. Although measurement of the I&O may be requested by the physician, the nurse should initiate it as a nursing order whenever it is indicated. Nursing units, such as critical care or surgical units, which have a high percentage of patients at risk for fluid and electrolyte imbalance routinely measure I&O on all their patients. To ensure an accurate, and thus, useful record, liquid intake and output should be measured exactly whenever possible and all unmeasured volumes estimated and noted. The I&O record should include the following:

Intake

Oral fluids: Ice chips must be included and recorded as fluids at approximately one half their volume. Include all foods that are liquid at room temperature.

Parenteral fluids: Parenteral fluid containers often are overfilled. The excess should be discarded during setup, or the exact amount given should be recorded.

Tube feedings: Often a 30 to 50 ml water flush is given at the end of intermittent tube feedings or periodically during continuous tube feedings. This flush needs to be included in the intake record.

Catheter irrigants: If the catheter is irrigated or lavaged and an equal amount is not withdrawn and discarded, the extra irrigant should be added to the intake record.

Medications: Parenteral and liquid oral medications often are overlooked yet can be significant sources of fluids. For example, mannitol and the antibiotic metronidazole (Flagyl) usually are both administered in 50 ml volumes.

Output

Urine output: Ideally, it is measured hourly using a urine meter (Figure 5-1).

Liquid feces: For individuals who are incontinent of liquid stool or urine, determine volume of loss by weighing soiled pads. First weigh a dry incontinence pad or diaper (using a gram scale) and document, then weigh all soiled pads. Each increase in weight by 1 g reflects approximately a 1 ml of fluid lost by the patient.

Vomitus

Nasogastric drainage

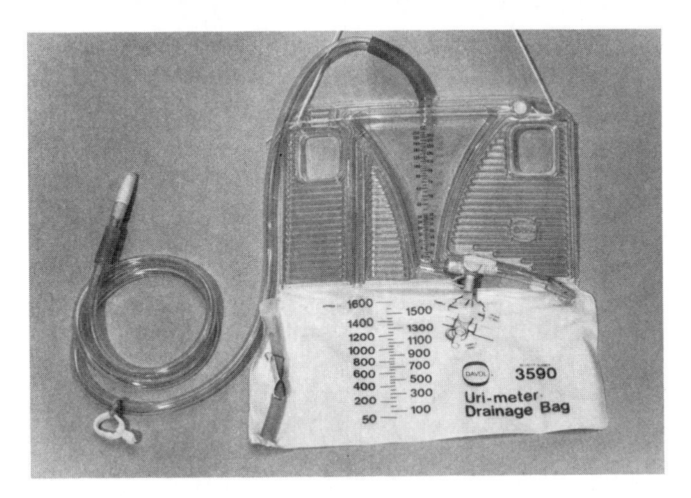

Figure 5-1 Urine meter.

(From Swearingen P: Addison-Wesley photo-atlas of nursing procedures, ed 2, Redwood City, Calif, 1991, Addison-Wesley Publishing Co, Inc.)

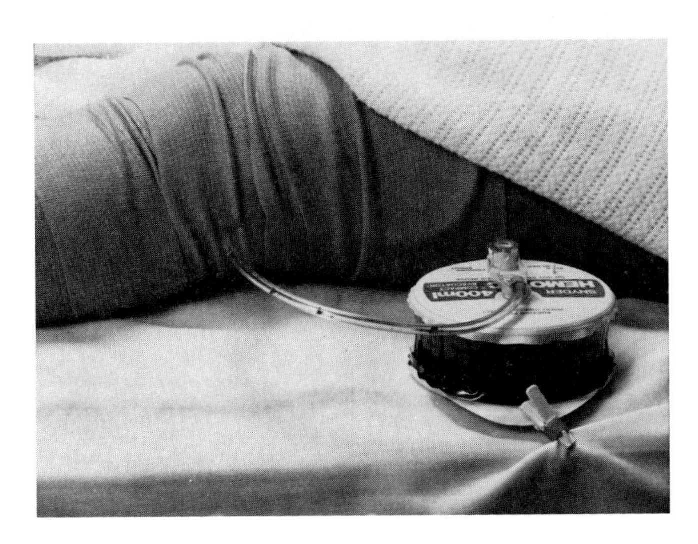

Figure 5-2 Portable wound drainage system.

(From Swearingen P: Addison-Wesley photo-atlas of nursing procedures, ed 2, Redwood City, Calif, 1991, Addison-Wesley Publishing Co, Inc.)

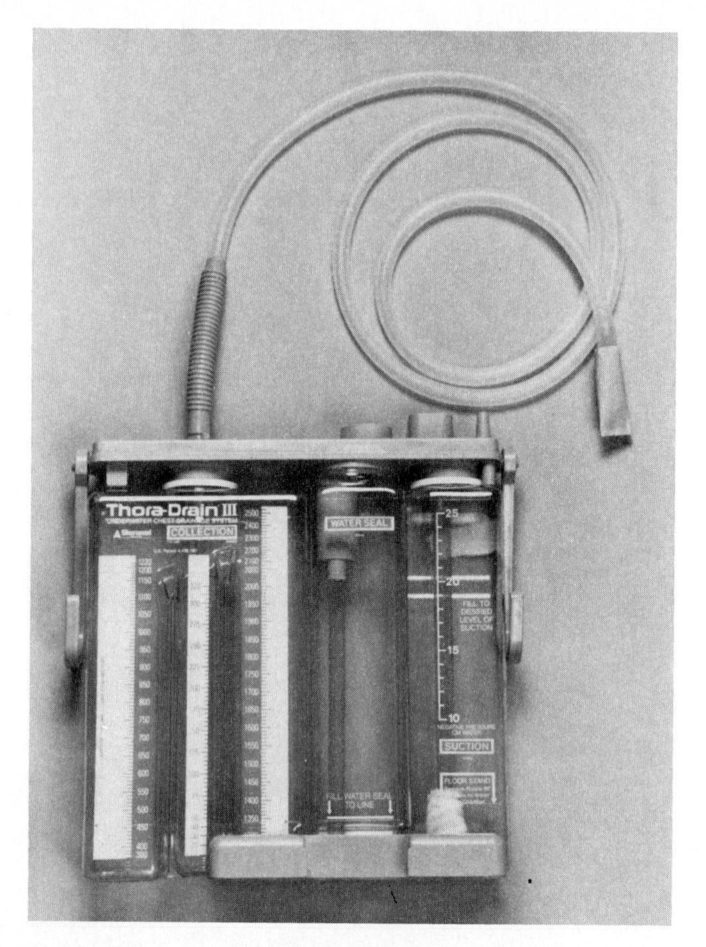

Figure 5-3 Disposable closed-chest drainage system.

(From Swearingen P: Addison-Wesley photo-atlas of nursing procedures, ed 2, Redwood City, Calif, 1991, Addison-Wesley Publishing Co, Inc.)

Excessive diaphoresis: May be documented either via a rating system (1+ for noticeable sweating to 4+ for profuse sweating) or by documenting the amount of linen saturated with sweat.

Wound drainage: May be documented by noting the type and number of dressings saturated, weighing dressings, or direct measurement of drainage contained in a gravity or vacuum drainage device for wounds (Figure 5-2) or chest (Figure 5-3).

Draining fistulas: If possible, collect drainage in a stoma bag or document amount of dressings or linen saturated.

Increased depth of respirations: Will increase a patient's insensible fluid loss but usually is not clinically significant.

Table 5-1 Hemodynamic evaluation of fluid volume abnormalities

Clinical Reading	Potential Cause
CVP <2 mm Hg or <5 cm H_2O PAP <20/8 mm Hg	Decreased effective circulating volume owing to true volume depletion (e.g., bleeding), shift of fluid out of the vascular space (e.g., burns), or vasodilatation (e.g., after administration of certain antihypertensive medications).
CVP >6 mm Hg or >12 cm H_2O	Fluid overload, poor right ventricular function, or constriction of the pulmonary vascular bed.
PAP >30/15 mm Hg	Increases in fluid volume or in pulmonary vascular resistance.

From Horne M and Swearingen PL: Pocket guide to fluids and electrolytes, St Louis, 1989, The CV Mosby Co.

Hemodynamic Monitoring

Invasive hemodynamic monitoring can be very useful in evaluating fluid volume abnormalities. It involves insertion of catheters into specific vessels for the purpose of measuring pressure or flow of blood. The information obtained can be used to differentiate between true intravascular volume changes and altered cardiac function (Table 5-1).

Central venous pressure (CVP)

Central venous pressure (CVP) is the measurement of mean right atrial pressure by means of a catheter that is inserted in or near the right atrium. It reflects right ventricular end-diastolic pressure. A normal reading is 2 to 6 mm Hg or 5 to 12 cm H_2O. Whether using a fluid-filled manometer or an electronically read pressure transducer, the monitoring equipment must be level with the right atrium to provide accurate and consistent measurements. The standard reference point for the right atrium is the fourth intercostal space half way between the anterior and posterior chest (Figures 5-4 and 5-5). Ideally, the patient is placed in the same position for all readings, usually supine with the head of the bed flat. Electronic equipment should be calibrated at least every 8 hours.

Pulmonary artery pressure (PAP)

Pulmonary artery pressure (PAP) is measured by means of a catheter that is passed through the right heart and into the pulmonary artery (PA), with the tip positioned in the pulmonary capillary bed. Normal PAP is 20-30/8-15 mm Hg. PA diastolic pressures may be used to estimate left ventricular end-diastolic pressure and

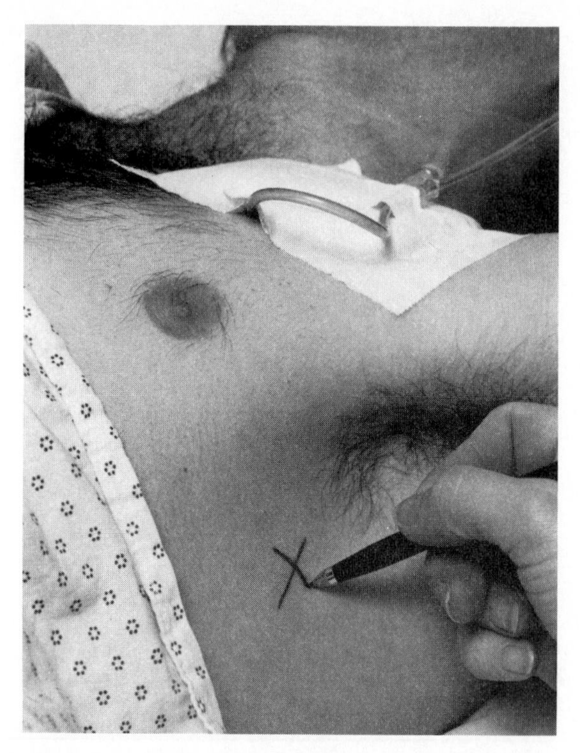

Figure 5-4 Marking the level of the right atrium.

(From Swearingen P: Addison-Wesley photo-atlas of nursing procedures, ed 2, Redwood City, Calif, 1991, Addison-Wesley Publishing Co, Inc.)

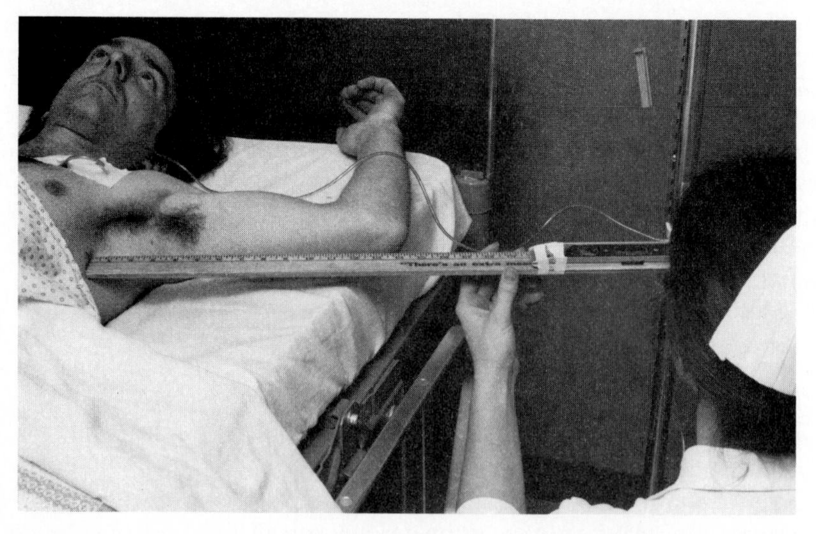

Figure 5-5 Measuring the right atrium.

(From Swearingen P: Addison-Wesley photo-atlas of nursing procedures, ed 2, Redwood City, Calif, 1991, Addison-Wesley Publishing Co, Inc.)

thus evaluate cardiac performance. As with CVP readings, the pressure transducer must be level with the right atrium. Ideally, readings are obtained with the individual in the same position each time.

VITAL SIGNS

After reviewing the nursing history and clinical indicators, physical assessment is the next step in the evaluation of the individual with or at risk for fluid and electrolyte disturbances. A practical way to begin physical assessment is with measurement of vital signs. The following are examples of changes in vital signs that may signal fluid or electrolyte imbalance:

Body Temperature

Elevations in body temperature may lead to fluid and electrolyte losses due to increased sweating. Changes in body temperature also may be a symptom of fluid and electrolyte imbalance. Hypernatremic (elevated sodium) dehydration may cause an elevation in temperature, while decreases in body temperature may occur with hypovolemia. In severe fluid volume deficit, for example, the rectal temperature may drop to as low as 35° C (95° F).

Respiratory Pattern

Increases in respiratory depth increase insensible fluid loss and may contribute to the development of volume depletion, although, as stated, this loss usually is minimal. Increased production of respiratory secretions also will contribute to the development of fluid volume deficit. Individuals on mechanical ventilation actually may gain fluid via the lungs due to the administration of humidified gases within a closed system (Cox, 1987). Shortness of breath (SOB), crackles (rales), or rhonchi may signal fluid buildup in the lungs owing to fluid volume excess. Rapid, deep respirations may be a compensatory mechanism for metabolic acidosis as the lungs attempt to normalize pH by blowing off more carbon dioxide. Slow, shallow respirations may result in the retention of carbon dioxide and acidosis. Refer to Chapters 13-16 for a discussion of the relationship between ventilation and acid-base balance.

Heart Rate

An increased heart rate (HR) may occur with fluid volume deficit as a compensatory mechanism for maintaining cardiac output. An increased HR also may occur with fluid volume excess that is poorly tolerated by the heart. Electrolyte imbalances involving potassium, calcium, or magnesium may affect heart rate and regularity. See Chapters 9, 10, and 12 for more information.

Blood Pressure

Blood pressure (BP) is determined by multiplying cardiac output (CO) by systemic vascular resistance (SVR). CO, in turn, is the product of heart rate times stroke volume (the amount of blood moved with each contraction of the left ventricle). Thus changes in stroke volume, heart rate, or vascular resistance may result in changes in BP. A decreased BP may signal intravascular fluid volume deficit as a result of a reduction in stroke volume. Electrolyte imbalances that cause dysrhythmias also may decrease BP if either heart rate or stroke volume is affected. Likewise, a decrease in vascular resistance, for example, with increased magnesium levels, may cause a drop in BP. An elevated BP may be indicative of fluid volume excess because of an increase in stroke volume.

Orthostatic (postural) changes in BP or heart rate may provide an early clue to the presence of volume deficit. A drop of 15 mm Hg in either the systolic or diastolic BP or an increase of 20 beats per minute or greater in the heart rate when the individual changes position from lying to sitting or standing indicates a significant volume deficit. A decrease in the pulse pressure (the difference between the systolic and diastolic BP) also may occur with volume deficit. Normal pulse pressure is 30 to 40 mm Hg.

PHYSICAL ASSESSMENT

The following are some examples of changes noted on physical assessment that may be indicative of fluid or electrolyte imbalance. See individual fluid, electrolyte, and acid-base disorders (Chapters 8 to 18) for additional information.

Integument

Skin moisture, color, and temperature

Skin that is flushed and dry may signal fluid volume deficit as in the individual with diabetic ketoacidosis. The individual in shock may have skin that is cool, clammy, and pale.

Skin turgor

Skin turgor changes may reflect changes in interstitial fluid volume. Turgor may be assessed by pinching skin over the forearm, sternum, or dorsum of the hand (Figure 5-6, *A*). With adequate hydration, the pinched skin returns quickly to its original position when released (Figure 5-6, *B*). With fluid volume deficit,

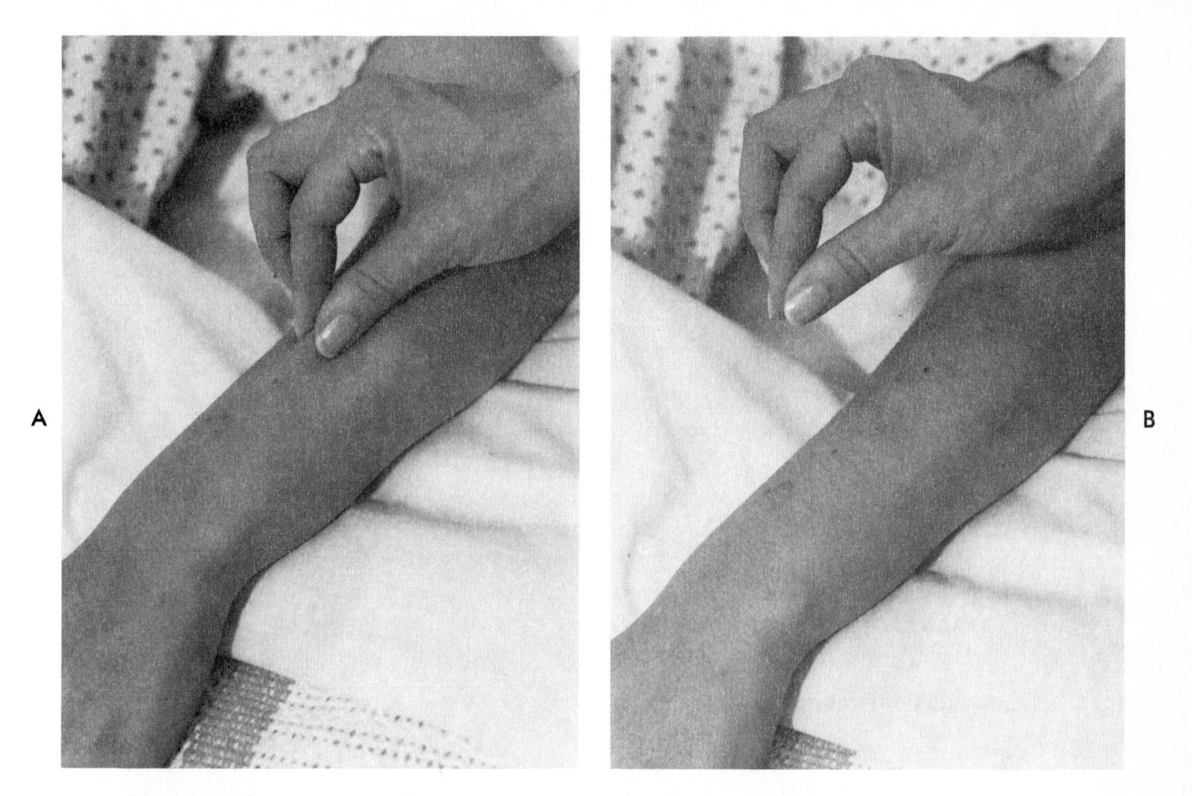

Figure 5-6 **A,** Assessing skin turgor. **B,** Example of normal skin turgor.
(From Swearingen P: Addison-Wesley photo-atlas of nursing procedures, ed 2, Redwood City, Calif, 1991, Addison-Wesley Publishing Co, Inc.)

the pinched skin remains elevated for several seconds. This is a less reliable indicator in the older adult, owing to the skin's decreased elasticity, or in individuals who recently have lost weight.

Edema

Edema is indicative of an expanded interstitial volume. It may be localized, which is usually the result of inflammation, or generalized secondary to altered capillary hemodynamics and the retention of excess sodium and water. Edema usually is most evident in dependent areas (i.e., legs and ankles in the ambulatory individual, and the sacrum and back in the individual requiring bedrest). Pitting should be assessed over a bony surface such as the tibia or sacrum and rated according to severity (i.e., 1+ for barely detectable edema to 4+ for deep, persistent pitting). See Chapter 6 for additional information.

Oral cavity

Increases in the longitudinal furrows of the tongue indicate a reduction in tissue volume of the tongue and are suggestive of fluid volume deficit. The mucous membrane between the cheek and gums is normally moist. In the presence of fluid volume deficit it may become sticky or dry due to decreased production of oral secretions. However, a sticky and dry oral mucous membrane is an unreliable indicator of fluid volume deficit in individuals who are mouth breathers.

Cardiovascular System

Jugular vein assessment

The degree of jugular venous distention provides an estimate of central venous pressure. With the HOB at a 30- to 45-degree angle, measure the distance between the level of the sternal angle (angle of Louis) and the point at which the internal and external jugular veins collapse. Optimally, this distance should be 3 cm or less (Figure 5-7). Visualization of distention above the level of 3 cm suggests intravascular volume excess or decreased cardiac function. Markedly decreased jugular venous distention (e.g., the absence of distention when the individual is fully reclined) indicates intravascular fluid volume deficit.

Hand vein assessment

Assessment of the hand veins also may be used to assess fluid volume status. Normally, elevating the hand above the level of the heart will collapse the veins in 3 to 5 seconds, and lowering the hand below the level of the heart will refill them in 3 to 5 seconds. With fluid volume deficit, the veins of the lowered hand require more than 3 to 5 seconds to fill. With fluid volume excess, the veins of the elevated hand require more than 3 to 5 seconds to empty.

Capillary refill

Arterial perfusion may be evaluated by testing capillary refill. To test capillary refill, apply pressure over a toenail or fingernail. The digit should blanch with pressure and then become pink again 2 to 4 seconds after pressure is removed. Delayed refill (greater than 4 to 6 seconds to return to pink) suggests poor arterial perfusion and may be indicative of decreased effective circulating volume. The older adult may have decreased capillary refill secondary to local peripheral vascular disease.

Pulse strength and volume

The strength and volume of the pulse is dependent on the volume of blood ejected by the left ventricle and the strength of the left ventricular contraction. Thus a bounding pulse may signal fluid volume excess, and a weak, thready pulse may signal a reduction in intravascular volume. Pulse irregularities may occur with potassium,

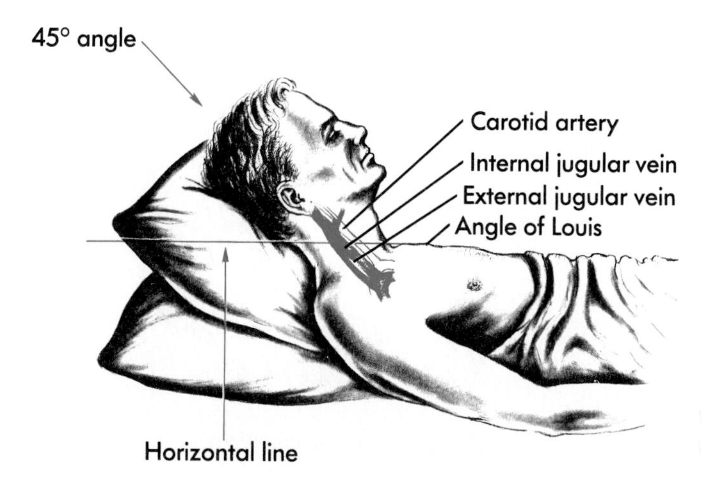

Figure 5-7 Inspection of external jugular venous pressure.
(From Thompson J and others: Clinical nursing, St Louis, 1986, The CV Mosby Co.)

calcium, and magnesium abnormalities (see individual diagnostic studies sections for a discussion of specific EKG changes).

Neurologic System

Sensorium

Changes in awareness, orientation, and level of consciousness (LOC) occur with changes in serum osmolality (e.g., changes in serum sodium) or severe acid-base imbalances. Severity of the symptoms will depend on the rate and degree of change. Individuals are more likely to become comatose, for example, if their serum sodium drops suddenly than if it decreases gradually. Gradual changes in the serum sodium level allow time for the cells to adjust their osmolality, thereby minimizing fluid shifts (Geheb, 1987). Restlessness and confusion may occur with fluid volume deficit, owing to decreased cerebral circulation or with acid-base imbalance due to altered cerebral cellular function. Markedly elevated calcium levels also can affect sensorium, as evidenced by agitation, confusion, and even psychosis.

Neuromuscular excitability

Abnormalities in neuromuscular excitability occur with calcium and magnesium changes. Calcium and magnesium deficits enhance neuromuscular excitability (e.g., hyperactive reflexes), while calcium and magnesium excesses depress neuromuscular function (e.g., diminished reflexes). The same neuromotor symptoms that occur with hypocalcemia also may develop in metabolic alkalosis. See Figures 5-8 and 5-9 for a depiction of evaluating deep tendon reflexes.

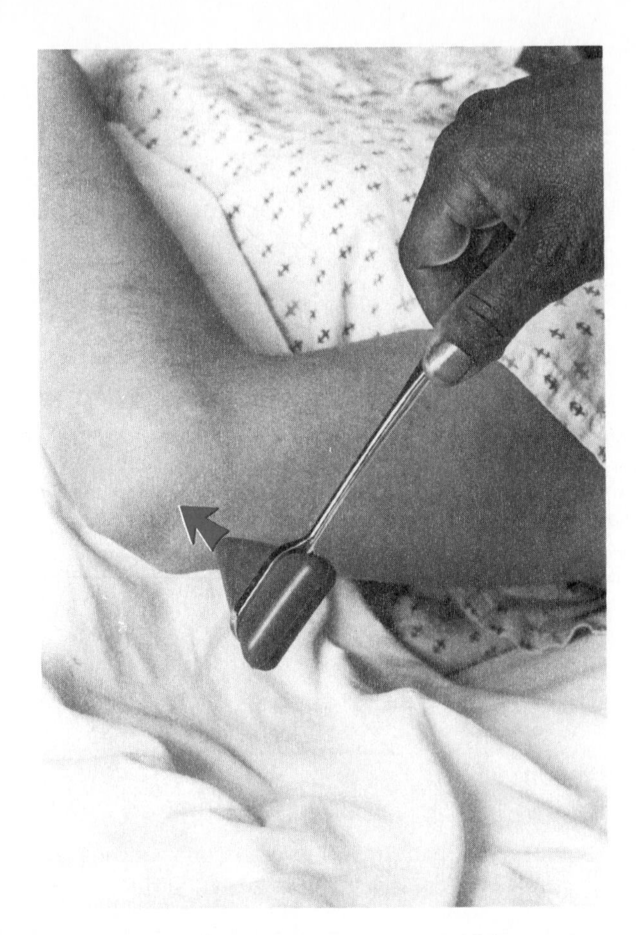

Figure 5-8 Assessing triceps reflex.

(From Swearingen P: Addison-Wesley photo-atlas of nursing procedures, ed 2, Redwood City, Calif, 1991, Addison-Wesley Publishing Co, Inc.)

Trousseau's and Chvostek's signs

Positive **Trousseau's and Chvostek's signs** can occur with hypocalcemia and hypomagnesemia. They are described and elicited as follows:

Positive Trousseau's sign: Ischemia-induced carpal spasm. It is elicited by applying a BP cuff to the upper arm and inflating it past systolic BP for 2 minutes.

Positive Chvostek's sign: Unilateral contraction of the facial and eyelid muscles. It is elicited when irritating the facial nerve by percussing the face just in front of the ear.

Abnormalities in potassium levels may also cause neuromotor symptoms. Tingling, paresthesias, weakness, and flaccid paralysis may occur with hyperkalemia. Weakness, cramps, tetany, and paralysis may occur with hypokalemia.

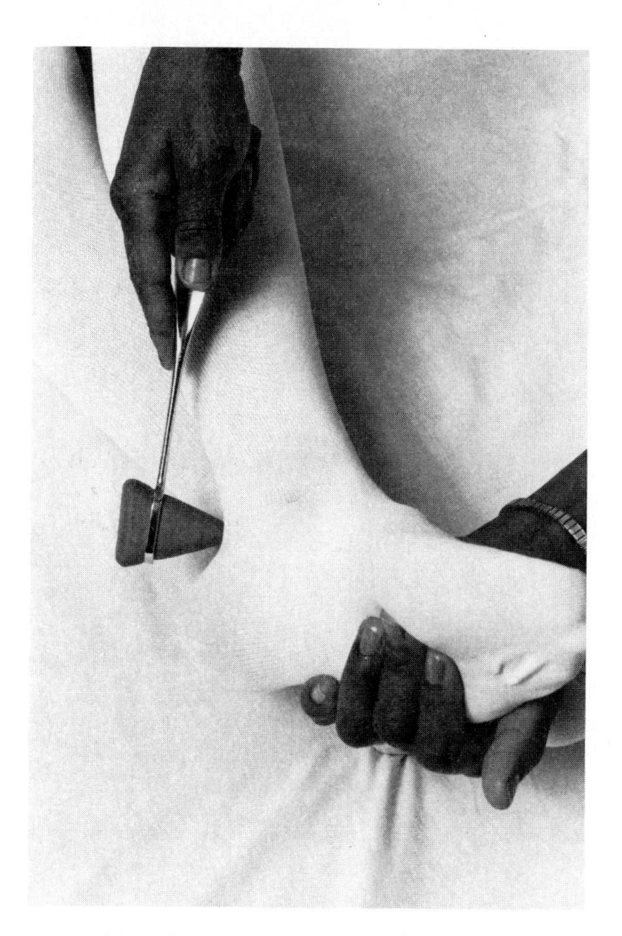

Figure 5-9 Assessing ankle reflex.

(From Swearingen P: Addison-Wesley photo-atlas of nursing procedures, ed 2, Redwood City, Calif, 1991, Addison-Wesley Publishing Co, Inc.)

Gastrointestinal System

GI disturbances

Anorexia, nausea, and vomiting may occur with acute fluid volume deficit or fluid volume excess. They occur in fluid volume deficit due to decreased mesenteric blood flow. These same symptoms develop in fluid volume excess secondary to edema of the GI tract. Alterations in bowel motility resulting in either diarrhea or constipation may develop with potassium or calcium imbalances.

Thirst

Thirst may be symptomatic of increased osmolality or fluid volume deficit. As discussed in Chapter 3, thirst provides the primary protection against the development

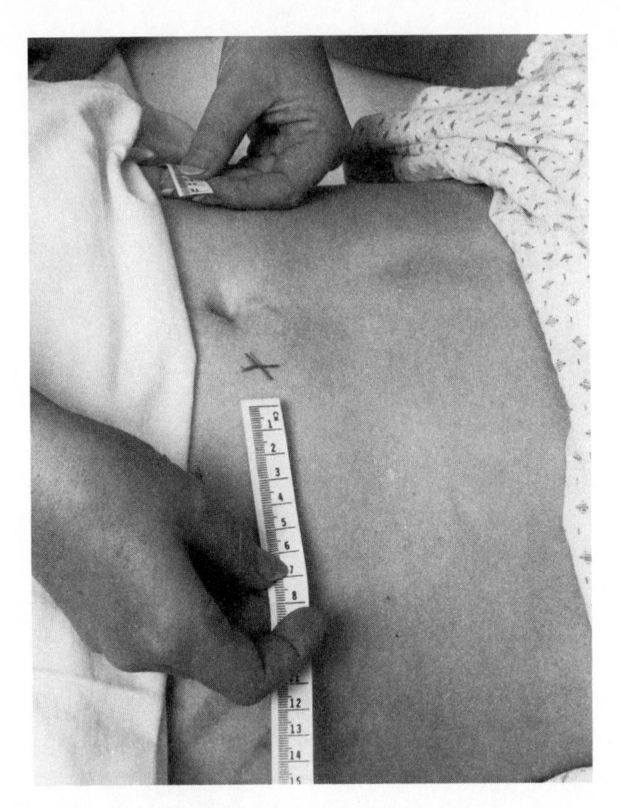

Figure 5-10 Measuring abdominal girth.

(From Swearingen P: Addison-Wesley photo-atlas of nursing procedures, ed 2, Redwood City, Calif, 1991, Addison-Wesley Publishing Co, Inc.)

of plasma hyperosmolality. Individuals with an intact thirst mechanism and adequate availability of water are able to maintain their plasma osmolality by satisfying their thirst. Although thirst is a less sensitive symptom of fluid volume deficit, it may provide a clue to the presence of an otherwise unsuspected fluid volume deficit. Remember, however, that thirst also is affected by advancing age and social and emotional factors.

Changes in abdominal girth

Increases in abdominal girth may indicate the retention of fluid in the GI tract or in the peritoneal cavity (e.g., ascites). Daily measurement of abdominal girth should be obtained in any individual with abdominal distention. Abdominal girth is measured at the same time(s) each day, using a nonstretchable tape measure. To ensure day-to-day consistency and accuracy, a mark should be made on the abdomen with indelible ink to identify the measurement site (Figure 5-10).

Renal System

Changes in urinary output

Although urine output was considered earlier in this chapter under the discussion of intake and output, it is also important as an aspect of physical assessment. A drop in urine output may be an early clue to fluid volume deficit. Or it may signal renal failure, which places the individual at risk for fluid volume excess.

An increased urine output may be a cause or contributing factor in the development of volume deficit. The polyuria that occurs with diabetes insipidus, during the diuretic phase of acute renal failure, or after the release of a urinary tract obstruction all may result in a volume deficit. Overuse of diuretics also can result in the development of volume deficit.

Changes in urine output also may contribute to the development of certain electrolyte imbalances. Individuals producing large volumes of urine, for example, are at increased risk for developing hypokalemia (a low plasma potassium level) due to increased loss of potassium in the urine. Individuals producing very small volumes of urine (anuria or oliguria) are at increased risk for hyperkalemia (an elevated plasma potassium level) due to the decreased volume of urine available for excretion of potassium (see Chapter 9).

LABORATORY ASSESSMENT OF FLUID AND ELECTROLYTE BALANCE

Laboratory tests are vital in the early identification and continuous monitoring of fluid, electrolyte, and acid-base imbalances. Consideration of laboratory results should be included in the nursing assessment of patients at risk for fluid and electrolyte disturbances. The laboratory values shown in Table 5-2 are applicable to the adult and are subdivided according to evaluation for fluid status and evaluation for acid-base balance.

Table 5-2 Laboratory tests that evaluate fluid status and acid-base balance

Tests for Fluid Status	Tests for Acid-Base Balance
Serium osmolality	Arterial blood gas values
Hematocrit	CO_2 content
Urea nitrogen	Anion gap
Urine osmolality	Urine pH
Urine specific gravity	Lactic acid
Urine sodium	

Tests to Evaluate Fluid Status

Serum osmolality

The normal range for serum osmolality is 280 to 300 mOsm/kg. Serum osmolality measures the solute concentration of the blood. It primarily reflects the concentration of sodium and its accompanying anions. Recall from Chapters 1 and 3 that the serum osmolality may be estimated by doubling the serum sodium level. It is also affected by the serum glucose, urea nitrogen (BUN) levels, and certain medications and ingestions. See the accompanying box for factors that may alter serum osmolality values.

Hematocrit

Normal ranges for hematocrit are 40% to 50% for males and 37% to 47% for females. Hematocrit measures the volume (percentage) of whole blood that is composed of red blood cells. Since hematocrit measures the volume of cells in relation

Factors that May Alter Serum Osmolality Values

Factors that may increase serum osmolality:

Serum osmolality will increase because of increases in serum sodium (secondary to a loss of water or a gain in sodium), serum glucose, or BUN; with ingestion of ethanol, methanol, or ethylene glycol; or with administration of mannitol.

1. Free water loss: For example, insensible water loss (See Chapter 4 for a discussion of insensible water loss.)
2. Diabetes insipidus: Due to loss of large volumes of dilute urine (See Case Study 3-2.)
3. Sodium overload: For example, with excessive administration of sodium bicarbonate ($NaHCO_3$), which increases the serum sodium level
4. Isotonic fluid loss that is replaced with water or hypotonic fluids: For example, vomiting of isotonic gastric contents with water replacement
5. Hyperglycemia: See Case Study 3-1.
6. Uremia due to the retention of uremia

Factors that may decrease serum osmolality:

Serum osmolality decreases due to a decrease in plasma sodium level (hyponatremia). Thus the factors that decrease serum osmolality are the same as those that will cause hyponatremia, i.e., retention of water or the loss of sodium. (See Chapter 8 for additional information.)

1. Syndrome of inappropriate antidiuretic hormone (SIADH): Occurs because of abnormal retention of water owing to excess ADH. (See Chapter 3 and Case Study 5-1 for additional information.)
2. Use of diuretics: Due to the loss of sodium
3. Adrenal insufficiency: Occurs because of loss of sodium in the urine secondary to a lack of aldosterone.
4. Renal failure: Due to retention of excess water

to plasma, it will be affected by changes in plasma volume. Thus the hematocrit will increase with dehydration and decrease with overhydration. The hematocrit may remain normal immediately following an acute hemorrhage (the concentration of RBCs to plasma has not changed), but over a period of hours there is a shift of fluid from the interstitial fluid (ISF) to the plasma and the hematocrit drops. The kidneys compensate for the loss of volume by retaining sodium and water. The hematocrit will decrease in the presence of anemia and increase with polycythemia.

Urea nitrogen

The normal range for blood urea nitrogen (BUN) is 6 to 20 mg/dl. Urea is produced by the body as a by-product of hepatic protein metabolism. Its only means of removal from the body is excretion by the kidneys. Urea production occurs at a fairly steady rate so that an increased BUN usually reflects a reduction in renal function. Urea synthesis and excretion can be affected, however, by such additional factors as hydration, protein intake, and tissue catabolism, thereby limiting the usefulness of BUN as an indicator of renal function. See the accompanying box for factors that may alter BUN values.

Urine osmolality

Physiologic range for urine osmolality is approximately 50 to 1,400 mOsm/kg; a typical 24-hour specimen is approximately 300 to 900 mOsm/kg. This is a measure of the solute concentration of the urine. Unlike plasma, the primary determinants of

Factors that May Alter BUN Values

Factors that may increase BUN values:

1. Decreased renal function: If the increase in BUN is solely the result of reduced renal function, the serum creatinine level will increase at approximately the same rate (creatinine to BUN ratio will be 1:10-15).
2. Excessive protein intake
3. GI bleeding: Owing to digestion of blood in the gut
4. Increased tissue catabolism (breakdown): For example, with fever, sepsis, antianabolic steroid use
5. Dehydration: Urea excretion varies with water excretion. In dehydration, decreased water excretion causes decreased urea excretion.

Factors that may decrease BUN values:

1. Low-protein diet
2. Severe liver disease: Due to decreased hepatic synthesis of urea
3. Volume expansion: Occurs because of the dilutional effect. For example, overhydration with IV fluids, pregnancy

Factors that May Alter Urine Osmolality Values

Factors that may increase urine osmolality:

1. Fluid volume deficit
2. SIADH: Urine osmolality will be inappropriately high, given the serum osmolality. (See Case Study 5-1.)

Factors that may decrease urine osmolality:

1. Fluid volume excess
2. Diabetes insipidus: Due to increased excretion of water secondary to a lack of ADH

urinary osmolality are nitrogenous wastes (e.g., urea, creatinine, uric acid). The kidney is capable of excreting a concentrated yet sodium-free urine.

Urinary values for concentration and composition are normal or abnormal only in relation to what is occurring in the blood. A patient with severe diaphoresis, for example, would be expected to have a relatively high urine osmolality (the kidneys should be compensating for the hypotonic fluid loss by retaining water). In contrast, a patient who has been overhydrated with IV D_5W would be expected to have a relatively low urine osmolality (the kidneys should be compensating for the excess water intake by excreting a dilute urine). In this case, a relatively high urine osmolality would be abnormal. (See the accompanying box for factors that may alter urine osmolality values.)

Urine specific gravity

The physiologic range for urine specific gravity is 1.001 to 1.040. A random specimen with normal fluid intake is approximately 1.010 to 1.020. **Specific gravity** measures the weight of a solution in relation to water (water = 1.000). Urine specific gravity evaluates the kidneys' ability to conserve or excrete water. It is a less reliable indicator of concentration than urine osmolality, since specific gravity is affected both by the weight and number of solutes. The presence in the urine of a few large solutes such as glucose or protein may cause a deceptively high specific gravity. Advantages of the test are that it may be performed quickly, easily, and inexpensively at the bedside by the nursing staff.

Factors that increase and decrease urine specific gravity are the same as those that affect urine osmolality (Table 5-3). Some substances that may give a false high specific gravity include glucose, protein, dextran, radiographic contrast material, and medications such as carbenicillin disodium.

Urine sodium

Normal range for urine sodium from a random specimen is 50 to 130 mEq/L. Urine sodium levels vary with sodium intake (e.g., increased intake results in increased excretion) and volume status (e.g., sodium is conserved in the presence of a decreased

Table 5-3 Relationship of urine osmolality to urine specific gravity

Osmolality	Specific Gravity
350 mOsm/kg	≈ 1.010
700 mOsm/kg	≈ 1.020
1050 mOsm/kg	≈ 1.030
1400 mOsm/kg	≈ 1.040 (physiologic maximum for urinary concentration)

In the absence of abnormal substances, e.g., sugar, protein, or medications

effective circulating volume). Levels may be measured from 24-hour specimens or from random spot specimens. Diuretics and advanced renal failure may increase urine sodium levels. Clinically, urine sodium levels are used for evaluation of volume status and for differential diagnoses of hyponatremia (decreased serum sodium) and acute renal failure.

Urine sodium, osmolality, and specific gravity may be helpful in differentiating between oliguria caused by decreased effective circulating volume (ECV) and oliguria secondary to acute tubular necrosis (ATN). In volume depletion or decreased ECV, the kidneys are able to respond appropriately by conserving sodium and concentrating urine. Thus urine sodium will be minimal, urine osmolality will exceed plasma osmolality, and urine specific gravity will be greater than 1.015. In ATN (a type of acute renal failure), the kidneys lose their ability to conserve sodium and concentrate urine appropriately. The urine osmolality will remain fixed at less than 350 mOsm/kg, the specific gravity will be fixed at approximately 1.010, and urine sodium typically will be greater than 20 to 40 mEq/L.

Tests to Evaluate Electrolyte Balance

Refer to Chapters 8-12 for individual discussions of each of the electrolytes. See Table 5-4 for normal values for serum electrolytes.

Tests to Evaluate Acid-base Balance

Refer to Chapter 13 for a review of acid-base physiology, Chapter 14 for a step by step discussion of arterial blood gas interpretation, and Chapters 15-19 for an in-depth discussion of acid-base disorders.

Arterial blood gases (ABGs)

Arterial blood gases (ABGs) measure the pH, carbon dioxide (CO_2) tension, and oxygen (O_2) tension of *arterial* blood, and oxygen saturation of hemoglobin. A bicarbonate level of arterial blood also is included in an ABG test and may be measured

Table 5-4 Serum electrolytes: normal values

Sodium (Na^+)	137-147 mEq/L
Chloride (Cl^-)	95-108 mEq/L
Potassium (K^+)	3.5-5.0 mEq/L
Bicarbonate (HCO_3^-)	22-26 mEq/L
Calcium (Ca^{2+})	8.5-10.5 mg/dl
	4.3-5.3 mEq/L
Magnesium (Mg^{2+})	1.8-3.0 mg/dl
	1.5-2.5 mEq/L
Phosphorus (PO_4^-)	2.5-4.5 mg/dl
	1.7-2.6 mEq/L

Table 5-5 Arterial blood gases: normal values

pH	7.35-7.45
$Paco_2$	35-45 mm Hg
Pao_2	80-95 mm Hg
O_2 saturation	95%-99%
HCO_3^- (bicarbonate)	22-26 mEq/L

directly or calculated from pH and $Paco_2$. ABGs evaluate acid-base balance and pulmonary function. See Table 5-5 for normal values. See Chapter 14 for additional information, including a step-by-step guide to ABG analysis.

CO_2 content or total CO_2

The normal range for CO_2 content is 22 to 28 mEq/L. Using a *venous* blood sample, this test measures carbon dioxide content in all its chemical forms: dissolved CO_2 ($Paco_2$), bicarbonate (HCO_3^-), and carbonic acid (H_2CO_3). Carbonic acid exists only briefly, and therefore, its concentration is negligible. Because dissolved CO_2 accounts for only 1.2 mEq/L, the CO_2 content of venous blood primarily reflects the bicarbonate level. It will increase in metabolic alkalosis and decrease in metabolic acidosis. Note that CO_2 content is different from the $Paco_2$ measurement included in an arterial blood gas analysis.

Anion gap

The normal range for an anion gap is 12 (+ or − 2) mEq/L. Anion gap reflects the normally unmeasured anions (e.g., phosphates, sulfates, and proteins) in the plasma. Anion gap = $Na^+ - (Cl^- + HCO_3^-)$. Measurement of the anion gap (Figure 5-11) may be helpful in the differential diagnosis of metabolic acidosis or in identifying hidden metabolic acidosis in certain mixed acid-base disorders.

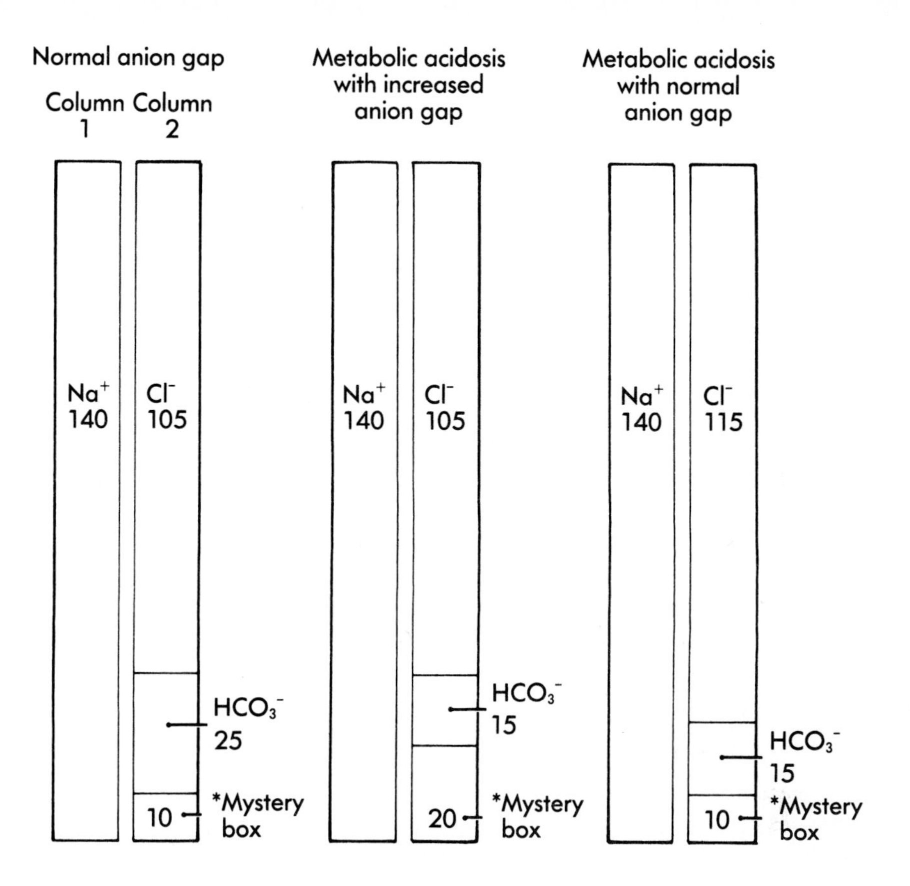

Figure 5-11 Anion gap and metabolic acidosis.
(From Horne M and Swearingen P: Pocket guide to fluids and electrolytes, St Louis, 1989, The CV Mosby Co.)

To understand anion gap, think of the extracellular fluid (ECF) as two equal-sized columns—one containing cations (positively charged ions) and the other containing anions (negatively charged ions). Since electroneutrality (an equal number of negative and positive charges) is maintained at all times within the body, the number of cations and anions (or the size of the two columns, as shown in Figure 5-11) always must be equal. Because sodium is the body's primary cation, the overall size of Column 1 may be determined by measuring serum sodium. The two primary anions of the ECF are chloride and bicarbonate. Their sum (Cl^- mEq/L plus HCO_3^- mEq/L) does not completely fill Column 2. The remaining portion of Column 2 (mystery box) represents the normally unmeasured anions of the ECF. This mystery box (anion gap) may be determined by obtaining a serum electrolyte panel and using the following formula:

$$\text{Anion gap (mystery box)} = Na^+ - (Cl^- + HCO_3^-)$$

The size of the anion gap is significant since the causes of metabolic acidosis fall into two categories: (1) those that cause an increase in unmeasured anions and thus increase the mystery box/anion gap, and (2) those that result from a loss of bicarbonate or an ingestion or administration of acidifying salts and do not alter the mystery box/anion gap. See the accompanying boxes and Chapters 13 and 17 for additional information.

Causes of Metabolic Acidosis with a Normal Anion Gap

GI Loss of HCO_3^-

- Diarrhea
- Lower GI fistulas

Renal Loss of HCO_3^-

- Renal tubular acidosis (type II)
- Acetazolamide (Diamox)

Inability of the Kidneys to Excrete H^+

- Early renal insufficiency
- Renal tubular acidosis (types I and IV)
- Diuretics: Triamterene (Dyrenium), spironolactone (Aldactone), Amilioride

Addition of Acidifying Salts

- Ammonium chloride
- Hyperalimentation fluids without adequate bicarbonate or bicarbonate-producing solutes (e.g., lactate or acetate)

Causes of Metabolic Acidosis with an Increased Anion Gap

Retention of Acids

- Renal failure

Abnormal Production of Acids

- Ketoacidosis
- Lactic acidosis
- Massive rhabdomyolysis

Ingestion of Acids

- Salicylates
- Methanol
- Ethylene glycol

Urine pH

The normal range of urine pH for a random specimen is 4.6 to 8. The kidneys play a critical role in the regulation of acid-base balance by excreting a portion of the hydrogen ions (H^+) produced each day. In spite of a **buffering system** within the renal tubule, which allows maximal excretion of H^+ with minimal decrease in urinary pH, the **pH** of the urine usually is markedly acidic (averaging approximately 6.0). Measurement of urine pH may be useful in determining if the kidneys are responding appropriately to metabolic acid-base imbalances. Urine pH should decrease in metabolic **acidosis** and increase in metabolic alkalosis. An inappropriately high urine pH in the presence of metabolic acidosis, for example, suggests renal tubular acidosis (a group of disorders that inhibit renal excretion of H^+). An inappropriately low pH in the presence of metabolic alkalosis may signal volume depletion (i.e., sodium bicarbonate is retained as the kidneys attempt to correct the volume deficit by conserving all filtered sodium). Urinary tract infections with pathogens that produce urease cause an alkaline urine due to excess ammonia production. Urine pH should be measured within 1 to 2 hours of collection.

Lactic acid

The normal arterial value for lactic acid is 0.5 to 1.6 mEq/L; the venous value is 1.5 to 2.2 mEq/L. Lactic acid is a by-product of the anerobic metabolism of glucose. Normally, the small quantity of lactic acid that is produced daily is immediately buffered by bicarbonate and lactate is generated. This lactate is then converted to CO_2 and H_2O or glucose by the liver and kidneys and HCO_3^- is regenerated. Any time there is an excess production of lactic acid (e.g., when there is a decreased oxygen delivery to the tissue) or decreased utilization of lactate, dangerous lactic acidosis may develop. (See the accompanying box.)

Factors Affecting Lactic Acid Values

Factors that may lead to the development of lactic acidosis:

- Increased production
  - Strenuous exercise
  - Shock
  - Cardiac arrest
  - Carbon monoxide poisoning
  - Hypoxemia

Factors that may lead to decreased utilization of lactate:

- Liver disease
- Severe acidosis

Related Tests

Creatinine

The normal range for creatinine is 0.6 to 1.5 mg/dl. Creatinine is a metabolic waste product produced by the breakdown of muscle creatine. The serum creatinine level reflects the balance between its production and excretion by the kidneys. Since it is produced at a steady rate dependent on muscle mass and is not affected by diet, hydration, or tissue catabolism, the creatinine level is a more accurate indicator of renal function than BUN. The serum creatinine level will increase as renal function decreases.

Creatinine clearance

The normal range for creatinine clearance is 107 to 141 ml/min for the adult male and 87 to 132 ml/min for the adult female. A creatinine clearance test provides an estimate of the glomerular filtration rate, and thus renal function. It measures the volume of blood completely cleared of creatinine over a specific period of time, usually 24 hours. It is expressed in ml/min and factored to consider body surface area. The value is determined by comparing the plasma creatinine level drawn sometime during the test period with the amount of creatinine excreted in the urine during the full test period, using the following formula:

$$\frac{\text{Urine creatinine} \times \text{Urine volume}}{\text{Plasma creatinine}}$$

Changes in creatinine clearance reflect similar changes in renal function. A drop of 75% in creatinine clearance, for example, indicates a 75% reduction in renal function. Since renal function decreases with advancing age, creatinine clearance drops approximately 10% with each decade past the age of 50. The accuracy of the test depends on complete collection of all the urine passed during the test period. Timing of the test period should begin after the individual has voided (this initial sample is discarded). Collection continues for the full time period (usually 24 hours). The individual empties his or her bladder (this specimen is saved) just prior to the end of the test.

Serum albumin

The normal range for serum albumin is 3.5 to 5.5 g/dl. Albumin is a small plasma protein produced by the liver that acts osmotically to help hold the intravascular

Factors that May Decrease Serum Albumin

- Decreased protein intake (e.g., protein malnutrition)
- Decreased hepatic synthesis (e.g., cirrhosis)
- Abnormal urinary loss (e.g., nephrotic syndrome)

volume in the vascular space. Decreased serum albumin (hypoalbuminemia) may lead to the development of edema secondary to a reduction in oncotic pressure, causing the movement of water out of the vascular space and into the interstitial space. The edema seen in protein malnutrition occurs due to decreased albumin production. (See the accompanying box on p. 92 for factors that may decrease serum albumin.)

Case Study 5-1

J.K. is a 54-year-old man just admitted to the oncology floor with a primary diagnosis of oat cell carcinoma of the lungs. He is a patient of the medical house staff, and the resident caring for him tells the nurse in charge "With his history and diagnoses, we will have to monitor him closely for volume and electrolyte problems." Unfortunately, before she had a chance to elaborate, the resident was paged to the emergency department. Her written orders for J.K. were as follows:

1. Admit to medical floor.
2. Diagnoses: CA of the lung, COPD, R/O gastroenteritis, and R/O SIADH.
3. Regular diet if tolerated.
4. Activity: up ad lib.
5. Insert heparin lock for IV meds.
   If diet not tolerated, begin IV of 5% dextrose in 0.9% NaCl at 50 cc/hr.
7. Stat labs: serum electrolytes, serum osmolality, arterial blood gases, urine osmolality, and random urine sodium. Call results to physician.
8. Serum electrolytes, BUN, creatinine, and thyroid panel in the AM.
9. Stool culture including ova and parasites.
10. Chest x-ray now.
    Meds: furosemide (Lasix) 20 mgm IV bid, diphenoxylate HCl with atropine (Lomotil) 2.5 mg PO qid, ranitidine (Zantac) 50 mg IV q8h.
12. Oxygen at 2 L/min via nasal cannula.

1. **Question:** Considering the physician's concern that J.K. might experience problems with fluid and electrolyte balance, what nursing orders should be included in this patient's care?

Answer:

a. Daily weights
b. Measurement and recording of all intake and output

Before obtaining a nursing history and performing a complete physical assessment, the nurse caring for J.K. does a quick mental review of the medical diagnoses and their possible effects on fluid and electrolyte balance. SIADH sometimes occurs with pulmonary tumors due to ectopic production of an ADH-like substance. Since ADH acts on the distal tubule and collecting duct in the kidney to cause the retention of water, abnormal production will lead to abnormal retention of water. Abnormal retention of water, in turn, may cause symptomatic hyponatremia (a low plasma sodium) and hypoosmolality. Medical management may include fluid restriction, a high-sodium, high-protein diet (to increase solute excretion, thereby increasing the water that must be excreted with it), medications such as lithium or demeclocycline (Declomycin), which suppress the action of ADH, and in severe cases hypertonic NaCl (3%) combined with furosemide. J.K.'s lung cancer and COPD (chronic obstructive pulmonary disease) may adversely affect oxygenation and excretion of carbon dioxide by the lungs, which if severe enough may result in hypoxemia and respiratory acidosis. Gastroenteritis may lead to (1) fluid and electrolyte imbalance secondary to decreased intake of food and fluids due to nausea and (2) increased GI losses due to vomiting and diarrhea. The medication furosemide (a potent diuretic) will cause increased urinary losses of water and electrolytes.

2. Question: What are the five dimensions of the health history that this nurse must include in assessing the patient?

Answer:

a. Physiologic
b. Developmental
c. Psychologic
d. Spiritual
e. Sociocultural

3. Question: J.K. states that today he has vomited four times, passed two loose, watery stools, and currently is nauseated and having abdominal cramps. Which fluid and electrolyte imbalances might occur with the loss of upper GI fluids due to vomiting? (Review Chapter 4, if necessary.)

Answer:

a. Fluid volume deficit
b. Decreased serum sodium (hyponatremia)
c. Decreased serum potassium (hypokalemia)
d. Alkalosis secondary to the loss of acidic gastric contents

The development of alkalosis secondary to the loss of gastric contents will be discussed further in Chapter 18.

4. Question: What are some nursing interventions that might help to relieve J.K.'s nausea and vomiting?

Answer:

a. Maintain a calm and quiet atmosphere.
b. Avoid sudden or jerking motions when repositioning.
c. Promptly remove any soiled linen and minimize any strong or noxious odors.
d. Provide mouth care regularly and after each emesis.
e. Offer cool compresses to the forehead.
f. Offer sips of tea with lemon or gingerale.
g. Obtain prescription for antiemetic medication.

5. Question: Although the electrolyte composition of diarrhea is different from that of plasma, it is isotonic in concentration. What might you expect to happen to J.K.'s serum osmolality, effective circulating volume, and urinary sodium if his diarrhea continues?

Answer: Initially the serum osmolality would remain unchanged because the loss is isotonic. The effective circulating volume would be decreased secondary to the loss of lower GI fluids. If J.K. were to drink water to replace the loss, his serum osmolality eventually would decrease. ADH-induced water retention also would contribute to the development of hypoosmolality. The urinary sodium level would be expected to decrease secondary to a hypovolemia-induced increase in the release of aldosterone.

6. Question: In reviewing the medications that he takes at home, the nurse determines that J.K. is not currently taking any diuretics. Before administering the first dose of IV furosemide (Lasix), the nurse discusses its indications and actions with J.K. and obtains the laboratory specimen for the urine sodium and osmolality. Why is it important that the urine sodium and osmolality be obtained prior to administering the furosemide?

Answer: Most diuretics, including furosemide, work by decreasing the renal reabsorption of sodium and water. Thus they will affect urine sodium and urine osmolality values.

7. Question: If J.K. does have SIADH, how will that affect his plasma and urine osmolality? Remember that osmolality is a measurement of the concentration of body fluids and that SIADH causes an abnormal retention of water.

Answer: The serum osmolality will be decreased, reflecting a dilute serum. The normal physiologic response to hypoosmolality is to decrease the production of ADH, thereby increasing the excretion of water. But in this case there is an abnormal production of ADH by the pulmonary tumor causing the kidneys to save water inappropriately; therefore the urine osmolality will be inappropriately high.

8. Question: As discussed in this chapter, what changes in the neurologic system might this nurse observe that suggest altered serum sodium and serum osmolality?

Answer: Changes in level of consciousness. Additional signs and symptoms of hyponatremia will be discussed in Chapter 8.

9. Question: Which of the tests ordered for J.K. will evaluate his pulmonary function and acid-base balance?

Answer: The arterial blood gases (ABGs).

10. Question: Which of the tests ordered for J.K. will evaluate his renal function?

Answer: The BUN and creatinine levels.

References

Adams F: How much do elders drink? Geriatr Nurs 9(4):218-221, 1988.

Baas L: Cardiovascular dysfunctions. In Swearingen PL and others: Manual of critical care: applying nursing diagnoses to adult critical illness, St Louis, 1988, The CV Mosby Co.

Beare PG and others: Nursing implications of diagnostic tests, ed 2, Philadelphia, 1985, JB Lippincott Co.

Corbett JV: Laboratory tests and diagnostic procedures with nursing diagnoses, ed 2, Norwalk, Conn and San Mateo, Calif, 1987, Appleton & Lange.

Cox P: Insensible water loss and its assessment in adult patients: a review. Acta Anaesthesiol Scand 31:771-776, 1987.

De Growin RL: De Growin and De Growin's bedside diagnostic examination, ed 5, New York, 1987, Macmillan Publishing Co.

Ellis K: One piece of paper we can't do without, RN 51(8):110, 1988.

Folk-Lighty M: Solving the puzzles of the patient's fluid imbalance, Nursing 84 14(2):34-41, 1984.

Gaspar PM: What determines how much patients drink? Geriatr Nurs 9(4):221-224, 1988.

Geheb MA: Clinical approach to the hyperosmolar patient, Critical Care Clinics 5:797-815, 1987.

Guyton AC: Textbook of medical physiology, ed 7, Philadelphia, 1986, WB Saunders Co.

Hazinski MF: Understanding fluid balance in the seriously ill child, Pediatr Nurs 14(3):231-236, 1988.

Holloway NM: Nursing the critically ill adult, ed 3, Menlo Park, Calif, 1988, Addison-Wesley Publishing Co, Inc.

Horne M and Swearingen PL: Pocket guide to fluids and electrolytes, St Louis, 1989, The CV Mosby Co.

Lattanzi WE and Siegel NJ: A practical guide to fluid and electrolyte therapy, Curr Probl Pediatr 16(1):1-43, 1986.

Metheny NM: Fluid and electrolyte balance: nursing considerations, Philadelphia, 1987, JB Lippincott Co.

Potter PA: Pocket nurse guide to physical assessment, St Louis, 1986, The CV Mosby Co.

Potter PA and Perry AG: Clinical nursing skills and techniques, St Louis, 1986, The CV Mosby Co.

Swearingen PL: The Addison-Wesley photo-atlas of nursing procedures, ed 2, Redwood City, Calif, 1991, Addison-Wesley Publishing Co, Inc.

Thompson JM and others: Clinical nursing, ed 2, St Louis, 1989, The CV Mosby Co.

Chapter

6

Disorders of Fluid Balance

FLUID VOLUME DEFICIT

A reduction in the volume of the extracellular fluid (ECF) is termed fluid volume deficit or **hypovolemia.** Depending on the cause, hypovolemia may be accompanied by acid-base, osmolar (concentration), or electrolyte imbalances. Volume disturbances associated with changes in serum sodium (and thus changes in concentration) will be discussed in Chapter 8. Compensatory mechanisms in hypovolemia include increased sympathetic nervous system stimulation (i.e., increased heart rate, increased inotropy—cardiac contraction, increased vascular resistance), increased thirst, increased release of antidiuretic hormone (ADH), and increased release of aldosterone. Severe ECF volume deficit can lead to hypovolemic shock. Prolonged hypovolemia may lead to the development of acute renal failure.

Fluid volume deficit may develop because of decreased intake, abnormal fluid losses, or a combination of the two. Abnormal fluid loss may occur as a result of increased skin, GI (gastrointestinal), or renal losses; bleeding; or movement of fluid into a nonequilibrating third space.

History and Risk Factors

Abnormal GI losses

As discussed in Chapter 4, the gastrointestinal tract is the most common site of abnormal fluid and electrolyte loss. Because of the large volume of fluid that moves through the GI tract each day, abnormal losses may lead to significant fluid loss. Causes of abnormal GI losses include vomiting, nasogastric (NG) suctioning, diarrhea, intestinal fistulas, and intestinal drainage. The GI tract is also one of the most common sites for third-space fluid loss. Bowel surgery or bowel obstruction may result in third-space fluid shifts caused by sequestering of fluid within the bowel lumen. In addition, abuse of laxatives or enemas may result in fluid volume deficit. In individuals undergoing diagnostic studies of the colon, volume deficit also may develop because of the combination of restricted intake and bowel cleansing with purgatives and enemas.

Abnormal skin losses

Insensible fluid loss from the skin may increase because of fever or use of radiant warmers or phototherapy in infants. Sensible fluid loss (sweat) increases with activity and increased ambient temperature. Sensible fluid helps to dissipate the body heat that is generated by oxidative metabolism and to prevent the body from gaining heat when the ambient temperature is greater than that of the body. Human survival in hot climates depends on the ability to dissipate body heat effectively through sweating. This requires adequate circulation to the skin. The combination of hypovolemia and hyperthermia may be lethal because of the body's limited ability to dissipate body heat. Sensible fluid loss is variable and can be as great as 2 liters an hour. Marathon runners, for example, can lose as much as 6% to 7% of their body weight by sweating (Better, 1987). Intact skin is an important barrier to abnormal fluid loss; thus burns and large wounds will contribute to the development of fluid volume deficit.

Abnormal renal losses

Normally the kidneys adjust urine output to match intake, thus helping to maintain fluid volume balance. Certain disorders and medications, however, affect the kidneys' ability to customize the urine output to the body's needs, thereby causing or contributing to the development of fluid volume deficit. Diuretics inhibit the reabsorption of sodium and water by the renal tubules and when used to excess, may lead to fluid volume deficit. The risk of hypovolemia is greatest with such diuretics as furosemide and ethacrynic acid, which work in the loop of Henle, since they affect the largest percentage of filtered sodium. Certain substances, such as mannitol, glucose, or contrast media (dye), will act as osmotic diuretics, inhibiting the reabsorption of water.

Polyuric forms of renal failure will result in an abnormal loss of fluids because of the diseased kidneys' inability to reabsorb sodium and water appropriately. Decreased production of or decreased kidney responsiveness to ADH will result in the production of a large amount of dilute urine (diabetes insipidus—refer to Case Study 3-2 for a clinical example). Adrenal insufficiency will cause an increased production of urine secondary to decreased production of aldosterone. Remember that aldosterone is a hormone produced by the cortical portion of the adrenal gland that acts on the renal tubule to increase the reabsorption of sodium (and thus water).

Third-space fluid shifts

Third-space fluid shifts, as described in Chapter 4, will result in a decrease in the volume of blood available to perfuse the tissues (see Table 4-5 for a description of third-space fluid shifts). Although there has not been an actual loss of fluid from the body, there is a reduction in the effective circulating volume. Examples of third-space fluid shifts include internal bleeding, ascites, accumulation of fluid in the peritoneal cavity with acute peritonitis, or plasma-to-interstitial fluid shift with tissue trauma or burns.

Hemorrhage

Uncontrolled bleeding may occur from trauma, surgery, childbirth, or clotting disorders. Significant blood loss also may occur with accidental disconnection of IV lines or vascular catheters. Whenever possible, all blood loss should be measured or estimated. Saturated dressings or bed linen may be weighed to estimate loss.

Decreased intake

Decreased fluid intake may occur because of unavailability of fluids (for example, the individual lost in the desert), altered thirst mechanism, or inability to express thirst. As discussed in previous sections, older adults are at increased risk for fluid volume deficit in part due to a less sensitive thirst mechanism. They are less inclined to seek fluids in response to a volume deficit. This may be aggravated by problems with mobility. Confused or comatose individuals are unable to express their thirst and seek adequate oral intake. Likewise, infants have a limited ability to express their thirst and are dependent on their caregivers to interpret their cries correctly.

Inadequate intake is more likely to lead to hypovolemia in the infant and the older adult because of greater obligatory fluid losses. In addition, infants have a proportionately greater body surface area than adults. Thus they have a proportionately greater daily fluid loss from the skin. Each day infants lose and must replace a volume of fluid equal to one third their ECF volume. In contrast, adults normally lose and replace a volume of fluid equal to only one sixth of their ECF volume.

Assessment

The signs and symptoms of fluid volume deficit will depend on the severity of the deficit and how quickly it develops. Mild deficits that develop gradually, as might occur with diarrhea, will cause few symptoms. In contrast, large volume losses that develop quickly, as in acute hemorrhage, will cause the most dramatic changes. Signs and symptoms of fluid volume deficit occur as a result of changes in both the intravascular and interstitial fluid volumes as well as overall volume loss. See Table 6-1 for quick assessment guidelines for individuals with fluid volume deficit.

Intravascular deficit

Flattened jugular veins

A reduction in vascular volume will cause a decrease in the amount of blood in the venous circulation and thus a reduction in venous pressure. Clinically the venous pressure may be assessed by measuring the level of distention in the jugular veins.

Table 6-1 Quick assessment guidelines for individuals with fluid volume deficit

Signs and Symptoms	Physical Assessment	Hemodynamic Measurements
Dizziness, weakness, fatigue, syncope, anorexia, nausea, vomiting, thirst, confusion, constipation, oliguria	↓ BP, especially when standing (orthostatic hypotension); ↑ HR; poor skin turgor; dry, furrowed tongue; sunken eyeballs; depressed fontanel and absent tearing in the infant Flattened neck veins; ↑ temperature; weight loss (except with third spacing); pallor; diaphoresis	↓ CVP, ↓ PAP, ↓ CO, ↓ MAP, ↑ SVR

History and Risk Factors

Abnormal GI losses: Vomiting, NG suctioning, diarrhea, intestinal drainage
Abnormal skin losses: Insensible losses secondary to fever; excessive diaphoresis secondary to ↑ ambient temperature or exercise; burns
Abnormal renal losses: Diuretic therapy, diabetes insipidus, renal disease (polyuric forms), adrenal insufficiency, osmotic diuresis (e.g., uncontrolled diabetes mellitus, postdye study)
Third spacing or plasma-to-interstitial fluid shift: Peritonitis, intestinal obstruction, burns, ascites
Hemorrhage: Major trauma, GI bleeding, obstetric complications
Altered intake: Coma, fluid deprivation

With the individual reclining at a 30- to 45-degree angle (see Figure 5-7), the distance between the level of the sternal angle (angle of Louis) and the point at which the internal and external jugular veins collapse is measured. Normally this distance should be about 2 to 3 cm. In fluid volume deficit, however, the venous pressure may be so low that there is no measurable vein distention with the individual in this position. When the head is lowered, neck vein distention may then be noted. Flat neck veins, when the individual is fully reclined, is strongly suggestive of ECF volume deficit.

Decreased cardiac output (CO)

A reduction in the volume of blood returning to the heart will lead to a drop in cardiac output (CO) (the volume of blood pumped by the heart each minute) secondary to a drop in stroke volume. Since the systemic blood pressure is determined in part by cardiac output, a drop in cardiac output will result in a drop in blood pressure. As discussed in Chapter 3, a drop in blood pressure activates the sympathetic nervous system (SNS). Stimulation of the SNS causes an increase in heart rate, cardiac contractility, and vascular resistance. Each of these changes results in rapid restoration of normal blood pressure. Thus early symptoms of fluid volume deficit are the result of SNS-induced compensatory mechanisms.

In addition to direct action on the heart and blood vessels, stimulation of the SNS increases the release of renin, thus activating the renin-angiotensin-aldosterone system, resulting in the retention of sodium and water by the kidneys. Fluid volume deficit also increases the release of ADH secondary to decreased atrial stretch. ADH acts on the renal collecting ducts to enhance water conservation.

Hypotension

In a mild volume deficit (less than 10% of the blood volume), blood pressure may return to normal or near normal with SNS compensation. The only symptoms may be a slight increase in heart rate and decreased jugular vein distention. With increasing volume deficit, the blood pressure may be maintained only while the individual is recumbent. Hypotension and increasing tachycardia may develop when the individual suddenly assumes a sitting or standing position. This is termed *orthostatic or postural hypotension* and may be an early clue to the presence of fluid volume deficit. A drop of 15 mm Hg in either the systolic or diastolic BP or an increase of 20 beats per minute or greater in the heart rate is considered significant. In moderate to severe volume deficit, SNS compensation is no longer adequate to maintain blood pressure. Hypotension then will be present when the individual is lying flat. The pulse pressure usually narrows as a result of a drop in systolic pressure. The diastolic pressure remains normal or may even rise secondary to SNS-induced vasoconstriction. Hemodynamic monitoring will reveal decreased central venous pressure (CVP), decreased pulmonary artery pressure (PAP), decreased cardiac output (CO), decreased **mean arterial pressure (MAP),** and increased systemic vascular resistance (SVR).

Hypotension will cause additional symptoms related to decreased tissue perfusion. Muscle weakness and fatigue are caused by decreased nutrient delivery to the muscles. A reduction in cerebral perfusion may cause dizziness, lethargy, confusion, and syncope. Chest pain or the development of dysrhythmias suggests coronary ischemia, while anorexia, nausea, vomiting, or abdominal pain suggests decreased mesenteric blood flow. Capillary refill will be delayed (greater than 4 seconds).

Decreased urinary output

A drop in urine output may be an early clue to fluid volume deficit. Initially, urine output decreases as the kidneys attempt to compensate for volume depletion by retaining sodium and water (via the actions of aldosterone and ADH). If hypotension is severe or prolonged, urine output drops because of decreased glomerular filtration. Remember that urine formation begins with glomerular filtration and that glomerular filtration is dependent on the hydraulic pressure (i.e., blood pressure) within the glomerulus. Also remember that the kidneys' ability to autoregulate blood flow is decreased in the presence of SNS stimulation and absent when the mean arterial pressure drops below 70 mm Hg. An hourly urine output of less than 30 ml suggests hypovolemia. Keep in mind, though, that factors other than volume deficit may alter renal perfusion and the production of urine.

Thirst

Thirst may provide another important clue to the presence of fluid volume deficit. Increased thirst develops from release of angiotensin II and from dryness of the mouth secondary to decreased production of oral secretions. Thirst also may be caused by increased or unreplaced loss of hypotonic fluid (e.g., sweat and insensible fluid). Recall that thirst is stimulated by an increase in the concentration of body fluids, as might occur with the loss of hypotonic fluids. Thirst usually does not develop until at least 10% of the intravascular volume has been lost or the serum osmolality exceeds normal.

Shock

Shock may develop when there is a loss of greater than 20% to 25% of the intravascular volume. In addition to the above symptoms, the individual in shock will appear pale and diaphoretic with cool, clammy extremities and a markedly decreased urine output (less than 15 ml/hr). In severe shock, the systolic BP will drop below 80 mm Hg. The risk of acute renal failure secondary to decreased renal perfusion is great.

Interstitial deficit

A reduction in the volume of the interstitial space will be evidenced by a decrease in the turgor or tone of body tissues. Skin turgor may be assessed by pinching skin over the forearm, sternum, or dorsum of the hand. With adequate hydration the pinched skin returns quickly to its original position when released. In fluid volume deficit the skin will flatten more slowly and will tend to remain *tented.* Remember, though, that skin turgor is a less reliable indicator of fluid volume deficit in the older adult because of decreased skin elasticity. Changes in skin turgor also may be a less reliable sign in the infant or small child due to changes in subcutaneous fat. Skin turgor may appear normal in the obese infant with hypovolemia owing to increased subcutaneous fat, or abnormal in the adequately hydrated but undernourished infant.

Decreased interstitial fluid also will cause a reduction in the size of the tongue and an increase in its longitudinal furrows. The skin and oral cavity will be dry from decreased secretions. The eyeballs may be soft and sunken due to decreased fluid tension. The loss of tearing in the older infant or child also may occur with fluid volume deficit. The anterior fontanel, which is normally soft and flat, will be depressed in moderate to severe volume depletion.

Overall fluid deficit

With the exception of fluid volume deficit secondary to third-space fluid shifts or plasma-to-interstitial fluid shifts, body weight will decrease in hypovolemia. As discussed in Chapter 5, acute changes in body weight *usually* reflect changes in fluid

Table 6-2 Weight loss as an indicator of ECF deficit in adults and children

Acute Weight Loss (%)	Severity of Deficit
2 to 5	Mild
5 to 10	Moderate
10 to 15	Severe
15 to 20	Fatal

volume (Table 6-2). In the individual who is being fed, a decrease in daily weight of a kilogram, for example, is equal to the loss of a liter of fluid. Keep in mind, though, that the adult who is not eating or is not being maintained through enteral or parenteral nutrition, will lose one quarter of a kilogram of weight daily because of the actual loss of body tissue. This weight loss does not reflect volume loss.

Diagnostic Studies

With the exception of hemodynamic monitoring (see Table 6-1), there are no diagnostic tests specific to fluid volume deficit. Diagnosis depends largely on physical findings. The BUN and hematocrit, however, may provide clues to the presence of fluid volume deficit.

Blood urea nitrogen (BUN) and creatinine

In the absence of renal dysfunction, an elevated BUN usually indicates volume deficit. Remember that the normal renal response to hypovolemia is to conserve sodium and water. Since urea passively follows the water, increased reabsorption of water will result in increased reabsorption of urea. Thus urea excretion will decrease and serum levels will increase. Often the BUN and creatinine are compared to determine if an elevated BUN reflects decreased renal function, some other cause such as hypovolemia, or a combination of the two. If an elevated BUN is solely the result of decreased renal function, the serum creatinine will increase at approximately the same rate, and the creatinine-to-BUN ratio will be 1:10-15. If the BUN increases at a greater rate, then it suggests that some other factor, such as hypovolemia, is involved.

Hematocrit

Since the hematocrit measures the percentage of whole blood that is composed of red blood cells, it will be affected by changes in the volume of the plasma. Decreases in the plasma volume, as occur in hypovolemia, will increase the percentage of the blood that is composed of red blood cells (RBCs). Thus the hematocrit usually increases with fluid volume deficit (Figure 6-1). The one exception, of course, is hypovolemia secondary to bleeding. The hematocrit remains normal immediately

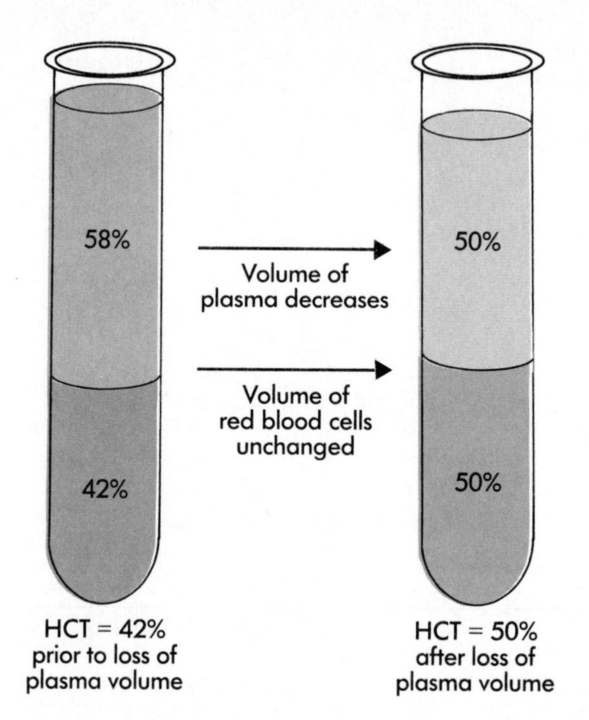

Figure 6-1 An example of the effect of plasma volume loss on hematocrit.

following acute hemorrhage because RBCs and plasma are lost in equal amounts. Over a period of hours, however, there will be a shift of fluid from the **interstitial fluid (ISF)** to the plasma and the hematocrit will drop.

Serum electrolyte levels

Serum electrolyte levels will vary depending on the type of fluid lost. Hypokalemia often occurs with abnormal GI or renal losses. Hyperkalemia often occurs with adrenal insufficiency. Hypernatremia may be seen with increased insensible or sweat losses and diabetes insipidus resulting from the loss of hypotonic fluid. Hyponatremia occurs in most types of hypovolemia because of increased thirst and ADH release, which lead to increased water intake and retention, thus diluting the serum sodium. See individual electrolyte disorders, Chapters 8 to 12 for additional information.

Acid-base changes

Changes in acid base balance also depend on the type of fluid that is lost and the severity of the deficit. Metabolic acidosis (pH <7.40 and HCO_3^- <22 mEq/L) may occur with lower GI losses, shock, or diabetic ketoacidosis. Metabolic alkalosis (pH >7.45 and HCO_3^- >26 mEq/L) may occur with upper GI losses or diuretic therapy. For a discussion of the relationship between shock and the development of metabolic acidosis, refer to Chapter 17. See Chapter 18 for a discussion of the relationship of volume depletion to the development and maintenance of metabolic alkalosis.

Urine specific gravity and osmolality

Urine specific gravity and urine osmolality both are increased in fluid volume deficit, except when the deficit is caused by abnormal renal losses. As the kidneys compensate for fluid volume deficit by retaining sodium and water, the urine becomes increasingly concentrated. Remember that unlike serum osmolality, urine osmolality (concentration) is determined by the concentration of nitrogenous wastes, not sodium.

Urine sodium

In hypovolemia due to causes other than renal, urine sodium will be low, demonstrating the kidneys' ability to conserve sodium in response to an increased aldosterone level. This will not be the case, however, when hypovolemia is the result of adrenal insufficiency or polyuric renal failure.

Serum osmolality

The serum osmolality will vary, depending on the type of fluid lost and the body's ability to compensate with thirst and ADH. As discussed, though, hyponatremia (and thus hypoosmolality) is common.

Medical Management

The primary goals of medical management are to treat or correct the precipitating event and to restore normal fluid volume, correcting any accompanying acid-base or electrolyte disturbances. The type of fluid used and the route of fluid replacement depend on the nature of the loss, severity of the deficit, serum electrolytes, serum osmolality, and acid-base status. Fluid may be replaced orally, enterally, or parenterally (see the accompanying box on p. 106). Parenteral fluid therapy is discussed in greater detail in Chapter 7.

Fluid challenge

Fluids should be administered rapidly enough and in sufficient quantity to maintain adequate tissue perfusion without overloading the cardiovascular system. Underlying cardiac and renal function will determine how well the individual will tolerate fluid

Commonly Prescribed IV Fluids

Crystalloids

- Dextrose and water solutions: Provide free water only and will be distributed evenly throughout both the ICF and ECF. These solutions are used to treat total body water deficits.
- Isotonic (*normal*) saline: Expands ECF only; does not enter ICF. Usually it is used as an intravascular volume expander or to replace abnormal losses.
- Mixed saline/electrolyte solutions: Provide additional electrolytes (e.g., potassium and calcium) and a buffer (lactate or acetate). Usually hypotonic solutions are used as maintenance fluids, while isotonic solutions are used as replacement fluids since most abnormal fluid losses are isotonic.

Colloids

- Blood, albumin, fresh frozen plasma: Expand only the intravascular portion of the ECF. Blood may be administered as whole blood (approximately 500 ml/unit, contains all blood components) or as packed RBCs (approximately 250 to 300 ml/unit, contains less plasma and no clotting factors). Albumin is produced in two concentrations: 5% and 25%. Five percent albumin is osmotically equivalent to plasma and will expand the plasma volume one ml for each ml given. In contrast, 25% albumin is hypertonic and will expand the plasma by 3 to 4 ml for each ml given (Rice, 1984).

replacement. Thus the rate of fluid administration should be based on both the severity of the loss and the individual's hemodynamic response to volume replacement. During a fluid challenge, volumes of IV fluid are administered at specific rates and intervals, and the patient's hemodynamic response is monitored and documented. A fluid challenge may be useful in determining the cause of oliguria. If oliguria is the result of hypovolemia, the urine output will increase with the administration of fluid. The urine output may remain unchanged when the primary problem is decreased renal or cardiac function. Hemodynamic parameters for the fluid challenge are prescribed by the MD or, depending on agency policy, determined by a fluid challenge protocol. A typical fluid challenge includes the following steps:

- Baseline VS, hemodynamic measurements (e.g., CVP, PAP, and CO), and clinical data (e.g., breath sounds, skin color and temperature, and sensorium) are obtained.
- Initial volume of fluid is administered as prescribed or per protocol (e.g., 100 to 200 ml normal saline [NS] over 10 minutes).
- At the end of the 10-minute infusion period the patient is reassessed.
- If the patient continues to demonstrate signs of hypovolemia (i.e., CVP and PAP remain low), additional fluid may be administered per MD or agency protocol.

- If the CVP or PAP increases too rapidly (e.g., >2 mm Hg for the CVP or >3 mm Hg for the PAP), fluid administration is discontinued and the patient is reassessed after 10 minutes. If after 10 minutes the CVP or PAP has dropped and the patient shows no signs of fluid overload, fluid administration is resumed.
- Fluid administration is continued until the desired hemodynamic parameters are achieved or a specific volume has been infused (e.g., 500 to 1000 ml). The fluid challenge should be discontinued anytime the patient shows signs of fluid volume excess (e.g., crackles/rales, increased HR, increased RR) or there is a rapid increase in CVP or PAP. Continued nursing assessment is essential during and after the fluid challenge.

Treatment of hypovolemic shock

Hypovolemic shock develops when the intravascular volume has decreased to the point that compensatory mechanisms are unable to maintain adequate tissue perfusion and normal cellular metabolism. Decreased delivery of nutrients to the tissues results in acidosis, cardiac depression, intravascular coagulation, increased capillary permeability, and release of toxins. If shock is allowed to progress too long without treatment, or with inadequate treatment, it may become irreversible and no amount of intervention will prevent death. Hence the treatment of hypovolemic shock involves rapid and vigorous volume replacement. The type of fluid recommended for treatment remains controversial. The controversy centers on whether to administer crystalloids alone or a combination of **crystalloids** and **colloids.** One of the concerns with rapid volume replacement is the loss of fluid into lung tissue (pulmonary edema). Some authorities believe that this may be minimized by the administration of colloids, while others feel that colloid administration may actually increase the risk of pulmonary edema (Strum and Wisner, 1985). Regardless of the choice of fluids, an important nursing function during the treatment of hypovolemic shock is continuous assessment of pulmonary status.

Initially, hypovolemic shock, secondary to hemorrhage, is treated with an isotonic electrolyte solution and blood is given as the hematocrit level drops. A balanced electrolyte solution such as Ringer's lactate is recommended since isotonic saline (0.9% NaCl) contains excessive amounts of sodium and chloride (see Chapter 7, the section on fluid therapy, p. 137). Medications that cause increased cardiac contractility, such as dopamine, or that cause vasoconstriction, such as levophed, are used only when hypovolemic shock does not improve with volume replacement.

Oral rehydration in pediatric diarrhea

As discussed in Chapter 4, diarrhea is a common source of abnormal fluid loss in infancy and early childhood. By the age of 3, most children have experienced one to three episodes of acute diarrhea (Tucker and others, 1987). Oral rehydration

Table 6-3 Composition of oral rehydration solutions

Formula	Na^+ (mEq/L)	K^+ (mEq/L)	Cl^- (mEq/L)	Base (mEq/L)	Carbohydrate (g/L)
Lytren (Mead-Johnson)	50	25	45	30 (citrate)	20 (dextrose, corn syrup, solids)
Pedialyte	45	20	35	30 (citrate)	25 (dextrose)
Rehydrolyte	75	20	65	30 (citrate)	25 (dextrose)
Infalyte powder (Pennwalt)	50	20	40	30 (bicarbonate)	20 (glucose)
WHO (World Health Organization)	90	20	80	30 (bicarbonate)	18 (dextrose)

From Whaley L and Wong D: Nursing care of infants and children, ed 3, St Louis, 1987, The C.V. Mosby Co, p. 1192.

solutions have been developed to treat this common health problem. Various types of solutions have been studied for use in the treatment of mild to moderate dehydration. The major advantage of these solutions is that they can be administered by parents under nursing or medical supervision, thus eliminating the need for hospitalization (Candy, 1987). This is particularly important in underdeveloped countries where infantile diarrhea poses a serious health threat and medical resources are limited.

Acute diarrhea results in the loss of water, sodium, potassium, and bicarbonate. Because oral rehydration solutions were first developed for use in countries where diseases such as cholera are a common cause of diarrhea, the original formulas contained relatively high concentrations of sodium (90 mEq/L). Concern over the possibility of hypernatremia has led to the development of commercially available solutions that contain far less sodium (30 to 75 mEq/L) for use in developed countries. These solutions also contain varying amounts of glucose, potassium, and some form of buffer, either bicarbonate or citrate (Table 6-3). The glucose concentration of oral rehydration solutions has received a great deal of consideration because of the relationship between glucose and sodium absorption in the gut. Maximal sodium and water absorption is believed to occur with glucose concentrations of 10 to 25 g/L (Lattanzi and Siegal, 1986). Higher concentrations of glucose decrease the absorption of sodium and water and contribute to increased diarrhea secondary to the presence of unabsorbed glucose in the gut. Commonly available fluids such as cola drinks, ginger ale, and apple juice are often prescribed to replace fluids lost in mild episodes of diarrhea. However, these are poor choices for fluid replacement in prolonged diarrhea because of their high glucose and low electrolyte concentrations.

Nursing Diagnoses and Interventions

Fluid volume deficit related to abnormal loss of body fluids or reduced intake

Desired outcomes: Patient attains adequate intake of fluid and electrolytes as evidenced by urine output ≥30 ml/hr, stable weights, specific gravity 1.010 to 1.030, no clinical evidence of hypovolemia, BP within patient's normal range, CVP 2 to 6 mm Hg (or 5 to 12 cm H_2O), and HR 60 to 100 bpm. Serum sodium is 137 to 147 mEq/L and hematocrit and BUN are within patient's normal range. For patients in critical care the following are attained: PAP 20-30/8-15 mm Hg and CO 4 to 7 L/min.

1. Monitor I&O hourly. Initially, intake should exceed output during therapy. Alert MD to urine output <30 ml/hr for 2 consecutive hours. Measure urine specific gravity q8h. Expect it to decrease with therapy.
2. Monitor VS and hemodynamic pressures for signs of continued hypovolemia. Be alert to decreased BP and CVP and increased HR and SVR. For critical care patients, also be alert to decreased PAP, CO, and MAP and to increased SVR.
3. Weigh patient daily. Weigh patient at the same time of day (preferably before breakfast) on a balanced scale, with patient wearing approximately the same clothing. Document type of scale used (i.e., standing, bed, chair).
4. Administer PO and IV fluids as prescribed. Document response to fluid therapy. Monitor for signs and symptoms of fluid overload or too rapid fluid administration: crackles (rales), SOB, tachypnea, tachycardia, increased CVP, increased PAP, neck vein distention, and edema. If the patient shows any of the above signs and symptoms, follow agency protocol for fluid challenge.
5. Monitor patient for hidden fluid losses. For example, measure and document abdominal girth or limb size, if indicated.
6. Notify MD of decreases in hematocrit that may signal bleeding. Remember that hematocrit will decrease in the dehydrated patient as he or she becomes rehydrated. Decreases in hematocrit associated with rehydration may be accompanied by decreases in serum sodium and BUN.
7. Place shock patient in a supine position with the legs elevated at 45 degrees to increase venous return. This position returns approximately 500 ml of blood pooled in the veins of the legs to the central circulation. Avoid the Trendelenburg position because it causes abdominal viscera to press on the diaphragm, thereby impairing ventilation.
8. Securely tape all non-Luer lock connections on IV lines to prevent accidental bleeding from disconnected lines. Luer lock-type connections *must* be used on arterial lines because of the high risk of hemorrhage and on central lines due to the additional risk of air embolus.

9. After evaluating individual learning needs, provide patient and significant others, as indicated, with verbal and written instructions for the following: signs and symptoms of hypovolemia, importance of maintaining adequate intake (especially in infants and older adults who are more likely to develop dehydration), and medications.

Altered cerebral, renal, and peripheral tissue perfusion related to decreased circulation secondary to hypovolemia

Desired outcome: Patient has adequate perfusion as evidenced by alertness, urinary output ≥30 ml/hr for 2 consecutive hours, warm and dry skin, BP within patient's normal range, HR <100 bpm, peripheral pulses >2+ on a 0-4+ scale, and brisk capillary refill (<3 seconds).

1. Monitor for signs of decreased cerebral perfusion: vertigo, syncope, confusion, restlessness, anxiety, agitation, excitability, weakness, nausea, and cool and clammy skin. Alert MD to worsening symptoms. Document response to fluid therapy.
2. Protect patients who are confused, dizzy, or weak. Keep siderails up and bed in lowest position with wheels locked. Assist with ambulation. Raise patient to sitting or standing positions slowly. Monitor for indicators of orthostatic hypotension: decreased BP, increased heart rate, dizziness, and diaphoresis. If symptoms occur, return patient to supine position.
3. To avoid unnecessary vasodilation, treat fevers promptly.
4. Reassure patient and significant others that sensorium changes will improve with therapy.
5. Monitor I&O and alert MD to urine output less than 30 ml/hr for 2 consecutive hours. Prolonged reduction in renal perfusion may result in ischemic damage to the kidneys with the development of acute renal failure.
6. Evaluate capillary refill, noting whether it is brisk (<3 seconds) or delayed (>5 seconds). Notify MD if refill is delayed.
7. Palpate peripheral pulses bilaterally in arms and legs (radial, brachial, dorsalis pedis, and posterior tibial). Use a Doppler stethoscope if unable to palpate pulses. Rate pulses on a 0 to 4+ scale or as weak or diminished, normal or present, and strong or bounding. Notify MD if pulses are absent or barely palpable.

   NOTE: Abnormal pulses also may be caused by a local vascular disorder.

FLUID VOLUME EXCESS

Expansion of ECF volume is termed fluid volume excess or **hypervolemia.** Hypervolemia may occur as the result of an increase in either the intravascular or interstitial fluid compartment or both. A shift of fluid from the interstitial to intravascular space also may be considered a hypervolemic state, although there has been no change in the overall volume of the ECF. If severe, hypervolemia can lead to heart failure and pulmonary edema, especially in the individual with cardiovascular dysfunction. Compensatory mechanisms for hypervolemia include the release of atrial natriuretic factor, leading to increased filtration and excretion of sodium and water by the kidneys, and decreased release of aldosterone and ADH.

History and Risk Factors

Fluid volume excess may develop whenever there is chronic or abnormal stimulus to the kidneys to save sodium or water, abnormal renal function with reduced excretion of sodium and water, excessive intake of sodium and water, or interstitial-to-plasma fluid shift. As with hypovolemia, hypervolemia may be associated with changes in the concentration of the ECF. Excessive retention of water, as occurs with increased production of ADH, will result in hypotonic volume expansion. In contrast, volume expansion that develops as the result of increased production of aldosterone will be isotonic due to equal retention of both sodium and water. Hypertonic volume expansion may occur with excessive administration of sodium or hypertonic sodium chloride solutions. See Chapter 8 for additional information concerning volume disturbances associated with sodium imbalances.

Increased stimulus to conserve sodium and water

Excess water may be retained any time there is an abnormal production of ADH or enhanced action of ADH. As discussed in Chapter 3, stress, pain, surgery, certain anesthetics, and medications such as morphine will increase the release of ADH. This is why fluid retention is common for the first few days following surgery. Postoperative fluid retention is further enhanced by increased release of aldosterone secondary to a stress-induced stimulus to the sympathetic nervous system.

Congestive heart failure, hepatic cirrhosis, and nephrotic syndrome result in ECF volume expansion due to chronic stimulus to the kidney to conserve sodium and water. In each of these conditions there is a reduction in the effective circulating volume (remember that this is the volume of blood available to perfuse the tissues). In congestive heart failure the effective circulating volume (ECV) is reduced because of the failing heart, which is unable to circulate the vascular volume adequately. In both cirrhosis and nephrotic syndrome, decreased plasma albumin levels result in a loss of intravascular volume into the interstitial space. Cirrhosis is further complicated by hepatic venous obstruction, which increases the movement of fluid into the peritoneal cavity. The kidneys respond to the reduction in ECV by increasing the release of renin. In turn, renin leads to an increased release of aldosterone, with retention of sodium and water. Yet in each of these conditions, the retained volume does not remain in the vascular space; instead, it expands the interstitial volume, resulting in the development of **edema.** Thus these individuals will exhibit the signs of both poor tissue perfusion and fluid volume excess. Excessive administration of adrenal cortical hormones, e.g., prednisone, can also cause fluid volume excess because of increased retention of sodium and water.

Renal failure

Decreased renal function may result in fluid volume excess resulting from the retention of fluid that is normally excreted by the kidneys. The daily intake of fluid (from liquid and solid food and oxidative metabolism) is approximately 2,700 ml/day The kidneys are responsible for excreting more than half of this volume. In addition, to maintain a balance between intake and output, the kidneys normally increase urine output in response to an increased intake of fluid. In renal failure, however, the kidneys lose their ability to adjust the urine output according to need. Hypervolemia may develop quickly in the presence of oliguria or anuria.

Excessive intake of sodium and water

Increased intake of oral fluids does not cause hypervolemia, except in individuals with cardiac or renal disease. Rapid administration of excess IV fluids, however, may result in symptomatic hypervolemia even in individuals with normal cardiac and renal function. Increased intake of sodium (in the form of salty foods or sodium-containing medications) may contribute to the development of hypervolemia due to water retention and increased thirst. For this reason sodium intake is often restricted in individuals at risk for developing fluid volume excess (e.g., those with congestive heart failure).

Interstitial-to-plasma fluid shifts

Intravascular volume excess may develop when there is a shift of fluid from the interstitial fluid space to the plasma. This may occur with the remobilization of fluid 2 days after a burn injury. Immediately after a burn injury there is a shift of fluid from the plasma to the interstitial space because of damage to the capillary membranes. During this initial phase the individual is at risk for intravascular fluid volume deficit, and treatment may include administration of large volumes of fluid in order to maintain adequate tissue perfusion. Two to three days following the burn injury, however, there is a shift of fluid from the interstitium back into the plasma. The individual is then at increased risk for intravascular fluid volume excess, especially if there has been aggressive fluid replacement during the initial phase.

Interstitial-to-plasma fluid shifts also may occur in individuals treated with hypertonic IV medications or fluids. Administration of IV sodium bicarbonate, for example, increases the tonicity of the plasma, resulting in a shift of fluid into the intravascular space.

Assessment

As with fluid volume deficit, the signs and symptoms that occur with fluid volume excess will increase with the severity of the imbalance and rapidity with which it develops. Sudden increases in the ECF, as might occur with too rapid an administration of large volumes of IV fluids, will result in more dramatic symptoms

Table 6-4 Quick assessment guidelines for individuals with fluid volume excess

Signs and Symptoms	Physical Assessment	Hemodynamic Measurements
Shortness of breath, orthopnea	Edema, weight gain, ↑ BP (with ↓ BP as the heart fails), bounding pulses, ascites, crackles, rhonchi, wheezes, distended neck veins, moist skin, tachycardia, gallop rhythm	↑ CVP, PAP, and MAP

History and Risk Factors

Retention of sodium and water: Heart failure, cirrhosis, nephrotic syndrome, excessive administration of adrenal cortical hormones

Abnormal renal function: Acute or chronic renal failure with oliguria

Excessive administration of IV fluids

Interstitial-to-plasma fluid shift: Remobilization of fluid after treatment of burns, excessive administration of hypertonic solutions (e.g., mannitol, hypertonic saline) or colloid oncotic solutions (e.g., albumin)

than would occur with a gradual fluid gain. Symptoms also will vary with the location of the excess fluid and whether both the interstitial and intravascular spaces are expanded. An increase in the volume of the interstitial fluid space, termed *edema,* is less likely to disrupt body functions than an increase in the intravascular fluid. The one exception to this generalization is an increase in the volume of the pulmonary interstitium (pulmonary edema), which can be life threatening. As discussed above, individuals with congestive heart failure, hepatic cirrhosis, and nephrotic syndrome may exhibit some of the signs and symptoms of intravascular deficit while suffering from overall expansion of the ECF. See Table 6-4 for quick assessment guidelines for individuals with fluid volume excess.

Increased intravascular volume

Increases in the volume of the intravascular space will be evidenced by increased blood pressure, bounding pulses, and distended neck veins. In severe hypervolemia the neck veins may remain distended even when the individual is sitting fully upright. If the increase in vascular volume is so great that the heart is unable to pump the blood forward adequately, the blood pressure may drop and the heart rate increase. The presence of an extra heart sound (gallop rhythm) also may signal fluid volume excess due to diastolic overloading of the ventricles. Hemodynamic monitoring will reveal an increased MAP, CVP, and PAP.

Increased intravascular volume leads to an increase in the hydraulic pressure (also termed *hydrostatic pressure*) within the capillaries. This in turn may cause a shift of fluid into the interstitial space, with the development of edema (see discussion below). One of the greatest dangers of an expanded vascular volume is a shift of fluid into the pulmonary interstitium, resulting in pulmonary edema.

Peripheral edema

Peripheral edema is defined as a localized or generalized palpable swelling of the interstitial space. Localized edema occurs in a specific area secondary to local tissue trauma, e.g., ankle edema that develops after a sprain. In contrast, generalized edema occurs nonspecifically throughout the body and is most apparent in dependent areas. Generalized edema is common in cardiac and renal failure. Severe generalized edema is termed **anascara.**

Edema may develop any time there is an alteration in capillary hemodynamics favoring either increased formation or decreased removal of interstitial fluid (Figure 6-2). Increased capillary hydraulic pressure from volume expansion or venous obstruction or increased capillary permeability owing to burns, allergy, or infection causes an increase in interstitial fluid volume. Decreased removal of interstitial fluid occurs when there is an obstruction to lymphatic outflow or a decrease in plasma

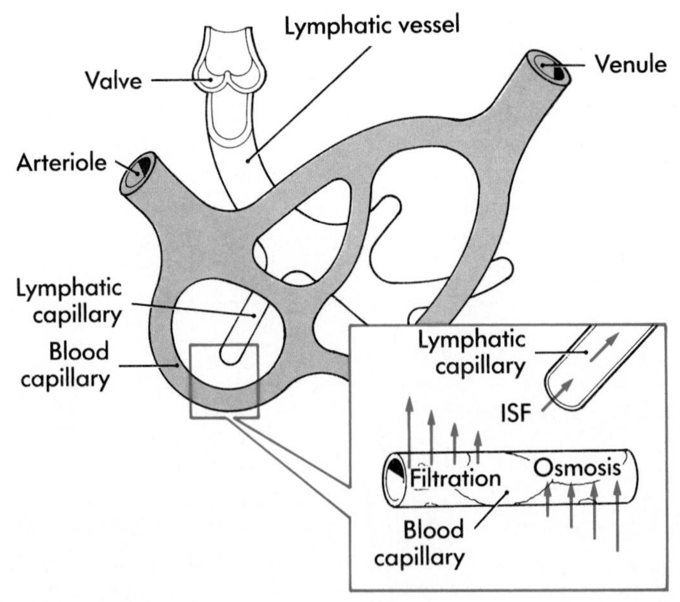

Figure 6-2 Capillary diagram.

(From Horne M and Swearingen P: Pocket guide to fluids and electrolytes, St Louis, 1989, The CV Mosby Co.)

oncotic pressure (recall that the plasma proteins help to hold the vascular volume in the vascular space). Furthermore, retention of sodium and water by the kidneys enhances and maintains edema. This may result from a decreased ability to excrete sodium and water, as in renal failure, or an increased stimulus to conserve sodium and water, as in heart failure. Heart failure leads to a reduction in cardiac output, with a drop in ECV, which in turn stimulates the kidneys to conserve sodium and water via the renin-angiotensin system. Since the diseased heart is unable to circulate this increased volume, the pressure within the venous circuit increases, capillary hydrostatic pressure increases, and edema is formed.

Ascites

The edema seen in nephrotic syndrome and hepatic cirrhosis is the result of both an increased stimulus to conserve sodium and water and a decreased ability to excrete it. Ascites, an accumulation of fluid in the abdominal cavity, is common in both of these conditions. Severe ascites, usually associated with advanced hepatic disease, may impair cardiac and respiratory function due to increased intraabdominal pressure. Pressure on the diaphragm results in shortness of breath. Poor right ventricular filling (secondary to increased intraabdominal pressure) may lead to a drop in cardiac output and decreased tissue perfusion.

Pitting edema

As stated above, generalized edema usually is most evident in dependent areas. The ambulatory patient will exhibit pretibial or ankle edema, while the patient restricted to bed rest will exhibit sacral edema. Generalized edema also may manifest itself around the eyes (periorbital) or in the scrotal sac due to the low tissue pressures in these areas. Sacral edema may be identified by pressing the index finger firmly into the sacral tissue and maintaining the pressure for several seconds. If a pit remains after the finger has been withdrawn, edema is present. Pitting also may be assessed over the tibia or ankle. Pitting may be rated according to severity (Figure 6-3). A notation of trace or +1 edema indicates edema that is barely detectable, while +4 indicates edema that leaves a deep, persistent pit.

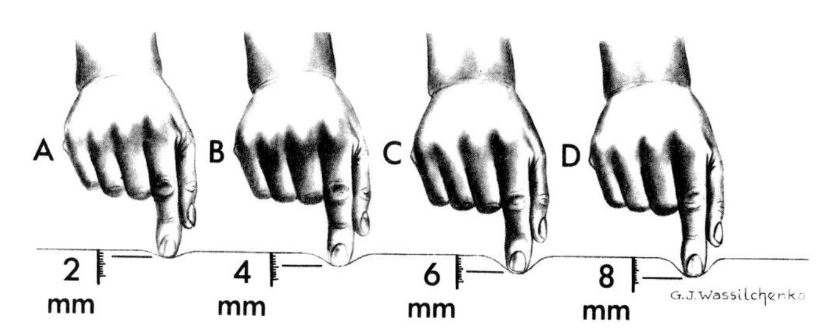

Figure 6-3 **Assessment of pitting edema. A, +1; B, +2; C, +3; D, +4.**
(From Bobak IM and Jensen MD: Essentials of maternity nursing, ed 2, St Louis, 1986, The CV Mosby Co.)

Pulmonary edema

Pulmonary edema occurs when there is an increase in the volume of the pulmonary interstitium. This increase in fluid decreases the diffusion of oxygen from the alveoli to the pulmonary capillaries and may cause a shift of fluid into the alveoli. This leads to shortness of breath, especially when the individual is recumbent (orthopnea), and increased respiratory rate (tachypnea). Auscultation of the lungs may reveal crackles (rales), rhonchi, and wheezes created by the sound of fluid within the airways. In severe pulmonary edema the individual will appear anxious, diaphoretic, and markedly short of breath, and will produce frothy, pink sputum. Rapid correction of this condition is necessary to prevent a fatal outcome.

Overall fluid gain

Body weight will increase with an increase in ECF volume unless the increase is the result of fluid shifts between compartments. Increased body weight may provide an early clue to the development of fluid volume excess that is gradual in progression. It also may provide a means of quantifying fluid gains. In the individual with chronic renal failure, for example, the need for fluid removal during dialysis may be determined by the interdialytic weight gain.

Diagnostic Studies

Laboratory findings in hypervolemia are variable and will depend on the underlying etiology. As with hypovolemia, diagnosis usually is based on a combination of physical findings and history. Laboratory tests, however, may be helpful in determining or confirming a suspected cause.

Hematocrit

A decreased hematocrit is one finding that is common to most types of volume excess. It occurs as the result of hemodilution. As the plasma volume expands because of the retention of fluid, the percentage of blood that is composed of red blood cells decreases.

Arterial blood gases (ABGs) and chest x-ray

In the presence of pulmonary edema, arterial blood gases may reveal hypoxemia (decreased Pao_2) and respiratory alkalosis (increased pH and decreased $Paco_2$), and the chest x-ray will show signs of pulmonary vascular congestion.

Serum sodium and serum osmolality

Serum sodium and serum osmolality will decrease if hypervolemia occurs as a result of excessive retention of water (e.g., in chronic renal failure).

BUN and creatinine

An increased BUN and creatinine will be present in renal failure. The BUN also may increase in cardiac failure secondary to decreased renal perfusion.

Urine specific gravity

Urine specific gravity will decrease if the kidney is attempting to excrete excess water. In acute renal failure, though, it may be fixed at approximately 1.010 due to the kidneys' inability to modify what is filtered.

Urinary sodium

The urinary sodium will be decreased in conditions associated with an increased production of aldosterone (e.g., congestive heart failure, cirrhosis, nephrotic syndrome) since hypervolemia occurs secondary to a chronic stimulus to conserve sodium and water.

Medical Management

The goals of therapy are to treat the precipitating problem and return the ECF to normal. Treatment may include restriction of sodium and water (see the accompanying box for a list of foods that are high in sodium) and diuretics (see Chapter 7 for a discussion of diuretic therapy). Hidden sources of sodium such as medications should be avoided. Bed rest and supportive hose may be used to help mobilize edema. Abdominal paracentesis (needle aspiration of the fluid within the peritoneal cavity) may be performed for the treatment of severe ascites, which adversely affects cardiopulmonary functioning. Following paracentesis the individual must be monitored for the development of intravascular volume deficit secondary to rapid reaccumulation of ascites. Dialysis or continuous arterial-venous hemofiltration (CAVH) may be necessary for the treatment of renal failure or life-threatening fluid overload.

Foods High in Sodium

- Bouillon
- Celery
- Cheeses
- Dried fruits
- Frozen, canned, or packaged foods
- Mustard
- Olives
- Pickles
- Preserved meat
- Salad dressings and prepared sauces
- Sauerkraut
- Snack foods (e.g., crackers, chips, pretzels)
- Soy sauce

Preload reduction

In addition to goals already mentioned, the aim of therapy in acute pulmonary edema is to protect the individual from immediate danger by reducing the volume of blood returning to the heart and improving oxygenation. Medications such as morphine, furosemide, and nitrites (e.g., sublingual nitroglycerin) will reduce the volume of the blood returning to the heart by increasing the capacitance of the venous circulation. This type of therapy is termed *preload reduction.*

Afterload reduction

Medications that dilate the arterial vessels, such as hydralazine and captopril, also may be given to reduce the workload of the heart and myocardial oxygen consumption. This therapy is termed *afterload reduction.* Diuretics such as furosemide are used to reduce the overall vascular volume (see the section on diuretic therapy, Chapter 7, p. 129, for additional information). Oxygen is administered to help maintain tissue oxygenation.

Nursing Diagnoses and Interventions

Fluid volume excess related to excessive fluid or sodium intake or compromised regulatory mechanism

Desired outcomes: Patient is normovolemic as evidenced by adequate urinary output of at least 30 to 60 ml/hr, specific gravity of approximately 1.010 to 1.020, stable weights, and edema ≤ +1 on a 0 to +4 scale. BP is within patient's normal range, CVP is 2 to 6 mm Hg (5 to 12 cm H_2O), and HR is 60 to 100 bpm. In addition, for critical care patients PAP is 20 to 30/8 to 15 mm Hg, MAP is 70 to 105 mm Hg, and CO is 4 to 7 L/min.

1. Monitor I&O hourly. With the exception of oliguric renal failure, urine output should be >30 to 60 ml/hr. Measure urine specific gravity every shift. If patient is experiencing diuresis, specific gravity should be <1.010 to 1.020.
2. Observe for and document the presence of edema: pretibial, sacral, periorbital. Rate pitting on a 0 (no edema) to +4 scale (Figure 6-3).
3. Weigh patient daily. Daily weights are the single most important indicator of fluid status. For example, a 2 kg acute weight gain is indicative of a 2 L fluid gain. Weigh patient at the same time each day (preferably before breakfast) on a balanced scale, with patient wearing approximately the same clothing each day. Document type of scale used (i.e., standing, bed, chair).
4. Limit sodium intake as prescribed by MD (Table 6-4). Consider use of salt substitutes.

    NOTE: Some salt substitutes contain potassium and may be contraindicated in patients with renal failure or in patients receiving potassium-sparing diuretics (e.g., spironolactone, triamterene).

5. Limit fluids as prescribed. Offer a portion of allotted fluids as ice chips to minimize patient's thirst. Teach patient and significant others the importance of fluid restriction and how to measure fluid volume.
6. Provide oral hygiene at frequent intervals to keep oral mucous membrane moist and intact.
7. Document response to diuretic therapy. Many diuretics (e.g., furosemide, thiazides) cause hypokalemia. Observe for indicators of hypokalemia: muscle weakness, dysrhythmias (especially PVCs and EKG changes—flattened T wave, presence of U waves). (See Chapter 9, the section on hypokalemia, p. 174). Potassium-sparing diuretics (e.g., spironolactone, triamterene) may cause hyperkalemia: weakness, EKG changes (e.g., peaked T wave, prolonged PR interval, widened QRS). (See Chapter 9, the section on hyperkalemia, p. 181). Notify MD of significant findings.
8. Observe for physical indicators of overcorrection and dangerous volume depletion secondary to therapy: dizziness, weakness, syncope, thirst, confusion, poor skin turgor, flat neck veins, acute weight loss. Monitor VS and hemodynamic parameters for signs of volume depletion occurring with therapy: decreased BP, CVP, PAP, MAP, and CO; increased HR. Alert MD to significant changes or findings.
9. After evaluating individual learning needs, provide patient and significant others with verbal and written instructions for the following: signs and symptoms of hypervolemia; symptoms that necessitate MD notification after hospital discharge: SOB, chest pain, new pulse irregularity; low-sodium diet; fluid restriction; and medications.

Impaired gas exchange related to alveolar-capillary membrane changes secondary to pulmonary vascular congestion occurring with ECF expansion

Desired outcomes: Patient has adequate gas exchange as evidenced by RR ≤ 20 breaths/min, HR ≤ 100 bpm, and $Pao_2 \geq 80$ mm Hg. Patient does not exhibit crackles, gallops, or other clinical indicators of pulmonary edema. For patients in critical care, PAP is $\leq 30/15$ mm Hg.

1. Acute pulmonary edema is a potentially life-threatening complication of hypervolemia. Monitor patient for indicators of pulmonary edema including air hunger, anxiety, cough with production of frothy sputum, crackles (rales), rhonchi, tachypnea, tachycardia, gallop rhythm, and elevation of PAP.
2. Monitor ABGs for evidence of hypoxemia (decreased Pao_2) and respiratory alkalosis (increased pH and decreased $Paco_2$). Increased oxygen requirements are indicative of increasing pulmonary vascular congestion.
3. Keep patient in semi-Fowler's or position of comfort to minimize dyspnea. Avoid restrictive clothing.
4. Administer oxygen according to unit protocol or MD prescription.

Potential for impaired skin and tissue integrity related to edema secondary to fluid volume excess

Desired outcome: Patient's skin and tissue remain intact.

1. Edema increases the risk of tissue breakdown secondary to altered nutrient delivery. Assess and document circulation to extremities at least every shift. Note color, temperature, capillary refill, and peripheral pulses. Determine whether capillary refill is brisk (<3 seconds) or delayed (>5 seconds). Palpate peripheral pulses bilaterally in arms and legs (radial, brachial, dorsalis pedis, and posterior tibial). Use Doppler stethoscope if unable to palpate pulses. Notify MD if capillary refill is delayed or pulses are absent.
2. Turn and reposition patient at least q2h to minimize tissue pressure.
3. Check tissue areas at risk with each position change (e.g., heels, sacrum, and other areas over bony prominences).
4. Use Eggcrate mattress or other device to minimize pressure.
5. Treat decubitus ulcers with occlusive dressings (e.g., Duoderm, Op-Site, Tegaderm) per unit protocol. Notify MD of the presence of sores, ulcers, or areas of tissue breakdown in patients who are at increased risk for infection (e.g., diabetics, immunosuppressed individuals, those with renal failure).

Case Study 6-1

B.K., a 32-year-old mother of three small children, delivered twins (her fourth and fifth children) 2 hours ago. Her labor and delivery were uneventful and her initial postpartum course was stable. Presently, her nurse is concerned because of B.K.'s increased vaginal bleeding. B.K. has saturated four peripads in the last half hour and when the nurse turned her she discovered a large blood stain in the bed.

1. Question: B.K. is at risk for developing what type of fluid imbalance?

Answer: Fluid volume deficit because of abnormal blood loss.

After the delivery of a baby, torn blood vessels in the lining of the uterus bleed. Normally, contraction of the uterine muscles is sufficient to compress the vessels and control the bleeding. If there are retained placental fragments or blood clots or if the uterine muscle has been overly stretched (as with multiple births), the uterus may be unable to contract completely and control the bleeding. Left untreated, this situation will result in postpartum hemorrhage. B.K.'s nurse massages the fundus (the upper portion of the uterus, just below the umbilicus) to stimulate uterine muscle contraction. Unfortunately, B.K.'s uterus does not become firm with massage but remains soft and *boggy,* indicating that the bleeding will continue. Another nurse in the postpartum unit pages B.K.'s physician, while B.K.'s nurse completes her assessment of B.K.'s condition.

2. Question: In addition to massaging the fundus, the nurse immediately lowers the head of B.K.'s bed to a fully reclined position. Why is this an important nursing action?

Answer: Hypotension, if present, may be minimized by lowering the head of the bed. Remember that if fluid volume deficit is not severe, compensatory mechanisms may be able to maintain the blood pressure when the individual is recumbent.

3. Question: A complete nursing assessment, as described in Chapter 5, is not practical in this situation. Based on the history, B.K.'s nurse will assess B.K. for the signs and symptoms of fluid volume deficit. What are they?

Answer:

a. *Decreased blood pressure (BP).* If the BP is normal with the patient lying flat, it should be checked with the head of the bed elevated. Orthostatic hypotension may be the first clue to symptomatic fluid volume deficit. Remember that a drop of 15 mm Hg in either the systolic or diastolic BP is considered significant.
b. *Increased heart rate.* The nurse should note whether the heart rate increases when the head is elevated (suggesting orthostatic hypotension). An increase in the heart rate of 20 beats per minute or greater is considered significant.
c. *Symptoms that suggest decreased tissue perfusion.* Weakness, dizziness, confusion, nausea, vomiting, decreased urine output, and changes in skin color and temperature.

B.K.'s blood pressure has dropped from 128/76 to 100/62 mm Hg and her heart rate has increased from 84 beats per minute to 106 beats per minute. These latest measurements were obtained with B.K. in a fully recumbent position. At this time B.K. complains of a vague feeling of weakness and nausea but denies dizziness and is fully oriented. Her skin is cool, dry, and pale. The vaginal bleeding remains excessive.

4. Question: While awaiting the physician's call or arrival, the nurse inserts a 16-gauge angiocatheter (according to unit policy) to provide a means of administering IV medications and fluids. Why is this an important nursing action?

Answer: Since B.K. shows signs of fluid volume deficit (decreased BP, increased heart rate, weakness, and nausea) and may be unable to tolerate oral fluids because of nausea, she will likely require treatment with IV fluids. If her bleeding continues, she could develop hypovolemic shock, which would necessitate rapid volume replacement. A 16-gauge angiocatheter is used to facilitate the administration of whole blood or packed red blood cells. If B.K.'s condition suddenly worsens, this simple nursing action could prove to be life saving.

5. **Question:** What signs and symptoms may signal that B.K. is in hypovolemic shock?

Answer: B.K. would appear pale and diaphoretic with cool, clammy extremities and her urine output would be markedly decreased. Her level of consciousness would be diminished. If the shock were severe, her systolic BP would drop below 80 mm Hg.

B.K.'s physician arrives on the unit and prescribes oxytocin (Pitocin) IV drip to increase B.K.'s uterine contractions and Ringer's lactate IV (a balanced electrolyte solution with an electrolyte composition similar to plasma) to restore vascular volume. She also requests a blood specimen for a stat hemoglobin and hematocrit to help determine the extent of bleeding and for a type and crossmatch so that compatible units of blood will be available if B.K. needs a blood transfusion.

Like ADH, oxytocin is produced in the hypothalamus and released into the general circulation by the posterior pituitary gland. It stimulates the contractile activity of uterine smooth muscle and plays a role in the initiation of labor. Oxytocin and ADH have similar chemical structures and some overlapping actions. Thus oxytocin preparations have mild antidiuretic (ADH) effects. Pitocin is used therapeutically to increase the strength of contractions during labor or decrease postpartum bleeding caused by uterine atony (as in B.K.'s case).

6. **Question:** Knowing that Pitocin has mild antidiuretic properties, what are some nursing considerations when caring for a woman receiving a Pitocin drip?

Answer:

a. *Monitor hourly urine output.* Remember that ADH increases the reabsorption (saving) of water by the renal tubule. Thus Pitocin may cause a decrease in urine output.
b. *Monitor I&O.*
c. *Monitor for the signs and symptoms of water retention.* Prolonged or excessive oxytocin administration, combined with the administration of hypotonic IV fluids, has led to fatal water intoxication (Danforth and Scott, 1986). The signs and symptoms of water intoxication will be discussed in the section on hyponatremia, Chapter 8.

B.K. has responded well to the Pitocin drip. Her uterus no longer feels boggy and soft and the bleeding has decreased. However, her BP has continued to drop and is now 90/60 mm Hg. The physician has decided to administer a transfusion to B.K. with 1 unit of blood and replace her remaining volume loss with the Ringer's lactate. She prescribed 250 ml of Ringer's lactate/hr until B.K.'s blood pressure and heart rate are normal.

7. **Question:** The nursing staff should now start monitoring B.K. for which potential fluid imbalance and why?

Answer: Fluid volume excess (hypervolemia) caused by rapid IV fluid replacement.

8. Question: What are some potential signs and symptoms of fluid volume excess?

Answer:

a. *Increased blood pressure (BP)* (greater than the individual's normal range).
b. *Bounding pulses.*
c. *Increased neck vein distention.*
d. *Edema.*
e. *Shortness of breath, orthopnea, adventitious breath sounds.* Crackles, rhonchi, wheezes.
f. *Increased heart rate.*

9. Question: How might B.K. appear if she were suddenly to develop pulmonary edema?

Answer: Anxious, diaphoretic, and markedly short of breath, possibly with production of frothy, pink sputum. Auscultation of her chest might reveal crackles throughout her lung fields.

During the next few hours B.K. receives 1,000 ml of IV fluid. Throughout this period the nurse assesses B.K.'s condition continuously. B.K.'s recumbent blood pressure is now normal and she is asking to get out of bed to void. The physician has indicated that B.K. may use the bedside commode if she is able.

10. Question: Given B.K.'s history of recent volume depletion, what are some of the necessary nursing assessments before getting her up to the commode?

Answer: B.K. should be assessed for orthostatic hypotension. Even though she has a normal recumbent blood pressure, her blood pressure may drop when she sits or stands up. The nurse should sit B.K. at the bedside and check her for dizziness or decreased blood pressure before allowing her to stand. B.K. should be returned to a recumbent position if at any time she becomes symptomatic of orthostatic hypotension.

References

Bastin JP: Action stat: postpartum hemorrhage, Nursing 89 19(2):33, 1989.

Better OS: Impaired fluid and electrolyte balance in hot climates, Kidney Int (suppl) 1:97-101, 1987.

Bowman-Perkins S and Kennally KM: The hidden danger of internal hemorrhage, Nursing 89 19(7):34-41, 1989.

Candy CE: Recent advances in the care of children with acute diarrhea: giving responsibility to the nurse and parents, J Adv Nurs 2(1):95-99, 1987.

Corbett JV: Laboratory tests and diagnostic procedures with nursing diagnoses, ed 2, San Mateo, Calif, 1987, Appleton & Lange.

Danforth DN and Scott JR (editors): Obstetrics and gynecology, ed 5, Philadelphia, 1986, JB Lippincott Co.

Donner C and Donner K: The critical difference: pulmonary edema, Am J Nurs 88:59, 1988.

Freeman LM and Bizek KS: A fluid challenge protocol, Crit Care Nurs 4:46-48, Jan/Feb 1984.

Horne M and Swearingen PL: Pocket guide to fluids and electrolytes, St Louis, 1989, The CV Mosby Co.

Goldberger E: A primer of water, electrolyte, and acid-base syndromes, ed 7, Philadelphia, 1986, Lea & Febiger.

Goodman LS and Gillman A: The pharmacological basis of therapeutics, ed 7, New York, 1985, Macmillan.

Guyton AC: Textbook of medical physiology, ed 7, Philadelphia, 1986, WB Saunders Co.

Jensen MD and Bobak IM: Maternity and gynecologic care: the nurse and the family, ed 3, St Louis, 1985, The CV Mosby Co.

Kokko JP and others: Fluids and electrolytes, Philadelphia, 1986, WB Saunders Co.

Lattanzi WE and Siegal NJ: A practical guide to fluid and electrolyte therapy, Curr Probl Pediatr 16(1):1-43, 1986.

Maxwell MH and others: Clinical disorders of fluid and electrolyte metabolism, ed 4, New York, 1987, McGraw-Hill, Inc.

Rice V: Shock management. I. Fluid volume replacement, Crit Care Nurse 69-82, Nov/Dec 1984.

Rose BD: Clinical physiology of acid-base and electrolyte disorders, ed 3, New York, 1989, McGraw-Hill, Inc.

Schrier RW: Renal and electrolyte disorders, ed 3, Boston, 1986, Little, Brown & Co., Inc.

Skidmore-Roth L: Mosby's 1989 nursing drug reference, St Louis, 1989, The CV Mosby Co.

Strum JA and Wisner DH: Fluid resuscitation of hypovolemia, Intensive Care Med 11:227, 1985.

Tucker JA and others: Treating acute diarrhea and dehydration with an oral rehydration solution, Pediatr Nurs 13(3):169-174, 1987.

Chapter

7

Diuretic and Intravenous Fluid Therapy

This chapter reviews the principles of diuretic and intravenous (IV) fluid therapy and identifies their nursing implications. Diuretic and IV fluid therapies have significant impact on fluid and electrolyte balance. Nurses are called upon almost daily to administer them and monitor their effects. Nearly all hospitalized patients receive continuous or intermittent IV therapy at some time during their hospital stay. IV fluids also may be administered in the home under the supervision of public health and home health nurses. Diuretics are one of the most commonly prescribed classes of medications in both acute and chronic care settings.

DIURETIC THERAPY

Diuretics are used primarily in the treatment of volume excess and edema. They reduce volume and edema by inhibiting the reabsorption of sodium and water by the kidneys. They may, however, also induce the loss of other important electrolytes and alter acid-base balance. For some individuals, a reduction in effective circulating volume (ECV) and alterations in electrolyte balance caused by diuretics may be detrimental. Individuals with hepatic cirrhosis, for example, may develop hepatic coma or hepatorenal syndrome with overuse of diuretics owing to diuretic-induced hypokalemia, metabolic alkalosis, and rapid fluid removal. Although retention of sodium and water by the kidneys is an important component in the development of edema, not all edematous states require treatment with diuretics. Treatment of mild edema, for example, may require restriction of dietary sodium intake only. The majority of edematous patients, however, may benefit from the judicious use of diuretics.

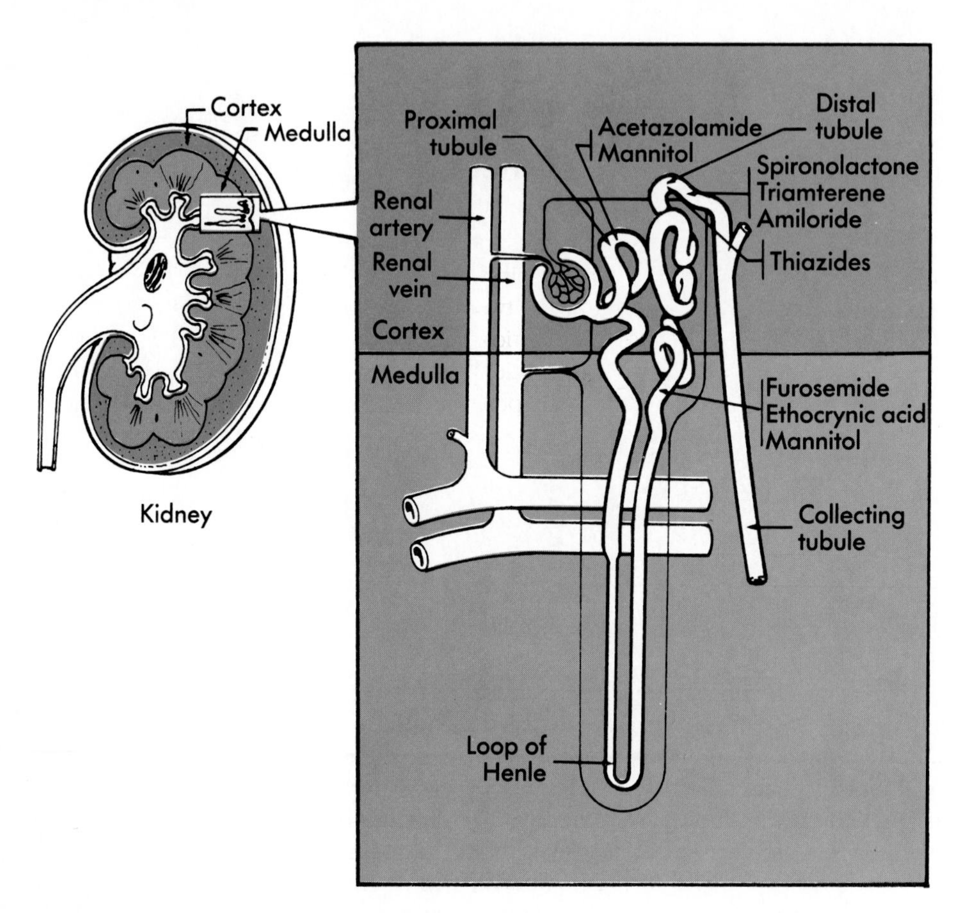

Figure 7-1 Sites of diuretic action.

(From Horne M and Swearingen P: Pocket guide to fluids and electrolytes, St Louis, 1989, The CV Mosby Co.)

Diuretics are also used in the treatment of hypertension, certain acid-base imbalances, glaucoma, hyponatremia, diabetes insipidus, hypercalcemia, renal stone disease, and the diagnosis and prevention of acute renal failure. These uses are, in some cases, unrelated to the diuretic action of the particular medication. These nondiuretic applications will be discussed in greater detail in subsequent sections.

As discussed in Chapter 2, active reabsorption of sodium by the renal tubule creates an osmotic (concentration) gradient that favors passive reabsorption of water. Most diuretics work by decreasing the reabsorption of sodium. Decreased sodium reabsorption leads to decreased reabsorption of water, and hence, to a diuresis. The quantity and characteristics of the diuresis varies, depending on the diuretic and its site of action within the renal tubule (see Figure 7-1 and Table 7-1). Although some complications of diuretic therapy are common to all or most diuretics (see the section on complications of diuretic therapy, p. 132), the specifics of nursing care will depend on the diuretic the individual is receiving (Table 7-1).

Sites of Action

Diuretics may be classified by their mechanism of action (e.g., osmotic diuretics, carbonic anhydrase inhibitors) or by their primary site of action within the renal tubule (e.g., proximal tubule, loop). These two classification systems overlap, since a particular diuretic's mechanism of action will often determine its site of action. Discussing diuretics by their sites of action has the practical advantage of enabling a review of the functions of the various portions of the renal tubule.

Proximal tubule

As discussed in Chapter 2, urine formation begins with the filtration by the glomerulus of as much as 180 L/day of fluid (filtrate). The filtrate is essentially protein-free plasma and has the same concentration of water and electrolytes (with the exception of protein-bound electrolytes) as the plasma. While in the proximal tubule, approximately 65% of the filtered sodium and water is reabsorbed and returned to the blood. The proximal tubule is also where most of the filtered bicarbonate is reabsorbed. There are no true proximal tubule diuretics, which means there are no diuretics that block this massive reabsorption of sodium and water. This is just as well, since a diuretic that would alter the reabsorption of such a great volume of sodium and water could induce dangerous volume depletion (although increased reabsorption of sodium and water in subsequent portions of the tubule would help blunt its effect). There are a group of diuretics, however, that do work in the proximal tubule. Their mechanism of action is that of limiting the reabsorption of bicarbonate. Normally, sodium and water are reabsorbed along with bicarbonate. When bicarbonate reabsorption is limited, so is that of sodium and water. These diuretics work by inhibiting the action of carbonic anhydrase.

Carbonic anhydrase is an enzyme that facilitates the reabsorption of bicarbonate (see Chapter 13 for a discussion of the role of carbonic anhydrase in the regulation of bicarbonate by the kidneys). The best known carbonic anhydrase inhibitor is acetazolamide (Diamox). Acetazolamide is a weak diuretic that is not typically prescribed for its diuretic action. More commonly, it is used to treat glaucoma, since carbonic anhydrase is also involved in the production of ocular fluid (aqueous humor). Glaucoma is characterized by excessive accumulation of fluid in the anterior chamber of the eye. Acetazolamide helps to prevent intraocular hypertension and damage by decreasing the secretion of aqueous humor.

Since acetazolamide increases the loss of bicarbonate in the urine, it is also used to treat metabolic alkalosis (excess bicarbonate). For the same reason, its use should be avoided in individuals at risk for developing metabolic acidosis (decreased

Table 7-1 Diuretic action

Diuretic (by site of action)	Potency	Characteristic of Diuresis
Proximal Tubule		
Acetazolamide (Diamox)[1]	Weak	$NaHCO_3^-$ diuresis with the loss of additional Na^+, Cl^-, and K^+
Proximal Tubule and Loop		
Mannitol[2]	Moderate	Osmotic diuresis with the loss of water in excess of Na^+ and Cl^-
Loop of Henle		
Furosemide (Lasix)[3]	Strong	Large diuresis (may affect 20%-30% of filtered load of sodium) with loss of Na^+, Cl^-, and K^+
Ethacrynic acid (Edecrin)[4]	Strong	Same as above
Bumetanide (Bumex)[5]	Strong	Same as above

Clinical indications and nursing considerations:

1. May be used in the treatment of metabolic alkalosis or in combination with other diuretics to treat refractory edema. It is most commonly used to treat glaucoma since it decreases the formation of aqueous humor. It is contraindicated in patients with acidosis. Monitor patients for hypokalemia (see Chapter 9).
2. May cause hyperosmolality and circulatory overload owing to the osmotic shift of fluid out of the cells and into the interstitium and intravascular space. Use with caution in patients with decreased cardiac function. Monitor for signs of circulatory overload (e.g., crackles, SOB, tachycardia). It may be used in the treatment of early acute renal failure.
3. Used to treat edema of congestive heart failure and advanced renal failure. It may be used alone or in combination with mannitol or dopamine to reverse early acute renal failure. It is effective in the treatment of acute pulmonary edema owing to its diuretic and direct venous vasodilatory actions. Do not give IV faster than recommended because of the risk of deafness. Stop the infusion and notify MD if the patient complains of ringing in the ears. Monitor for hypokalemia (see Chapter 9) and volume depletion (see Chapter 6).

bicarbonate). See Chapters 17 and 18 for additional information about acetazolamide.

Proximal tubule and descending portion of the loop

The osmotic diuretics work in the proximal tubule and the beginning (descending) portion of the loop of Henle (Figure 7-1). Remember from Chapter 1 that osmotic diuretics are substances that are filtered by the glomerulus but *not reabsorbed* (or, as in the case of glucose in uncontrolled diabetes, only partially reabsorbed) from the tubule. They remain within the tubule creating an osmotic gradient, which inhibits the reabsorption of some of the filtered water and sodium. Mannitol is the primary diuretic

Table 7-1 Diuretic action—cont'd

Diuretic (by site of action)	Potency	Characteristic of Diuresis
Early Distal Tubule		
Thiazides (Diuril, Hydrodiuril, Esidrix)[6]	Moderate	Diuresis affecting up to 5% of filtered load of sodium, with loss of Cl^- and K^+
Metolazone (Zaroxolyn)[7]	Similar to thiazides	
Chlorthalidone (Hygrotin)[8]	Similar to thiazides	
Late Distal Tubule—Potassium-sparing Diuretics		
Spironolactone (Aldactone)[9]	Weak	Blocks the action of aldosterone, with loss of Na^+ and Cl^- but not K^+
Triamterene (Dyrenium)[10]	Weak	Weak diuresis with loss of Na^+ and Cl^- but not K^+. Does not depend on the presence of aldosterone

4. Same as above.
5. Potent diuretic that may cause hypokalemia (see Chapter 9) and volume depletion (see Chapter 6). While furosemide may be used when renal function is markedly diminished, bumetanide should not.
6, 7, 8. Often used to treat hypertension owing to both diuretic action and direct antihypertensive effect, which is unrelated to diuretic action. This group of diuretics has the most nondiuresis-related side effects (e.g., decreased release of insulin, hyperlipidemia, skin rashes). Monitor patient for hypokalemia (see Chapter 9), hyperglycemia, and volume depletion (see Chapter 6). These diuretics also may cause hyperuricemia and hypercalcemia (see Chapter 10).
9, 10. May be combined with the thiazide diuretics for increased diuretic action and decreased hypokalemia. Used alone, these diuretics may cause hyperkalemia (see Chapter 9).

in this class. Although sometimes administered for its diuretic action, it is most commonly used for its effect on the osmolality of the extracellular fluid (ECF). See the discussion of mannitol in Case Study 1-1.

Understanding the concept of an osmotic diuresis can be helpful in anticipating abnormal fluid loss in certain individuals. The IV contrast media used in many diagnostic x-ray studies acts as an osmotic diuretic, as does urine glucose in uncontrolled diabetes. In both these situations, nurses must be aware of the risk of fluid volume deficit and monitor for its signs and symptoms (e.g., hypotension and tachycardia).

Loop of Henle

The loop diuretics (furosemide, ethacrynic acid, bumetanide) work in the ascending limb of the loop of Henle. The loop diuretics are the most potent because they affect the reabsorption of such a large percentage of sodium and, over time, they limit the kidneys' ability to produce a concentrated urine.

The loop is second only to the proximal tubule for its reabsorption of sodium (approximately 20% to 30% of the filtered load is reabsorbed in the ascending limb). Unlike the proximal tubule, however, where water and solutes are reabsorbed in equal concentrations, the loop reabsorbs less water than sodium. The reabsorption of solutes in excess of water is a vital component in the mechanism that enables the kidneys to adjust urinary concentration. The physiology of urinary concentration is complex and beyond the scope of this book. It is necessary to know only that the ability to concentrate and dilute urine according to need depends on two factors: the net reabsorption of solutes in the loop and the availability of antidiuretic hormone (ADH). Any condition or medication that affects either of these two factors will affect the kidneys' ability to adjust urinary concentration. Loop diuretics decrease the reabsorption of sodium in the loop and thus decrease the kidneys' ability to form a concentrated urine.

Furosemide (Lasix), the most commonly prescribed loop diuretic, has the additional effect of increasing venous vasodilatation. This effect, combined with its strong diuretic action, makes it very useful in the treatment of acute pulmonary edema. The immediate goal in the treatment of pulmonary edema is to reduce the volume of blood returning to the heart. Intravenous furosemide does this within minutes by increasing venous capacitance and initiating a brisk diuresis. Furosemide also may be given with mannitol or dopamine (an IV drip medication that increases cardiac output and renal perfusion) to reverse early acute renal failure.

Although urinary excretion of calcium is regulated by the actions of parathyroid hormone, it is also affected by sodium and water excretion. Most of the calcium that is filtered by the glomerulus is passively reabsorbed in the proximal tubule and loop of Henle, following concentration gradients created by the reabsorption of sodium and water. If sodium and water reabsorption is reduced, as occurs with diuretic therapy, calcium reabsorption is also reduced. Thus administration of loop diuretics will increase the excretion of calcium in the urine. This is the rationale for treating elevated serum calcium levels (hypercalcemia) with saline and loop diuretics. This will be discussed again in Chapter 10.

Early distal tubule*

As discussed in Chapter 2, approximately 5% of the filtered load of sodium is reabsorbed in the early portion of the distal tubule. This is the site of action of the thiazide and thiazide-like diuretics. They are significantly weaker diuretics than the loop diuretics because they affect a much smaller percentage of the sodium and do not alter the kidneys' ability to produce a concentrated urine. Thiazides are the most

* There is inconsistancy among authors as to the exact terminology used to describe the distal portion of the renal tubule. The confusion centers on where (anatomically and physiologically) the distal tubule ends and the collecting duct begins. This text will use the traditional approach of separating the distal tubule into early and late portions followed by the collecting duct. The reader should keep in mind though, that the late distal tubule may be referred to as the initial portion of the collecting duct in other sources.

commonly prescribed diuretics, primarily because of their role in the early treatment of hypertension. They reduce blood pressure both by decreasing vascular volume and by direct antihypertensive action, which is unrelated to their diuretic effect.

Unlike the loop diuretics, which increase the excretion of calcium, the thiazide diuretics *reduce* the excretion of calcium, which potentially leads to increased serum calcium (hypercalcemia). The exact mechanism for decreased excretion of calcium is unclear, but may be potentiated by diuretic-induced volume depletion. This side effect is sometimes used to clinical advantage in the treatment of renal stone formation. Low-dose thiazides may decrease the incidence of calcium stones in the urine. Thiazide diuretics have another important side effect—decreased release of insulin. Hyperglycemia may develop in individuals who have no history of diabetes or in diabetics who still have some endogenous insulin production (e.g., those whose diabetes is controlled by diet or oral medication). Individuals receiving thiazide diuretics should be instructed to observe for and report the signs and symptoms of hyperglycemia (e.g., increased hunger, thirst, and urine output).

Late distal tubule/cortical collecting tubule

The late distal tubule and cortical collecting tubule (the portion of the collecting tubule that lies within the cortex—see Figure 7-1) is the site of aldosterone action and potassium regulation. All diuretics that have been discussed so far increase the excretion of potassium and may result in significant hypokalemia (decreased plasma potassium level). The greater the diuresis, the greater the loss of potassium in the urine. Thus the loop diuretics cause the greatest loss of potassium. Potassium excretion by the kidneys (and thus overall potassium regulation by the body) is determined by the amount of potassium secreted in the distal portion of the nephron. Potassium secretion is increased by several factors, including increased fluid delivery to the distal tubule, increased sodium reabsorption, and elevated aldosterone levels. Additional factors affecting potassium excretion, such as increased intake and changes in body pH, will be discussed in Chapter 9. The potassium-wasting diuretics (those discussed so far) increase the passive secretion of potassium by increasing the amount of fluid in the distal tubule. This increased fluid simply *washes out* extra potassium. Increased sodium presented to the distal tubule also enhances potassium secretion; as more sodium is reabsorbed, the excretion of H^+ and K^+ increases. Volume depletion induced by overuse of diuretics further increases potassium loss owing to increased secretion of aldosterone.

The diuretics that work in the distal portion of the tubule do not increase the secretion and excretion of potassium and are termed *potassium-sparing diuretics.* In the proximal portion of the nephron, chloride is reabsorbed with sodium, thus maintaining electrical neutrality. The distal tubule, however, is less permeable to chloride, and thus electrical neutrality is maintained by the excretion of potassium and hydrogen. (Remember that sodium, potassium, and hydrogen are positive ions, while chloride is a negative ion). Therefore, decreased sodium reabsorption in the distal tubule (as with the administration of potassium-sparing diuretics) will result in decreased potassium and hydrogen secretion and excretion. Triamterene (Dyrenium) is one diuretic that acts in the distal nephron and decreases the secretion of potassium.

Spironolactone, another diuretic that acts in the distal nephron, works by blocking the action of aldosterone. Recall that aldosterone increases the reabsorption of sodium and the secretion of potassium and hydrogen. Thus a medication that blocks the affects of aldosterone will increase the excretion of sodium and decrease the excretion of potassium and hydrogen. Spironolactone is most effective in conditions that result in an increase in the release of aldosterone. Since only 1% to 2% of the filtered sodium is reabsorbed in this portion of the tubule, the potassium-sparing diuretics are relatively weak. They may be combined with other diuretics (usually thiazides) to increase the overall diuretic action and to counteract the potassium-wasting effects of the other diuretic.

Complications of Diuretic Therapy: Nursing Considerations

Volume abnormalities

Volume depletion, the most common complication of diuretic therapy, occurs as a result of overdiuresis. Nursing responsibilities for individuals undergoing diuretic therapy include monitoring for signs of fluid volume deficit: dizziness, weakness, fatigue, and postural hypotension. Nurses should initiate intake and output (I&O) measurement on all acute care patients started on diuretics. Diuretics that are given in one daily dose should be administered in the morning to minimize the need to void during the night (nocturia), thus preventing sleep pattern disturbance. Outpatients should be taught to monitor for signs and symptoms of fluid volume deficit and report them to a health care professional should they occur.

Electrolyte disturbances

Hypokalemia

The most common electrolyte disturbance that occurs with diuretic therapy is hypokalemia. It may develop with any of the diuretics except those that work in the late distal tubule. Hypokalemia occurs due to increased secretion and excretion of potassium in the distal tubule. Hypokalemia can be avoided by also administering a potassium-sparing diuretic (Table 7-1) or a potassium supplement. Individuals receiving potassium-wasting diuretics should be instructed to report indicators of hypokalemia (e.g., fatigue, muscle weakness, leg cramps, and irregular pulse). For additional indicators of hypokalemia, see Chapter 9.

Hyperkalemia

Hyperkalemia may develop with use of the potassium-sparing diuretics (the diuretics that work in the late distal tubule) secondary to the *decreased* secretion and excretion of potassium. They should not be given to individuals with decreased renal function or to those receiving potassium supplements because of the increased risk of hyperkalemia. Monitor individuals receiving potassium-sparing diuretics for indicators of hyperkalemia: irritability, anxiety, abdominal cramping, muscle weakness (especially in the lower extremities), and EKG changes. See Chapter 9 for additional indicators of hyperkalemia.

Hyponatremia

Hyponatremia may occur with diuretic therapy due to an increased stimulus to the release of ADH secondary to a reduction in effective circulating volume. Increased ADH results in increased retention of water, causing a dilutional drop in serum sodium. This is most likely to occur with use of thiazide diuretics because they decrease the kidneys' ability to excrete water in excess of solutes. Monitor patients for indicators of hyponatremia: irritability, apprehension, and dizziness. See Chapter 8 for additional indicators of hyponatremia.

Hypomagnesemia

Hypomagnesemia may develop secondary to decreased reabsorption and increased excretion of magnesium by the kidneys. This is most likely to occur with the loop diuretics since the loop is the primary site of magnesium reabsorption. Although magnesium loss may be significant, the serum magnesium level often remains normal because the loss occurs primarily from the intracellular space. Magnesium is vital to the function of the cellular sodium-potassium pump, and intracellular hypomagnesemia will contribute to the development of hypokalemia. Thus hypomagnesemia may be evidenced by persistant hypokalemia despite supplementation. Individuals should be monitored, however, for the following indicators of hypomagnesemia: lethargy, weakness, and dysrhythmias. For more information concerning the relationship of potassium and magnesium, see Chapter 12.

Hypocalcemia and hypercalcemia

As previously discussed, the loop diuretics increase urinary excretion of calcium. Typically, this does not result in the development of hypocalcemia because of a concomitant increase in intestinal absorption of calcium (Rose, 1989). Of greater concern, however, is the development of calcium stone formation in the urine. Individuals with a history of nephrolithiasis (kidney stones) should be monitored for indicators of stone formation, including flank or genital pain.

In contrast to loop diuretics, thiazide diuretics can decrease renal excretion of calcium, potentially resulting in hypercalcemia. This is most likely to occur in individuals already at risk for hypercalcemia, those with hyperparathyroidism, for example. Individuals at increased risk should be monitored for indicators of hypercalcemia (e.g., weakness, paresthesias, and altered mentation). Refer to Chapter 10 for additional indicators of hypercalcemia.

Acid-base disturbances

Metabolic alkalosis

Clinically, one of the most common causes of metabolic alkalosis is diuretic therapy. It may develop with use of the loop or thiazide-type diuretics secondary to an increased secretion and excretion of hydrogen by the kidneys and the contraction of the ECF around the existing bicarbonate (contraction alkalosis). Monitor patients for indicators of metabolic alkalosis: muscular weakness, dysrhythmias, apathy, and confusion. For additional indicators of metabolic alkalosis, see Chapter 18.

Metabolic acidosis

Metabolic acidosis may occur with acetazolamide (Diamox) use secondary to increased loss of bicarbonate in the urine. Indicators of metabolic acidosis include tachypnea, fatigue, and confusion. The potassium-sparing diuretics also may cause metabolic acidosis due to decreased secretion of H^+ and K^+ in the distal tubule. These medications should be given with caution to any individuals at risk for developing acidosis, such as those with renal failure. For more information concerning the relationship of diuretics to metabolic acidosis, see Chapter 17.

Other metabolic complications

Azotemia

Azotemia, the increased retention of metabolic wastes (e.g., urea and creatinine), may occur with diuretic therapy secondary to a reduction in effective circulating volume (ECV). A decrease in ECV will result in decreased kidney perfusion and decreased excretion of metabolic wastes. Alert the MD to changes in BUN and serum creatinine levels.

Hyperuricemia

Hyperuricemia may occur with diuretic therapy due to increased reabsorption and decreased excretion of uric acid by the kidneys. This condition usually is problematic only in individuals with preexisting gout. Instruct individuals with a history of gout to report increased episodes of gouty-type pain.

FLUID THERAPY

The goals of IV fluid therapy are to maintain or restore normal fluid volume and electrolyte balance and provide a means of administering medications quickly and efficiently. An additional goal is that of nutritional maintenance. Unfortunately, routine IV fluids (i.e., 5% dextrose solutions) contain only enough carbohydrates to reduce tissue breakdown and prevent total starvation. They do not provide the adequate calories and essential amino acids needed for tissue synthesis. For example, 5% dextrose solutions supply approximately 170 calories/L, while the average patient on bed rest requires a minimum of 1,500 calories/day. Individuals should not be maintained solely on 5% dextrose solutions for longer than a few days.

The type of IV fluid prescribed for volume replacement or maintenance depends on several factors, including the type of fluid lost and the patient's nutritional needs, serum electrolytes, serum osmolality, and acid-base balance.

Commonly Prescribed Fluids

IV fluids may be divided into two major categories: colloids and crystalloids. Colloids are solutions that contain such substances as proteins or synthetic macromolecules that do not readily cross the capillary membrane. These solutions remain within the vascular space and, depending on their concentration, may cause an osmotic shift of fluids from the interstitium to the intravascular space. Crystalloid solutions do not contain substances that are restricted to the intravascular space and will cross the capillary membrane.

Crystalloids

Dextrose in water solutions

Five percent dextrose in water (D_5W) is a solution that provides free (pure) water and a small number of calories. Pure water and other markedly hypotonic solutions cannot be administered intravenously since they cause swelling and hemolysis of red blood cells (RBCs) as they enter the vein (see Figure 3-4). This occurs because of an osmotic shift of water into the RBCs. Thus dextrose must be added to these solutions to allow safe administration. The addition of 5% dextrose to water renders it isotonic (i.e., its effective osmolality is similar to that of body fluids). The dextrose is quickly metabolized, though, and the net effect is the administration of pure water.

Water crosses all membranes and will equilibrate between the two major fluid compartments. Thus the water in D_5W will be distributed throughout the intracellular fluid (ICF) and extracellular fluid (ECF), expanding both fluid compartments. D_5W is best used to replace total body water deficits, treat hypernatremia caused by water loss, replace insensible losses, or maintain IVs on a *keep-open* basis. Although normal daily fluid losses are hypotonic (effective osmolality less than that of body fluids), D_5W is not an appropriate maintenance solution because it does not replace the sodium and other electrolytes lost in normal body fluids.

Dextrose solutions also are available in concentrations of 10%, 20%, 25%, and 50% to provide additional calories. These solutions are hypertonic (effective osmolality greater than that of body fluids) but may be administered safely if given slowly. If given too rapidly, they may cause an osmotic shift of water into the intravascular space or a transient hyperinsulinism with weakness, diaphoresis, and confusion. See the discussion of total parenteral nutrition that follows.

Sodium chloride solutions

Sodium chloride (saline) solutions may be hypotonic (e.g., 0.45% NaCl), isotonic (e.g., 0.9% NaCl), or hypertonic (e.g., 3% NaCl). These solutions may be administered alone or combined with dextrose. The more dilute saline solutions (e.g., 0.225% NaCl) must be administered with dextrose to prevent hemolysis of RBCs.

Isotonic NaCl (0.9% NaCl) is often referred to as *normal saline* because it has the same tonicity as plasma; however, it is not truly normal or physiologic. (See Figure 7-2 for a comparison of lactated Ringer's solution, 0.9% NaCl, and 0.45 NaCl to ECF.) Isotonic NaCl provides sodium and chloride in concentrations greater than the sodium

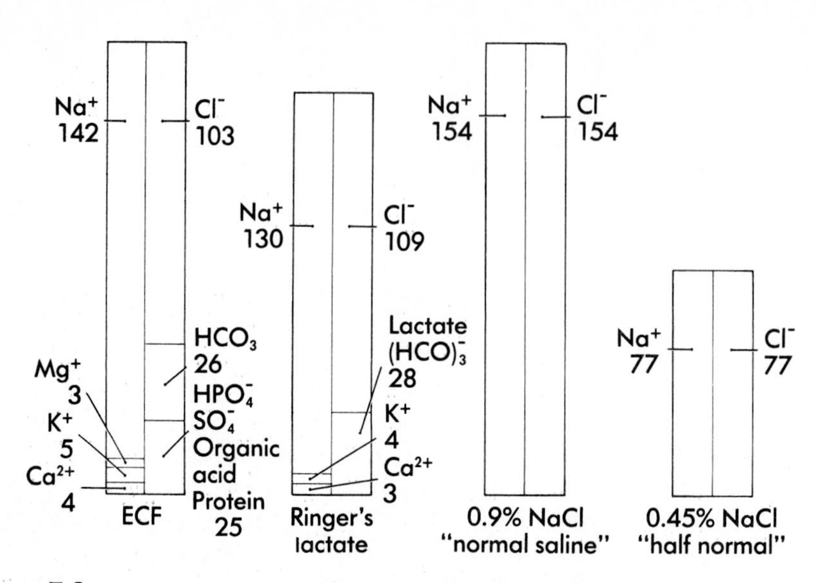

Figure 7-2 Composition of ECF and 3 common IV solutions.
(From Horne M and Swearingen P: Pocket guide to fluids and electrolytes, St Louis, 1989, The CV Mosby Co.)

and chloride concentrations in plasma. This is particularly true for chloride since the normal plasma chloride is only 100 mEq/L, while normal saline contains approximately 150 mEq/L of chloride. Isotonic NaCl expands only the ECF and does not enter the ICF (recall that sodium is restricted from the inside of the cell by the actions of the sodium-potassium pump). Isotonic NaCl is best suited for replacing ECF volume deficits (e.g., vomiting). It is not an appropriate solution for replacing primary water deficits or as a maintenance solution because it contains only NaCl and does not provide free water. The availability of free water is important in enabling the kidneys to excrete their daily solute load. Excessive administration of isotonic NaCl may cause fluid volume overload and hyperchloremic acidosis (see Chapter 17 for a discussion of hyperchloremic acidosis).

Hypotonic saline solutions provide both NaCl and free water. A 1,000 ml container of 0.45% NaCl (also referred to as *half normal saline* or just *half normal*) provides the equivalent of 500 ml of isotonic NaCl and 500 ml of pure (free) water. Thus 500 ml of the 0.45% NaCl solution will expand both the ICF and ECF, while the remainder will expand only the ECF. One-quarter normal saline (D_5/0.2% NaCl) provides an even greater volume of free water since it contains a smaller percentage of sodium and chloride. It may be used to replace hypotonic fluid losses, such as with an osmotic diuresis, or treat hypernatremia since it contains only a small amount of sodium.

Hypertonic NaCl solutions are used only to treat severe symptomatic hyponatremia and must be given with extreme caution due to the risk of intravascular fluid overload from an osmotic shift of water out of the cells and interstitium. A 3% NaCl solution contains 513 mEq/L each of sodium and chloride, and a 5% solution contains 755 mEq/L of each. Goldberger (1986) recommends that hypertonic NaCl be given at rates no greater than 100 ml/hr and that the individual be monitored carefully for signs of intravascular fluid excess (e.g., increased blood pressure, increased pulse, shortness of breath) for at least 30 minutes between 100 ml doses. The overall intake should not exceed 400 ml of 5% NaCl/day.

Ringer's solution

Ringer's solution is an isotonic NaCl solution that contains potassium and calcium in concentrations approximately equal to the potassium and calcium concentrations in plasma. Although the sodium content approaches the normal plasma level, the chloride content is significantly higher than that in plasma due to the lack of bicarbonate, the ECF's other major anion (Table 7-2). Lactated and acetated solutions were developed to correct this problem and provide buffer replacement. Both acetate and lactate are metabolized to produce bicarbonate. Lactated Ringer's (also known as Hartmann's solution) should not be given to patients with liver disease since lactate is converted by the liver. Also, it should not be given to patients with lactic acidosis because the pathway for metabolizing lactate would become overwhelmed. Acetate is preferred for these patients since it is metabolized by muscles and other peripheral tissues and is less dependent on the presence of adequate oxygen. Note that although the amount of potassium contained in Ringer's solution is physiologic, it is inadequate for replacing daily losses in patients who are not eating or who have abnormal potassium losses. Ringer's solution is often used instead of normal saline for replacement of ECF losses or ECF expansion. (See Figure 7-2 for a comparison of lactated Ringer's solution and ECF.)

Other multiple electrolyte solutions

Multiple electrolyte solutions were developed to replace the electrolytes and water lost with specific types of fluid loss. They may be either hypotonic or isotonic, depending on their NaCl content, and contain potassium, calcium, and magnesium (often in quantities greater than those in the plasma), acetate, or lactate. These solutions are usually combined with dextrose. Solutions containing less than plasma levels of sodium are used for the individual who cannot take fluids orally and who requires only a maintenance solution, since they provide the necessary free water. Solutions containing near normal plasma levels of sodium are used as replacement fluids since most abnormal fluid losses are isotonic (e.g., NG drainage, diarrhea). Some of these solutions have been specially formulated to replace specific types of fluid losses, for example, gastric fluid (contains ammonium—NH_4^+, which provides H^+, and a high Cl^-) or intestinal fluid (contains HCO_3^-). The ideal solution is one that has been customized to the patient's own combination of normal and abnormal losses. This may be accomplished by using the appropriate dextrose and NaCl solution and then adding each of the other electrolytes as needed. Fortunately, most patients are able to excrete excess water and electrolytes, thereby enabling the use of standard IV solutions that are less carefully formulated.

Table 7-2 Composition and use of commonly prescribed IV solutions

Solution	Glucose g/L	Electrolyte Composition mEq/L					Tonicity/ mOsm/L	Indications and Considerations*
		Na^+	K^+	Ca^{2+}	Cl^-	HCO_3^-		
Dextrose in water								
1. 5%	50	—	—	—	—	—	Isotonic/278	■ Provides free water necessary for renal excretion of solutes ■ Used to replace water losses and treat hypernatremia ■ Provides 170 Kcal/L ■ Does not provide any electroyltes
2. 10%	100	—	—	—	—	—	Hypertonic/556	■ Provides free water only, no electrolytes ■ Provides 340 Kcal/L
Saline								
3. 0.45%	—	77	—	—	77	—	Hypotonic/154	■ Provides free water in addition to Na^+ and Cl^- ■ Used to replace hypotonic fluid losses ■ Used as a maintenance solution although it does not replace daily losses of other electrolytes ■ Provides no calories

4. 0.9%	—	154	—	—	154	—	Isotonic/308	■ Used to expand intravascular volume and replace ECF losses ■ Only solution that may be administered with blood products ■ Contains Na^+ and Cl^- in excess of plasma levels ■ Does not provide free water, calories, or other electrolytes ■ May cause intravascular overload or hyperchloremic acidosis
5. 3.0%	—	513	—	—	513	—	Hypertonic/1026	■ Used to treat symptomatic hyponatremia ■ Must be administered slowly and with extreme caution because it may cause dangerous intravascular volume overload and pulmonary edema
Dextrose in saline								
6. 5% in 2.225%	50	38.5	—	—	38.5	—	Isotonic/355	■ Provides Na^+, Cl^-, and free water ■ Used to replace hypotonic losses and treat hypernatremia ■ Provides 170 Kcal/L
7. 5% in 0.45%	50	77	—	—	77	—	Hypertonic/432	■ Same as 0.45% NaCl except that it provides 170 Kcal/L
8. 5% in 0.9%	50	154	—	—	154	—	Hypertonic/586	■ Same as 0.9% NaCl except that it provides 170 Kcal/L

* Modified from Rose DB, Clinical pathology of acid-base and electrolyte disorders, ed, 3, New York, 1989, McGraw-Hill, Inc.

Continued.

Table 7-2 Composition and use of commonly prescribed IV solutions—cont'd

Solution	Glucose g/L	Electrolyte Composition mEq/L					Tonicity/ mOsm/L	Indications and Considerations
		Na^+	K^+	Ca^{2+}	Cl^-	HCO_3^-		
Multiple electrolyte solutions								
9. Ringer's	—	147	4	5	156	—	Isotonic/309	■ Similar in composition to plasma except that it has excess Cl^-, no Mg^{2+}, and no HCO_3^- ■ Does not provide free water or calories ■ Used to expand the intravascular volume and replace ECF losses
10. Lactated Ringer's (Hartmann's solution)	—	130	4	3	109	28†	Isotonic/274	■ Similar in composition to normal plasma except that it does not contain Mg^{2+} ■ Used to treat losses from burns and lower GI tract ■ May be used to treat mild metabolic acidosis but should not be used to treat lactic acidosis ■ Does not provide free water or calories

†In the form of lactate.

Colloids

Colloids are solutions such as blood, plasma, dextran, and albumin that remain in the vascular space and do not expand the interstitial or intracellular space. These solutions are useful in the treatment of shock when rapid expansion of the intravascular volume is desirable.

Blood and blood component therapies

Blood and blood component therapies are vital in the treatment of anemia, deficiencies in specific blood components, and volume deficit due to hemorrhage. Blood and most blood products require ABO and Rh typing to determine compatibility prior to administration (the reader is referred to a medical-surgical text for a discussion of the requirements for safe administration of blood and blood components).

Blood may be administered as whole blood or as packed red blood cells (PRBCs). Both whole blood and PRBCs therapies require typing and crossmatching prior to administration. Whole blood, administered in 500 ml units, is used to replace blood lost due to hemorrhage. Unless the blood is fresh (stored for less than 24 hours), it is a less than perfect replacement. As blood is stored, its composition is altered due to biochemical and metabolic changes. There is a reduction in the platelets and clotting factors, so that individuals who require large volumes of transfused blood may develop clotting abnormalities. Banked whole blood contains adenine, citrate, and ammonia, which may pose metabolic problems, especially for persons with decreased liver function.

Individuals requiring massive transfusions may develop symptoms of hypocalcemia (i.e., paresthesias, tetany, seizures) as the result of excessive binding of calcium with the citrate added to banked blood. Additionally, the older the unit of blood, the greater the potassium content due to the release of intracellular potassium that occurs with the death of RBCs. This may be of concern in the renal failure patient who is at risk for developing hyperkalemia due to decreased urinary excretion of potassium. Blood therapy also poses the risk of hepatitis, HIV infection, allergic reactions, and potentially fatal antigen-antibody reactions (occurs only with type and crossmatch errors).

Although not all volume loss that occurs with hemorrhage may be replaced with blood, blood replacement is usually indicated if the hematocrit drops below 25% to 30%. An adequate hematocrit is necessary for maintaining the oxygen-carrying capacity of the blood. The decision to transfuse will depend not only on an individual's absolute hematocrit, but also on his or her ability to tolerate a reduced hematocrit. The individual with heart disease, for example, may require a higher hematocrit to provide adequate oxygen delivery to the myocardium, while the person with chronic anemia may tolerate a very low hematocrit. The recipient's religious beliefs and concerns about the risks of transfusion also must be factored in the decision whether or not to administer blood products.*

* Certain religions prohibit the use of human blood products, Jehovah's Witnesses, for example.

A unit of PRBCs is produced by centrifuging a unit of whole blood to remove approximately 200 ml of plasma. This process reduces the volume of the unit to approximately 300 ml and increases the hematocrit to approximately 70% to 75%. The advantages of giving PRBCs are that they help to restore the oxygen-carrying capacity of the blood without providing excess volume and that they contain less plasma protein antigens, adenine, citrate, potassium, and ammonia than a unit of whole blood. Individuals with a history of transfusion reactions (onset of chills and fever within one hour of transfusion) may require leukocyte-poor PRBCs, which have nearly all the plasma, leukocytes, and platelets removed. The disadvantages of giving PRBCs are the relative lack of clotting factors as compared with whole blood and the increased viscosity of PRBCs, which slows their administration (although this limitation may be overcome by administering them with isotonic saline).

Individuals with clotting disorders or those receiving a large number of transfusions (either whole blood or PRBCs) may be given fresh frozen plasma to maintain their level of clotting factors. Platelet packs (usually given in groups of 10) are used to treat individuals with bleeding problems secondary to platelet deficiency (thrombocytopenia).

Albumin is the major protein of the plasma. As discussed in Chapter 1, it does not readily cross the capillary membrane and therefore helps to hold the vascular volume in the vascular space. Commercially available human serum albumin is used clinically to increase the plasma oncotic pressure and to expand the intravascular volume rapidly. It is often used in the treatment of shock to prevent the loss of fluid into the tissues secondary to rapid fluid replacement (this use, however, remains controversial). Albumin also may be used to restore serum levels in conditions associated with a decreased production of albumin (e.g., cirrhosis).

Albumin is obtained from pooled plasma, after which it is bottled and sterilized. It does not require typing and crossmatching, and it carries virtually no risk of hepatitis or HIV infection. It is available in two concentrations: 5% and 25%. Five percent albumin has an osmolality similar to normal plasma (300 mOsm/L) and will expand the vascular volume one ml for each ml administered (assuming there is no increase in capillary permeability with the loss of albumin into the interstitium). In contrast, 25% albumin has an osmolality of 1500 mOsm/L, and it will expand the intravascular volume 3 to 4 ml for each ml administered (Rice, 1984). Although currently both 5% and 25% albumin solutions contain equal concentrations of sodium (approximately 130 to 160 mEq/L), 25% albumin is often referred to as *salt poor* and nurses may see it prescribed that way. The term *salt poor* is a carryover from the time when 25% albumin was no longer prepared in a 1.8% salt solution.

Synthetic colloids

There are several commercially available synthetic colloids that act osmotically to expand the plasma volume. Mannitol, which was discussed in Case Study 1-1, is an example. Dextran, probably the most widely used, and hetastarch are two others. Both mannitol and dextran are types of sugars, while hetastarch is a form

Nursing Considerations for Delivering Fluid Therapy

1. Ensure that IV fluids infuse at the prescribed rate. Ideally, all IV fluids should be administered via an infusion pump or controller. IV fluids containing potentially dangerous medications or additives or TPN solutions must be on a pump. Pumps are preferred over controllers when the delivery of precise amounts is essential (e.g., when administering fluids to neonates or titrating medication in the critical care setting). Solutions containing potassium and calcium must not be administered at rates faster than recommended due to the risk of dangerous cardiac dysrhythmias. Pediatric volumetric control sets should be used when administering fluids to infants and small children.
2. Confirm that the correct solution is used. IV fluids must be checked against the MD's prescription. Label the IV solutions for content (when there is an additive), the time the infusion was initiated, the initials of the nurse who prepared and hung the infusion, and the infusion rate. Chart when new containers are hung, rates are changed, or medications are added.
3. Monitor the patient's response to fluid therapy. If the individual is being treated for fluid volume deficit, for example, does he or she continue to show signs of volume deficit? Use fluid challenge protocols for patients receiving rapid fluid replacement (see Chapter 6, p. 105).
4. Monitor for indicators of intravascular volume excess, especially when the individual is receiving colloids or has renal or cardiac disease.
5. Monitor the IV site for extravasation of IV fluids into tissues: swelling, coolness of skin, or discomfort (burning sensation or pain) at the site. If extravasation is suspected, stop the infusion and assess the adequacy of the IV site. If the infusion contains medications or additives that may damage tissue (e.g., high concentrations of calcium or sodium bicarbonate) alert the patient's physician or follow unit protocol.
6. Monitor for indicators of allergic reaction to the fluid (e.g., blood, hetastarch) or medication additives (e.g., antibiotics). Stop the infusion, keep the IV line open with 0.9% NaCl, and report any of the following to the patient's physician: sudden onset of chills, fever, rash, itching, shortness of breath, generalized edema.
7. Monitor for signs of phlebitis (i.e., tenderness or pain, erythema, warmth, or edema at site). Change IV site to a different vein if phlebitis is present.
8. Monitor for the following signs of IV-related infection (this is of particular concern with TPN solutions due to their high glucose content): unexplained fever or tachycardia, increase in WBC count, nausea, vomiting, or headache. Alert patient's physician to abnormal findings. Check all solution containers for cracks, leaks, or cloudy solution. Keep admixed TPN solutions refrigerated at 4° C (39.6° F) until 30 minutes before use. Use meticulous aseptic technique when manipulating TPN catheter, dressing, and lines. Change tubing and dressings on a routine basis according to agency protocol.

of starch. The synthetic colloids have the advantage of being relatively inexpensive as compared with albumin or plasma, and they do not carry the risk of bloodborne diseases. They do pose a small risk of allergic and anaphylactic reactions, and may affect clotting.

Total parenteral nutrition (TPN)

TPN is a form of IV fluid therapy aimed at meeting an individual's nutritional needs. Its use is indicated when oral and enteral nutrition are not feasible or adequate. TPN may be administered on a temporary basis using a peripheral venous access. Dextrose concentrations are then limited to only 10% concentration due to the potential for venous trauma. Long-term nutritional support with more concentrated dextrose solutions (e.g., 50%), requires the insertion of a special vascular catheter to provide access to a central vein. TPN solutions usually contain dextrose, amino acids, electrolytes, vitamins, trace elements, and sometimes insulin (to prevent the development of hyperglycemia). IV fat emulsions may be administered simultaneously with the TPN to provide additional energy via fatty acids. Carbon dioxide (CO_2) production may be reduced by providing a portion of the calories in the form of fat rather than carbohydrate. This may be helpful in conditions such as COPD, when CO_2 retention is a problem (see Chapter 15). (See the accompanying box on p. 143 for nursing considerations for delivering fluid therapy.)

Case Study 7-1

M.W. is a 67-year-old woman who arrives in the emergency room with complaints of generalized weakness and "palpitations" for the last 12 hours. The nurse obtains the following information and physical data:

History: negative for heart disease, diabetes, hypertension, neuromuscular disease, or central nervous system disorders. The patient denies any symptoms or problems other than weakness and "palpitations." Specifically, she denies chest pain, shortness of breath, and abnormal fluid loss.

Medications: None.

Allergies: None known.

Vital signs:

BP: 107/62 mm Hg sitting
    76/56 mm Hg standing

Apical/Radial pulses: 104/97 and irregular

Physical assessment: Skin cool and dry to touch, pallor, neck veins flat when fully reclined, breath sounds clear, abdomen soft and nontender with hypoactive bowel sounds, extremities cool with palpable pulses and without edema. In addition, the patient complains of dizziness and weakness that increase when standing.

1. **Question:** Although M.W.'s history does not suggest any fluid volume abnormality, do her physical findings suggest one? If so, which imbalance?

   Answer: Yes, fluid volume deficit. M.W. has significant orthostatic hypotension (a difference of greater than 20 mm Hg between her sitting and standing BP), weakness and dizziness that increase with standing, flat neck veins, an increased heart rate, and skin that is cool and pale.

Because the physical assessment findings are strongly suggestive of fluid volume deficit, the emergency room nurse discusses with her the importance of identifying any possible cause of the abnormal fluid loss. After some encouragement and support, M.W. admits that she has been on a "crash diet" and taking her husband's furosemide (a loop diuretic) to "help keep off the water weight." M.W. tearfully describes the stress she is under at work and how she feels pressured to "maintain" her appearance. She reluctantly admits to an unusually large urine output and thirst.

2. Question: What are potential nursing diagnoses for M.W.?

Answer:

a. **Fluid volume deficit** related to abnormal loss secondary to furosemide-induced diuresis
b. **Knowledge deficit:** Sound dieting practices and health hazards of medication abuse
c. **Ineffective individual coping** related to inability to deal effectively with the stress of employment
d. **Alteration in pattern of urinary elimination:** Diuresis

Based on the information obtained by the emergency room nurse, the physician requests the following tests: plasma electrolytes levels (to determine if the furosemide has resulted in any electrolyte imbalances; electrolyte imbalance is a possible cause of the irregular pulse and palpitations), urea nitrogen and plasma creatinine values (to determine if there is any evidence of renal disease, since an increased urine output could also indicate renal disease), blood glucose level (to rule out hyperglycemia—remember from Case Study 3-1 that an increased blood glucose may cause an increase in urine output and thirst), and an electrocardiogram ([EKG] to determine the cause of the irregular pulse and palpitations).

3. Question: Which electrolyte imbalances might occur with furosemide use?

Answer:

a. Hypokalemia due to increased secretion of potassium by the distal tubule.
b. Hyponatremia secondary to increased release of ADH and retention of water related to a diuresis-induced reduction in effective circulating volume. This is less likely to occur with the loop diuretics due to their effects on urinary concentration.
c. Hypomagnesemia secondary to decreased reabsorption of filtered magnesium in the renal tubule.
d. Hypocalcemia secondary to decreased reabsorption of calcium in the loop of Henle.

The results of the laboratory tests are as follows:

M.W.'s Test Values	Normal Values
Plasma Na^+ = 137mEq/L	137-147 mEq/L
K^+ = 2.7 mEq/L	3.5-5.0 mEq/L
Cl^- = 77 mEq/L	95-107 mEq/L
HCO_3^- = 35 mEq/L	22-26 mEq/L
Urea nitrogen = 40 mg/dl	6-20 mg/dl
Creatinine = 1.4 mg/dl	0.6-1.5 mg/dl
Glucose = 70 mg/dl	65-110 mg/dl
Mg^{2+} = 1.6 mEq/L	1.5-2.5 mEq/L

M.W. is, in fact, hypokalemic (has low plasma potassium). The EKG shows a normal sinus rhythm with occasional premature ventricular beats and changes consistant with hypokalemia.

4. Question: M.W.'s creatinine is normal but her urea nitrogen is elevated. Is this consistant with fluid volume deficit?

Answer: Yes. Remember from Chapter 5 that the urea nitrogen level increases with dehydration due to increased renal reabsorption of filtered urea. If the increased urea nitrogen were solely the result of renal insufficiency, the creatinine would have increased by an equal proportion.

5. Question: Acid-base imbalances may occur with diuretic therapy. Which imbalance may develop with the use of the loop diuretics? Does M.W. have laboratory evidence of that imbalance?

Answer: Metabolic alkalosis sometimes occurs with the use of loop diuretics. If present, this could contribute to M.W.'s feelings of weakness and possibly to her irregular pulse. The increased bicarbonate level on the electrolyte panel does reflect metabolic alkalosis. The relationship between metabolic alkalosis and diuretic therapy will be discussed in greater detail in Chapter 18.

M.W.'s test results are consistant with the diagnosis of diuretic abuse. The nurse assures her that fluid replacement, potassium supplementation, and discontinuation of the furosemide will resolve her weakness and palpitations. M.W. is to receive IV fluid and potassium replacement in the emergency room and then be followed by her private physician after discharge. The emergency room nurse provides her with information concerning medically-approved diet and stress reduction programs available in the community.

6. Question: What are some nursing considerations relevant to M.W.'s fluid replacement?

Answer:

a. Monitor and document her response to fluid therapy, including indicators of relief of the fluid volume deficit and for signs of fluid volume excess, especially if M.W. is to receive rapid fluid replacement.

b. Ensure that the IV potassium is not given at a rate faster than that recommended due to the risk of sudden hyperkalemia. Ensure that M.W. has an adequate urine output before administering potassium, since decreased renal function increases the risk of hyperkalemia. The risks of potassium administration will be discussed in Chapter 9.
c. Monitor the IV site for extravasation of IV fluids into tissues: swelling, coolness of skin, or discomfort (burning sensation or pain) at the site.

Case Study 7-2

T.O., a 34-year-old man brought to the emergency room with a gunshot wound to the abdomen, is bleeding profusely, and shows clinical indicators of hemorrhagic shock. His condition requires rapid administration of IV fluids to prevent the development of irreversible shock. The goals of therapy in the treatment of hypovolemic shock are to restore adequate intravascular volume and maintain nutrient delivery to the tissues.

1. **Question:** Would 5% dextrose in water (D_5W) be an appropriate choice of IV fluid?

   Answer: No. Administering D_5W is the physiologic equivalent to administering pure water. Water crosses all membranes and will equilibrate between all fluid compartments. Thus it will expand both the ICF and the ECF. Considering the relative size of the ECF as compared with the ICF, only one third of the water will remain in the ECF (see Figure 7-3). If 1 L (1,000 ml) of D_5W is administered IV, in theory only approximately 84 ml will remain in the intravascular space. Over time, 1,000 ml of D_5W will do little to maintain the volume of the intravascular space.

Isotonic NaCl solutions (either *normal saline* or Ringer's lactate) are used in combination with whole blood or packed red blood cells in the treatment of hemorrhagic shock.

2. **Question:** If 1 L of *normal saline* (NS) is administered IV, how much of the liter will expand the ECF?

   Answer: The full liter will expand the ECF. The sodium-potassium pump in the cell wall keeps sodium out of the cell. Thus, the sodium in NS will remain in the ECF, and because the concentration of NS is the same as that of the ECF (isotonic), the water also will remain outside of the cell (Figure 7-4). Assuming that the intravascular volume is one quarter of the ECF, one quarter of the liter (250 ml) should remain in the intravascular space. From this simple discussion it is easy to see that isotonic sodium chloride solutions are three times as effective at expanding the

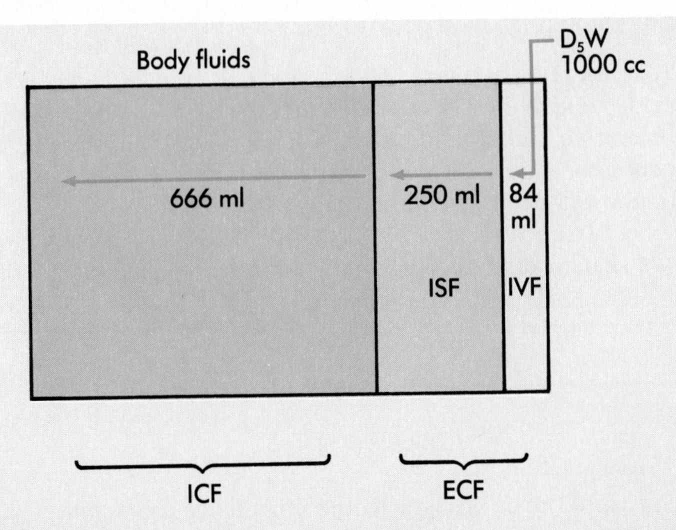

Figure 7-3 The effects of D_5W on body fluids.

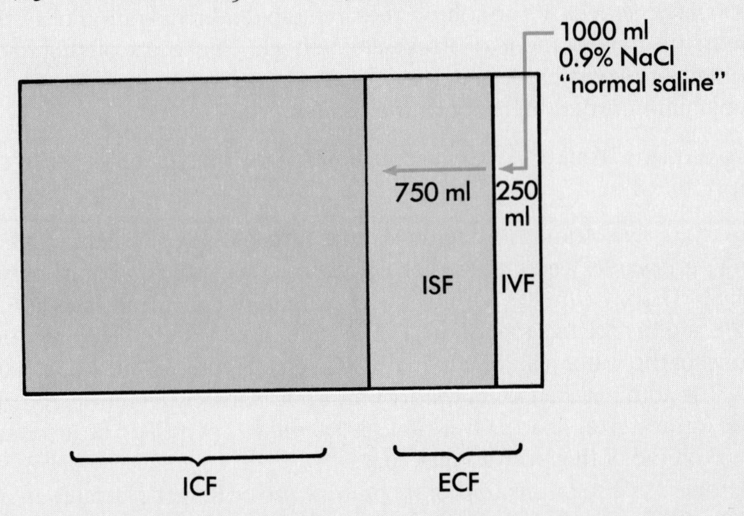

Figure 7-4 The effects of normal saline on body fluids.

intravascular volume as D_5W. This is why isotonic NaCl solutions pose a greater risk of intravascular volume overload and require careful nursing assessment during their administration.

3. **Question:** If 2 units of whole blood (1,000 ml total) are administered, what amount will expand the intravascular volume?

   Answer: The full liter will remain in the intravascular space because cells and plasma proteins do not cross the capillary membrane (Figure 7-5). This emphasizes the need to monitor individuals receiving blood for intravascular volume excess.

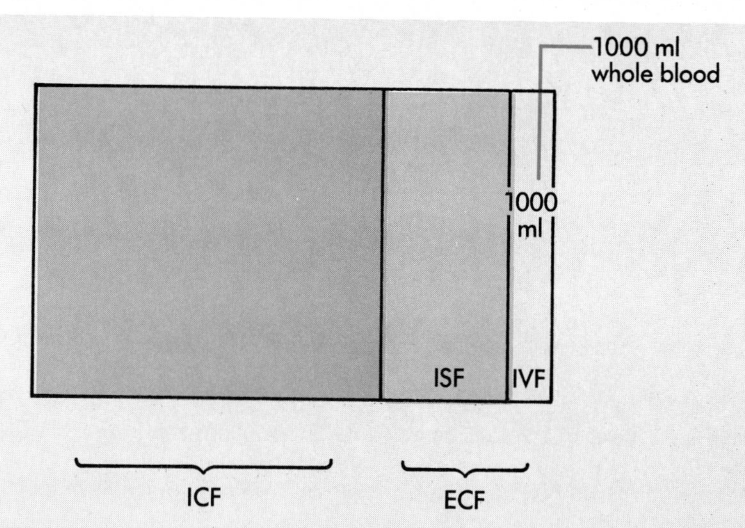

Figure 7-5 The effects of whole blood on body fluids.

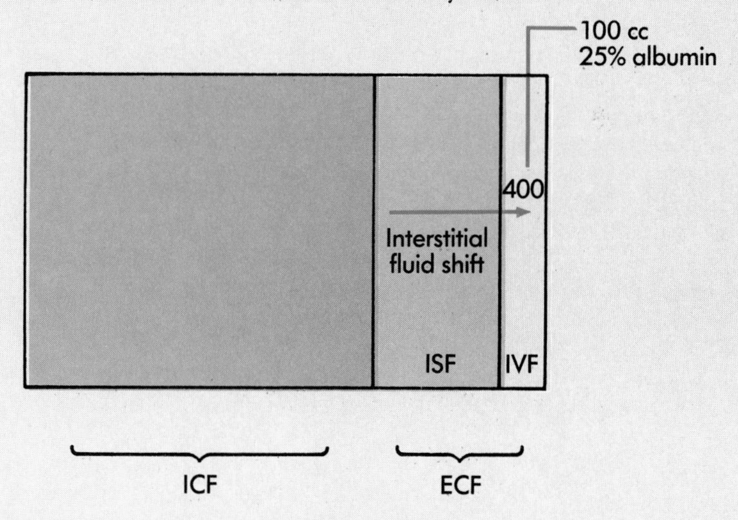

Figure 7-6 The effects of 25% albumin on body fluids.

4. **Question:** Concentrated albumin solutions (25%) are often used in the treatment of shock because they provide rapid volume expansion. If T.O. is given 100 ml of 25% albumin, will his intravascular volume increase by more or less than 100 ml?

Answer: More than 100 ml. Remember that 25% albumin has an osmolality of approximately 1,500 mOsm/L and will expand the intravascular volume by 3 to 4 ml for each ml administered (Figure 7-6). Concentrated albumin solutions may precipitate pulmonary edema in individuals with peripheral edema and heart failure or renal disease. This occurs because of the shift of fluid from the interstitial space into the intravascular space with an increase in pulmonary capillary hydrostatic pressure.

References

Borland C and others: Biochemical and clinical correlates of diuretic therapy in the elderly, Age Ageing 15:357-363, 1986.

Brater DC: Serum electrolyte abnormalities caused by drugs, Prog Drug Res 30:9-69, 1986.

Horne M and Swearingen PL: Pocket guide to fluids and electrolytes, St Louis, 1989, The CV Mosby Co.

Ghishan FK: The transport of electrolytes in the gut and the use of oral rehydration solutions, Pediatric Clin N Am 35(1):35-51, 1988.

Goldberger E: A primer of water, electrolytes, and acid-base syndromes, ed 7, Philadelphia, 1986, Lea & Febiger.

Goodman LS and Gillman A: The pharmacological basis of therapeutics, ed 7, New York, 1985, Macmillan.

In the market for an electronic infuser? Am J Nurs 88(9):1225-1227, 1988.

LaRocca JC and Otto SE: Pocket guide to intravenous therapy, St Louis, 1989, The CV Mosby Co.

Laski M: Diuretics: mechanism of action and therapy, Seminars in Nephrol 6(3):210-223, 1986.

Millam DA: Managing the complications of IV therapy, Nursing 88 17(3):34-42, 1988.

Rice V: Shock management Part I: fluid volume replacement, Crit Care Nurs 4:69-82, Nov/Dec 1984.

Rice V: The role of potassium in health and disease, Crit Care Nurs 2(3):54-73, 1982.

Rose BD: Clinical physiology of acid-base and electrolyte disorders, ed 3, New York, 1989, McGraw-Hill, Inc.

Smith K: Fluids and electrolytes: a conceptual approach, New York, 1980, Churchill Livingstone, Inc.

Chapter

8

Disorders of Sodium Balance

Sodium plays a vital role in maintaining the concentration and volume of extracellular fluid (ECF). It is the main cation of ECF and thus the major determinant of ECF osmolality. Under normal conditions, ECF osmolality can be estimated by doubling the serum sodium value. Sodium imbalances usually are associated with parallel changes in osmolality. Sodium also is important in maintaining irritability and conduction of nerve and muscle tissue and assisting in the maintenance of acid-base balance.

The average daily sodium intake far exceeds the body's normal daily requirements. The kidneys are responsible for excreting the excess and are capable of conserving sodium avidly during periods of extreme sodium restriction. Sodium concentration of the ECF is maintained via regulation of water intake and excretion. If serum sodium concentration is decreased (hyponatremia), the kidneys respond by excreting water. Conversely, if serum sodium concentration is increased (hypernatremia), serum osmolality increases, stimulating the thirst center and causing an increased release of antidiuretic hormone (ADH) by the posterior pituitary gland. ADH acts on the kidneys to conserve water. The adrenal cortical hormone, aldosterone, is an important regulator of sodium and ECF volume. The release of aldosterone causes the kidneys to conserve sodium and water, thereby increasing ECF volume.

Gains or losses of total body sodium are not necessarily reflected by the serum sodium level. If sodium and water are lost in the same concentration (e.g., with isotonic diarrhea), then the serum sodium level will remain unchanged despite an actual loss of sodium from the body. Likewise, increased sodium intake is typically accompanied by increased water intake (secondary to increased thirst), again maintaining the serum sodium level. Abnormal sodium levels most often reflect gains or losses in water. Normal serum sodium is 137 to 147 mEq/L.

HYPONATREMIA

Hyponatremia (serum sodium < 137 mEq/L) can occur because of a net gain of water or a loss of sodium-rich body fluids. Clinical indicators and treatment depend on the cause of hyponatremia and whether or not it is associated with a normal, decreased, or increased ECF volume.

History and Risk Factors

Decreased ECF volume

Losses of sodium-rich body fluids (due to abnormal GI, renal, or skin losses) alone will not result in serum hyponatremia since these losses are either isotonic or hypotonic. That is, sodium is lost with an equal or greater proportion of water. The physiologic response to this volume loss, however, will contribute to the development and maintenance of hyponatremia. Remember that volume deficit stimulates thirst and increases the release of ADH. This will result in the gain and conservation of water. Although volume loss also will stimulate the release of aldosterone causing increased retention of sodium and water, the net effect of all these factors will be a greater increase in water than sodium, leading to hyponatremia. As discussed in Chapter 7, hyponatremia may occur with diuretic therapy, especially with the use of thiazide diuretics. Adrenal insufficiency will result in hyponatremia due to the loss of sodium and water (secondary to the lack of aldosterone) combined with increased release of ADH secondary to volume depletion.

Normal/increased ECF volume

Syndrome of inappropriate antidiuretic hormone (SIADH), an excessive and inappropriate production of ADH, will result in dilutional hyponatremia due to abnormal retention of water. Total body sodium content has not changed; hyponatremia, in this situation, is solely the result of a water imbalance and is associated with a normal or increased fluid volume. Disorders that may result in SIADH are numerous and include neurologic disorders (e.g., meningitis, cerebral neoplasm, subarachnoid hemorrhage) and pulmonary diseases (e.g., pneumonia, tuberculosis, asthma). SIADH occurs with neurologic disorders that involve altered function of the hypothalamus or posterior pituitary, resulting in increased production of ADH by the hypothalamus or increased release of ADH by the posterior pituitary gland. The exact mechanism responsible for increased ADH levels in pulmonary disease remains unclear. Use of certain medications such as chlorpropamide (Diabinase—an antidiabetic medication) may lead to drug-induced SIADH. Again, the exact mechanism is unclear. Use of the hormone oxytocin to treat certain obstetrical conditions may result in hyponatremia due to its antidiuretic properties. Oat cell carcinoma and other tumors of the lung have been shown to produce an ADH-like hormone that will result in clinical hyponatremia. Postoperative patients may experience a period of inappropriate ADH secretion due to pain, stress, and medications, lasting 24 to 72 hours.

Hyponatremia also may develop in individuals with chronic psychiatric disorders that involve excessive water intake. This disorder is termed *psychogenic polydipsia.* Although this condition may be aggrevated by medications that cause a dry mouth (e.g., haloperidol), symptomatic hyponatremia is believed to develop because of a defect in water excretion. Excessive water intake alone does not result in hyponatremia since the kidneys normally are able to compensate with polyuria. Increased sensitivity of the hypothalamic osmoreceptors and increased renal sensitivity to ADH have been identified in these individuals; both of these conditions can result in decreased water excretion (Goldman and others, 1988). Development of symptomatic hyponatremia is often associated with worsening psychosis, although the relationship between the two remains unclear. Nurses must be on the alert for excessive fluid intake in psychotic individuals and monitor closely for signs and symptoms of hyponatremia.

Edematous states (e.g., congestive heart failure, cirrhosis, nephrotic syndrome) result in dilutional hyponatremia due to decreased effective circulating volume (ECV), which stimulates the release of ADH. Again, the problem is not one of sodium loss, but rather water gain. Hyponatremia also may develop with excessive or inappropriate administration of hypotonic IV fluids and oliguric renal failure due to the inability of the kidneys to excrete the daily water load.

Pseudohyponatremia

Hyperlipidemia, hyperproteinemia, and hyperglycemia may cause a pseudohyponatremia (false hyponatremia). Hyperlipidemia and hyperproteinemia reduce the total percentage of plasma that is sodium and water (a greater percentage of plasma in these conditions is lipids or protein). Sodium has not been lost and water has not been gained, thus the sodium-to-water ratio of the plasma does not change. The laboratory reports a decreased plasma sodium level, though, since there is a reduction in the percentage of sodium and water contained within the plasma. Pseudohyponatremia resulting from increased lipids or protein requires no treatment since there is no water or sodium imbalance.

With hyperglycemia, the osmotic action of the elevated glucose causes a shift of water out of the cells and into the ECF, thus diluting the existing sodium (see Case Study 3-1). For every 100 mg/dl glucose is elevated, sodium is diluted by 1.6 mEq/L. The hyponatremia associated with hyperglycemia will resolve with treatment of the elevated glucose.

Assessment

Decreased plasma osmolality

The signs and symptoms of hyponatremia are largely the result of a decrease in plasma osmolality, which causes the movement of water into the cells. As the concentration of sodium in the plasma decreases, a concentration gradient is created that favors the shift of water into the cells. An increase in the volume of brain cells leads to altered function and neurologic symptoms (e.g., irritability, lassitude, apathy, apprehension, personality changes, tremors, hyperreflexia, muscle spasms, convulsions, coma). Neurologic symptoms usually do not occur until the serum sodium level has dropped to approximately 120 to 125 mEq/L and are more likely to develop with a sudden decrease in plasma sodium than a gradual decrease. Early symptoms include nausea and malaise, progressing to headache, lethargy, and dulled sensorium if the sodium level drops to 115 to 120 mEq/L. Seizures, coma, and permanent neurologic damage may occur when the plasma sodium level is less than 115 mEq/L.

Decreased ECF volume

Some of the signs and symptoms associated with hyponatremia occur as the result of associated changes in vascular volume. The individual with hyponatremia secondary to the loss of body fluids also will exhibit the signs and symptoms of fluid volume deficit (e.g., dry mucous membranes, dizziness, postural hypotension, tachycardia).

Normal or expanded ECF volume

Hyponatremia resulting from water gain may be associated with the symptoms of fluid volume excess (e.g., edema, weight gain, elevated BP). Intracellular volume expansion may be evidenced by *fingerprint edema* (identified by the presence of a fingerprint mark after a finger is pressed firmly into tissue over the sternum or other bony surface). Changes in hemodynamic measurements will reflect ECF volume status (Table 8-1).

Diagnostic Studies

Both the plasma sodium and plasma osmolality will be decreased since sodium is the primary determinant of plasma osmolality. Hyponatremia combined with a normal or elevated osmolality indicates pseudohyponatremia or renal failure with the retention of urea (remember that urea also affects plasma osmolality). The sodium level will be less than 137 mEq/L, and if the individual is symptomatic, less than 125 mEq/L. In conditions with decreased ECV, the urine sodium will be decreased (usually <20 mEq/L) as the kidneys attempt to restore volume by retaining sodium. The two exceptions to this are adrenal insufficiency and salt-wasting renal disease in which the primary problem is the kidneys' inability to conserve sodium. In conditions associated with normal or increased ECV, such as SIADH, the urine will contain a greater than expected concentration of sodium since the increased volume *turns off* aldosterone.

Medical Management

The goal of therapy is to protect the patient from immediate danger (i.e., return sodium to >120 mEq/L) and then *gradually* return sodium to a normal level while restoring normal ECF volume.

Table 8-1 Quick assessment guidelines for individuals with hyponatremia

---

Signs and Symptoms

Hyponatremia with decreased ECF volume

Irritability, apprehension, dizziness, personality changes, postural hypotension, dry mucous membranes, tachycardia, tremors, seizures, coma

Hyponatremia with normal or increased ECF volume

Headache, lassitude, apathy, confusion, weakness, edema, weight gain, elevated BP, hyperreflexia, muscle spasms, convulsions, coma

NOTE: Neurologic symptoms usually do not occur until the serum sodium level has dropped to approximately 120-125 mEq/L.

Hemodynamic Measurements

Decreased ECF volume: Evidence of hypovolemia, including decreased CVP, PAP, CO, MAP; increased SVR

Increased ECF volume: Evidence of hypervolemia, including increased CVP, PAP, MAP

History and Risk Factors

Decreased ECF volume

- GI losses: Diarrhea, vomiting, fistulas, NG suction
- Renal losses: Diuretics, salt-wasting kidney disease, adrenal insufficiency
- Skin losses: Burns, wound drainage

Normal/increased ECF volume

- Syndrome of inappropriate antidiuretic hormone (SIADH): Excessive circulating antidiuretic hormone
- Edematous states: Congestive heart failure, cirrhosis, nephrotic syndrome
- Excessive administration of hypotonic IV fluids
- Oliguric renal failure

NOTE: Hyperlipidemia, hyperproteinemia, and hyperglycemia may cause a pseudohyponatremia. Hyperlipidemia and hyperproteinemia reduce the total percentage of plasma that is water. The sodium-to-water ratio of the plasma does not change, but the plasma sodium level is reduced since there is a reduction in plasma volume. With hyperglycemia, the osmotic action of the elevated glucose causes a shift of water out of the cells and into the ECF, thus diluting the existing sodium. For every 100 mg/dl glucose is elevated, sodium is diluted by 1.6 mEq/L.

Hyponatremia with reduced ECF volume

Treatment of hyponatremia associated with abnormal fluid loss includes fluid replacement with sodium-containing fluids. Adequate replacement of fluid volume is essential to *turn-off* the physiologic stimulus to ADH release and enable the kidneys to restore the balance between sodium and water. Treatment also includes replacement of other electrolyte losses incurred due to abnormal fluid loss. The individual with diarrhea, for example, may require replacement of potassium and bicarbonate. Hypertonic saline (3% to 5%) is used only if the plasma sodium is less than 110 to 115 mEq/L or the patient is very symptomatic.

Hyponatremia with normal or expanded ECF volume

Treatment of hyponatremia that results from water retention includes water restriction and the use of loop diuretics. Loop diuretics, such as furosemide, are used because of their effects on urinary concentration. They induce a fairly isotonic diuresis that results in volume loss without further hyponatremia. Thiazide diuretics are not used since they can cause a loss of sodium in excess of water, contributing further to the hyponatremia. Hypertonic NaCl is indicated only when hyponatremia is severe. It must be administered with extreme caution since it may induce dangerous intravascular volume overload due to its extreme hypertonicity. For this reason, hypertonic NaCl is usually given in combination with a loop diuretic.

Treatment of SIADH

The best treatment approach in SIADH is removal of the cause. For example, thoracotomy (to remove tumors of the lung) or craniotomy (to evacuate hematomas) are procedures that may remove the cause of this disorder. When the primary problem is not amenable to treatment, or when the condition is acute, management of SIADH may involve fluid restriction and diuretic therapy, as discussed previously. It also may include the use of medications that inhibit the action of ADH on the renal tubule (e.g., lithium and demeclocycline).

Nursing Diagnoses and Interventions

Potential fluid volume deficit or excess related to abnormal fluid loss, excessive intake of hypotonic solutions, or abnormal retention of water

Desired outcome: Patient is normovolemic as evidenced by HR 60 to 100 bpm, RR 12 to 20 breaths/min with normal depth and pattern (eupnea), BP within patient's normal range, and CVP 2 to 6 mm Hg (5 to 12 cm H_2O). For critical care patients, pulmonary artery pressure (PAP) is 20 to 30/8 to 15 mm Hg.

1. If patient is receiving hypertonic saline, assess carefully for signs of intravascular fluid overload: tachypnea, tachycardia, shortness of breath, crackles (rales), rhonchi, increased CVP, increased PAP, gallop rhythm, and increased BP. If given

too rapidly, hypertonic saline may cause red blood cells to shrivel in addition to causing an osmotic shift of fluid into the vascular space.

2. Explain the need for limiting fluids if a fluid restriction has been prescribed. Remind patient that a portion of fluid allotment can be taken as ice or popsicles to minimize thirst.
3. Review medications with patient and significant others. Include drug name, purpose, dosage, frequency, precautions, and potential side effects. Teach signs and symptoms of hypokalemia if patient is taking diuretics and provide examples of foods that are high in potassium (see the box on p. 169).
4. For other interventions, see Chapter 6, the section on hypovolemia, p. 109, for **Fluid volume deficit**; see Chapter 6, the section on hypervolemia, p. 118, for **Fluid volume excess**.

Sensory/perceptual alterations related to altered sensorium and LOC secondary to sodium level <120 to 125 mEq/L

Desired outcome: Patient does not exhibit signs of injury due to altered sensorium.

1. Assess and document LOC, orientation, and neurologic status with each VS check. Reorient patient as necessary. Alert MD to significant changes.
2. Inform patient and significant others that altered sensorium is temporary and will improve with treatment.
3. Keep side rails up and bed in lowest position, with wheels locked.
4. Use reality therapy, such as clocks, calendars, and familiar objects; keep these items at the bedside within patient's visual field and the call light within easy reach. Have patient wear glasses or hearing aid if needed.
5. If seizures are expected, pad side rails and keep an airway at the bedside.

HYPERNATREMIA

Hypernatremia (serum sodium level >147 mEq/L) may occur with water loss or sodium gain. Because sodium is the major determinant of ECF osmolality, hypernatremia always causes hypertonicity. In turn, hypertonicity causes a shift of water out of the cells, which leads to cellular dehydration and increased ECF volume.

As discussed in Chapter 3, the body's primary protection against the development of hypertonicity is thirst. As the plasma osmolality increases, the thirst center in the brain is stimulated and the individual seeks fluids. Although increased release of ADH is another important protective response to hypernatremia, thirst provides the ultimate defense. Significant hypernatremia due to water loss occurs only in individuals who have a diminished ADH response, do not have an intact thirst mechanism, or are unable to obtain fluids. Thus hypernatremia typically is seen in infants (due to their inability to adequately express their thirst), the elderly (due to a diminished thirst mechanism), those who are comatose, and individuals who do not have access to water.

History and Risk Factors

Water loss

Increased sensible and insensible fluid loss from the skin (e.g., diaphoresis, fever, burns) and insensible fluid loss from the lungs (e.g., respiratory tract infection) may result in the development of hypernatremia. Insensible fluid is almost electrolyte free; its loss from the body is nearly equivalent to the loss of pure water. Although sensible fluid does contain a significant amount of sodium, it is also hypoosmotic as compared to plasma and its loss will result in the loss of water in excess of sodium. Loss of water in excess of sodium results in the development of *hyper*natremia. Under the discussion of hyponatremia, it was pointed out that the loss of any body fluids, even hypotonic fluids, often leads to *hypo*natremia due to the body's normal response to volume loss (i.e., water gain secondary to increased thirst and release of ADH). Thus hypotonic fluid loss leading to hypernatremia is most likely to occur in the individual with impaired thirst, inability to access water, or decreased ADH action.

Osmotic diuresis

An osmotic diuresis also will result in the loss of water in excess of electrolytes and may lead to the development of hypernatremia. Any substance that is freely filtered by the kidneys but not reabsorbed may act as an osmotic diuretic. The conditions most commonly associated with an osmotic diuresis include an elevated blood glucose, radiographic studies using IV contrast media, and use of mannitol or dextran.

Diabetes insipidus (DI)

In addition to contributing to the development of hyponatremia, altered ADH also may be a primary cause of hypernatremia. A lack of functioning ADH is termed diabetes insipidus (DI) and may develop because of decreased production or release of ADH (central/neurogenic DI), or because of a decrease in the kidneys' responsiveness to it (nephrogenic DI). Central DI may occur secondary to head injury, brain tumors, infections such as meningitis or encephalitis, and conditions causing increased intracranial pressure or cerebral hypoxia (see Case Study 3-2). Typically, nephrogenic DI occurs as a complication of medication therapy or electrolyte imbalance. Lithium, which is used in the treatment of manic depression, and the antibiotic demeclocycline (Declomycin) are the two medications most commonly associated with this disorder. Hypercalcemia (increased plasma calcium) and hypokalemia (decreased plasma potassium) both alter the kidneys' responsiveness to ADH, causing nephrogenic DI.

Sodium gain

Sodium gain, a true increase in body sodium content, may occur with increased intake of sodium or decreased excretion of sodium. Examples of increased sodium intake include IV administration of hypertonic saline or sodium bicarbonate, use of medications such as sodium polystyrene sulfonate (Kayexalate), or increased oral intake of sodium, especially in infants and comatose persons, typically due to incorrectly prepared formulas or insufficiently diluted enteric tube feedings. Saltwater near-drowning also may result in hypernatremia due to the aspiration and ingestion of hypertonic sea water (the ocean has approximately three times the sodium content of plasma). Primary aldosteronism, which develops due to a tumor of the adrenal gland, results in excessive production of aldosterone with abnormal retention of sodium.

Assessment

As with hyponatremia, the signs and symptoms of hypernatremia occur largely as the result of changes in the plasma osmolality that lead to changes in the volume of cellular water. Hypernatremia (hypertonicity) creates an osmotic gradient that causes a shift of water out of the cells. Dehydration of cerebral cells leads to neurologic symptoms such as intense thirst, fatigue, weakness, lethargy, restlessness, agitation, seizures, and if severe, eventually to coma. Sodium excess also has a direct effect on irritability and conduction of nerve cells, causing them to be more easily excited. Changes noted on physical assessment may include low-grade fever, flushed skin, peripheral and pulmonary edema (sodium gain), or postural hypotension (water loss). The individual who has experienced a sodium gain will exhibit the accompanying signs of volume expansion, while the individual with water loss may show signs of fluid volume deficit. The volume effects of water loss, however, are minimized because the shift of water out of the cells helps to maintain the volume of the ECF. Hemodynamic measurements will vary, depending on volume status (increased central venous pressure [CVP] and pulmonary artery pressure [PAP] in sodium excess and decreased CVP and PAP in severe water loss). (See Table 8-2 for quick assessment guidelines for individuals with hypernatremia.)

Symptoms are most likely to develop with sudden increases in plasma sodium. After approximately 24 hours, brain cells adjust to ECF hypertonicity by increasing intracellular osmolality. The exact mechanism by which this occurs is unclear, but it is known that this increased osmolality helps to rehydrate the cell. Thus individuals with chronic hypernatremia may exhibit few symptoms. This adaptive mechanism has significance in the treatment of hypernatremia. If the plasma sodium level is reduced too quickly by the administration of water, there will be a rapid and dramatic movement of water into the cell, due to its increased osmolality, causing dangerous cerebral edema.

Diagnostic Studies

Determining the cause of hypernatremia requires a review of the patient's history and physical findings, along with laboratory results.

Serum sodium and osmolality

The diagnosis of hypernatremia is based on the finding of an elevated serum sodium level. The serum sodium will be >147 mEq/L and the serum osmolality will be >300 mOsm/kg.

Table 8-2 Quick assessment guidelines for individuals with hypernatremia

Signs and Symptoms

Intense thirst, fatigue, restlessness, agitation, coma. Symptomatic hypernatremia occurs only in individuals who have a diminished ADH response, do not have access to water, or who have an altered thirst mechanism (e.g., infants, the elderly, those who are comatose)

Physical Assessment

Low-grade fever, flushed skin, peripheral and pulmonary edema (sodium gain); postural hypotension (water loss)

Hemodynamic Measurements

Sodium excess: Increased CVP and PAP

Water loss: Decreased CVP and PAP

History and Risk Factors

Sodium gain: IV administration of hypertonic saline or sodium bicarbonate, increased oral intake, primary aldosteronism, saltwater near drowning, drugs such as sodium polystyrene sulfonate (Kayexalate)

Water loss: Increased insensible and sensible water loss (e.g., diaphoresis, respiratory infection), diabetes insipidus, or osmotic diuresis (e.g., hyperglycemia)

Urine specific gravity and osmolality

Urinary concentration may provide the first clue to the cause of hypernatremia. Urine specific gravity and osmolality will be increased in nonrenal causes of water loss and in sodium excess. The concentration of the urine is increased as the kidneys attempt to compensate for increased plasma osmolality by retaining water. Obviously this will not be the case in DI or water loss secondary to an osmotic diuresis, in which the primary problem is an inability to concentrate the urine. In DI, for example, the urine specific gravity is less than 1.007 (average normal is 1.010 to 1.020) and the urine osmolality is less than 200 mOsm/kg (average normal is 300 to 900 mOsm/kg).

Plasma ADH/vasopressin test

When DI is suspected, a plasma ADH level may be obtained or a vasopressin test performed. The vasopressin test involves administering vasopressin (Pitressin—an aqueous solution of the hormone ADH) and monitoring the urine for changes in output and concentration. Individuals with central DI should respond to the vasopressin test by decreasing their urine output and increasing urinary concentration (osmolality). In contrast, individuals with nephrogenic DI will not experience a significant change in either output or concentration since their kidneys have a diminished ability to respond to ADH.

Medical Management

Medical management of hypernatremia will depend largely on the etiology. Oral or IV water replacement is indicated for water loss. If the serum sodium level is >160 mEq/L, IV D_5W or hypotonic saline is given to replace the water deficit. Hypernatremia is corrected slowly, over approximately two days, to avoid too great a shift of water into brain cells, which could cause cerebral edema. An important nursing function during the treatment of hypernatremia is careful monitoring for the signs and symptoms of cerebral edema (increased BP, decreased heart rate, altered sensorium). Sodium intake is restricted in individuals with sodium gain. Additionally, diuretics may be given in combination with oral or IV water replacement.

Treatment of diabetes insipidus (DI)

Treatment of central DI includes administration of vasopressin (Pitressin) given IM or SC or dDAVP (desmopressin) by nasal spray to replace the ADH not produced or released by the brain. Central DI also may be treated with chlorpropamide or other medications that increase the release or action of ADH.

The usual approach in managing nephrogenic DI is to treat the primary problem (e.g., correct hypokalemia). Ironically, thiazide diuretics combined with sodium restriction is the most commonly used treatment when the primary problem is not reversible. While it seems odd to treat polyuria with a diuretic, recall that one of the side effects of the thiazide diuretics is hyponatremia. Hyponatremia occurs because the thiazides decrease the kidneys' ability to excrete water in excess of sodium *(free water)*. In DI, the problem is uncontrolled excretion of free water, and the thiazides actually decrease urine output in the presence of DI.

Nursing Diagnoses and Interventions

Potential for injury related to altered sensorium secondary to cerebral edema occurring with too rapid correction of hypernatremia

Desired outcome: Patient does not exhibit evidence of injury due to altered sensorium or seizures.

1. Cerebral edema may occur if hypernatremia is corrected too rapidly. Monitor serial serum sodium levels; notify MD of rapid decreases.

2. Assess patient for indicators of cerebral edema: lethargy, headache, nausea, vomiting, increased BP, widening pulse pressure (difference between the systolic and diastolic BP), decreased pulse rate, and seizures.
3. Assess and document LOC, orientation, and neurologic status with each VS check. Reorient patient as necessary. Alert MD to significant changes.
4. Inform patient and significant others that altered sensorium is temporary and will improve with treatment.
5. Keep side rails up and bed in lowest position, with wheels locked.
6. Use reality therapy, such as clocks, calendars, and familiar objects; keep these items at the bedside within patient's visual field. Keep the call light within easy reach. Have individual wear glasses or hearing aides if needed.
7. If seizures are anticipated, pad side rails and keep an airway at the bedside.

See the section on hypovolemia, p. 109, for **Fluid volume deficit** (applicable to hypernatremia caused by water loss); see the section on Hypervolemia, p. 118, for **Fluid volume excess** (applicable to hypernatremia caused by sodium gain).

Case Study 8-1

J.T. is an 84-year-old man admitted to the hospital because of increasing confusion and a 4-day history of bronchitis with fever and poor oral intake. The physician describes him as a pleasant but "senile gentleman," who is normally well cared for at home by his elderly wife. The nurse assigned to J.T. completes her nursing assessment, noting the following:

Confused as to person, place and time; restless; irritable
BP and pulse—sitting: 110/66 mm Hg; 94 bpm
BP and pulse—lying flat: 116/70 mm Hg; 92 bpm
Rectal temperature 39.2° C (102.6° F)
Auscultation of lungs reveals coarse rhonchi throughout lung fields
Skin is warm, dry, scaly, and pale. Skin turgor is poor.

Appearance indicates poor personal hygiene: dirty, matted hair; odor of urine about skin and clothes; ragged, crusted fingernails.

1. **Question:** Has J.T. experienced any unusual fluid losses?

   Answer: Yes, increased insensible fluid loss due to fever and respiratory infection.

2. **Question:** How would you describe J.T.'s fluid losses—isotonic, hypotonic, or hypertonic?

   Answer: Hypotonic. Insensible fluid is nearly equivalent to pure water.

3. Question: What would you expect to happen to his plasma sodium level if he is loosing extra water?

Answer: It should increase. An electrolyte panel drawn immediately after admission does in fact reveal an elevated plasma sodium level: 164 mEq/L (normal range = 137 to 147 mEq/L).

4. Question: How would this elevated sodium level affect J.T.'s plasma osmolality?

Answer: It would increase it. Remember that sodium is the primary determinant of the plasma osmolality. The osmolality of the ECF may be estimated by doubling the serum sodium level.

J.T.'s estimated osmolality = 2 × 164 = 328 mOsm/kg
(normal = 280-300 mOsm/kg).

5. Question: What are the normal physiologic mechanisms that help prevent the development of hypernatremia (hypertonicity)? Why were they not effective in preventing J.T.'s hypernatremia?

Answer: Thirst is the primary protection against the development of hypertonicity, although increased production and release of ADH is also important. In this case, J.T.'s ability to compensate with increased thirst is limited due to age (the thirst center is less sensitive to changes in osmolality in the older adult) and confusion. J.T.'s physical hygiene suggests that recently his wife has been unable to care for him adequately. Since his wife traditionally has provided all of his meals, it is possible that he simply did not consume food or fluids. Older adults are also at increased risk for hypertonicity due to the affects of advanced age on the capacity of the kidneys to concentrate the urine.

6. Question: Although J.T. has experienced abnormal fluid loss and decreased intake, he does not exhibit significant signs of fluid volume deficit. Why?

Answer: The symptoms of fluid loss are minimized by the osmotic shift of water out of the cells. This shift helps to maintain the volume of the ECF.

7. Question: The nurse has noted that J.T. has decreased skin turgor. Is this a reliable indicator of volume status in this situation?

Answer: No, because skin turgor typically is decreased in the older adult due to the loss of normal elasticity of the skin.

J.T. is to be treated with antibiotics and hypotonic IV fluids (e.g., D_5/0.2 NaCl).

8. Question: Identify an important nursing consideration when administering hypotonic IV fluids to an individual with hypernatremia.

Answer: Monitoring for the development of cerebral edema secondary to a rapid shift of fluid into the cerebral cells. This is of greatest concern in the individual who is chronically hypernatremic. The plasma sodium level should not be dropped at a rate greater than 2 mEq/hr. Rapid correction may cause cerebral edema, which in turn may result in permanent brain damage or death. The symptoms of acute hypernatremia occur as the result of a shift of water out of the brain cells. After approximately 24 hours, however, the brain cells compensate by increasing intracellular osmolality, which restores cell volume. Thus the individual with chronic hypernatremia may exhibit few symptoms. J.T. was confused prior to the development of hypernatremia, so his confusion does not necessarily indicate acute hypernatremia. It is likely that he has been hypernatremic for over 24 hours, placing him at increased risk for the development of cerebral edema secondary to fluid therapy.

9. Question: What are some potential nursing diagnoses for J.T.?

Answer:

a. **Potential for injury** related to confusion and the risk of cerebral edema
b. **Fluid volume deficit** related to decreased intake and increased fluid loss
c. **Alteration in nutrition:** Less than body requirements related to decreased intake
d. **Self-care deficit:** Hygiene/feeding/dressing/grooming related to confusion and debilitated state
e. **Altered health maintenance** related to confusion and debilitated state
f. **Impaired home maintenance management** related to confusion and debilitated state

References

Arieff AI and De Fronzo RA: Fluid, electrolyte, and acid-base disorders, New York, 1985, Churchill Livingstone, Inc.

Cowley AW and others: Osmoregulation during high salt intake: relative importance of drinking and vasopressin secretion, Am J Physiol 25:878-886, 1986.

Geheb MA: Clinical approach to the hyperosmolar patient, Crit Care Clin 5(4):797-815, 1987.

Goldberger E: A primer of water, electrolyte, and acid-base syndromes, ed 7, Philadelphia, 1986, Lea & Febiger.

Goldman MB and others: Mechanisms of altered water metabolism in psychotic patients with polydipsia and hyponatremia, N Engl J Med 318(7):397-403, 1988.

Hartshorn J and Hartshorn E: Vasopressin in the treatment of diabetes insipidus, J Neurosci Nur 20(1):58-59, 1988.

Horne M and Swearingen PL: Pocket guide to fluids and electrolytes, St Louis, 1989, The CV Mosby Co.

Kokko JP and others: Fluids and electrolytes, Philadelphia, 1986, WB Saunders Co.

Maxwell MH and others: Clinical disorders of fluid and electrolyte metabolism, ed 4, New York, 1987, McGraw-Hill, Inc.

Metheny NM: Fluid and electrolyte balance: nursing considerations, Philadelphia, 1987, JB Lippincott Co.

Rose BD: Clinical physiology of acid-base and electrolyte disorders, ed 3, New York, 1989, McGraw-Hill, Inc.

Schrier RW: Renal and electrolyte disorders, ed 3, Boston, 1986, Little, Brown & Co.

Vokes TJ: Disorders of antidiuretic hormone, Endocrinol Metabolism Clin N Am 17(2):281-299, 1988.

Chapter

9

Disorders of Potassium Balance

Potassium is the primary intracellular cation. It plays a vital role in cell metabolism and maintenance of the resting potential of nerve and cardiac cells. A relatively small amount (approximately 2%) of the body's potassium is located within the extracellular fluid (ECF) and is kept within a narrow range. Normal serum potassium is 3.5 to 5.0 mEq/L, and it reflects the potassium concentration of the ECF as a whole. The vast majority of the body's potassium is located within the cells. The potassium concentration of intracellular fluid (ICF) is approximately 150 mEq/L.

There is a constant shift of potassium into and out of all body cells. Potassium passively diffuses out of the cells due to the large concentration gradient between the ICF and ECF. At the same time, there is passive diffusion of sodium into the cells. If this simple diffusion of potassium and sodium were left unchecked, the concentration of these electrolytes would reach an equilibrium between the two compartments, and the unique compositions of the ICF and ECF would not be maintained. Equilibrium between these two compartments is prevented by the action of the sodium-potassium pump. As discussed in Chapter 1, a pump exists in the membrane of all body cells that actively moves potassium into the cells and sodium out. A steady supply of energy and sufficient quantities of the electrolyte magnesium are necessary for this pump to work.

Distribution of potassium between ECF and ICF is also affected by ECF pH, glucose and protein metabolism, and several hormones, including insulin and epinephrine. In acidosis, for example, excess hydrogen ions (H^+) move into the cells to be buffered. In order to maintain electroneutrality (equal number of negative and positive charges) within the cells, another positive ion (e.g., potassium [K^+]) must move out. In alkalosis the reverse occurs. H^+ shift out of the cells and K^+ shift in to replace them. Most acute changes in serum pH are accompanied by reciprocal changes in serum potassium concentration. Additionally, potassium is released when cells are broken down or destroyed (catabolism), and potassium moves into the cells when new tissue is formed (anabolism). Both insulin and epinephrine promote cellular uptake of potassium and help the body compensate for an acute potassium load by facilitating the movement of potassium into the cells. Additional factors affecting the shift of potassium into and out of the cells will be discussed in greater detail in the section on history and risk factors, which follows.

The body gains potassium through foods (primarily meats, fruits, and vegetables) and medications. In addition, the ECF gains potassium any time there is a breakdown of cells or movement of potassium out of the cells. However, an elevated serum potassium level usually does not occur unless there is a concomitant reduction in renal function. Potassium is lost from the body through the kidneys, GI tract, and skin. Potassium may be lost from the ECF because of an intracellular shift. Some form of potassium intake, whether from food or medication, is necessary each day to maintain balance. It may be necessary to increase intake when there are abnormal losses.

The kidneys are the primary regulators of potassium balance. They regulate potassium balance by adjusting the amount of potassium that is excreted in the urine. As the serum potassium level rises after a potassium load, so does the level in the renal tubular cells. This increased potassium level creates a concentration gradient favoring the movement of potassium into the renal tubule, with loss of potassium in the urine. An increase in the plasma concentration of potassium also will stimulate the release of aldosterone. Aldosterone increases the reabsorption of sodium and the secretion of potassium into the lumen of the renal tubule. Although sodium reabsorption and potassium secretion in the distal tubule are interrelated, there is *not* an ion-for-ion exchange of sodium and potassium. In most conditions associated with increased aldosterone levels (e.g., postsurgical stress) there will be an increase in urinary excretion of potassium. The kidneys are unable to conserve potassium as avidly as they do sodium, and a significant amount of potassium may be lost in the urine in the presence of potassium depletion.

HYPOKALEMIA

Hypokalemia occurs because of a loss of potassium from the body, movement of potassium into the cells, and inadequate intake (rare). Changes in serum potassium levels reflect changes in ECF potassium, not necessarily changes in total body levels.

History and Risk Factors

Increased urinary loss

Diuretic therapy

The most common cause of clinically significant potassium loss occurs secondary to diuretic therapy. All diuretics, except those that work in the late distal tubule, increase the loss of potassium in the urine. The greater the diuresis, the greater the loss of potassium. Thus the loop diuretics (furosemide and ethacrynic acid) and the thiazide diuretics cause the greatest hypokalemia. Increased potassium loss may occur with any condition that causes a diuresis, such as the osmotic diuresis that occurs with uncontrolled diabetes mellitus. Although the clinical significance of mild diuretic-induced hypokalemia remains controversial, an increased incidence of sudden cardiac death has been observed in hypertensive individuals treated with diuretics, possibly related to hypokalemia- or hypomagnesemia-induced dysrhythmias (Rose, 1989).

Diuretic-induced hypokalemia is also of concern in individuals with hepatic cirrhosis due to increased risk of hepatic coma. As potassium moves out of the cells in response to hypokalemia, positive hydrogen ions (H^+) move into the cells in order to maintain electroneutrality within the cells. This causes a relative intracellular acidosis. The renal tubular cell responds by increasing renal production of ammonia. This increased production of ammonia may be enough to precipitate hepatic coma in individuals with severe hepatic disease (see Chapter 13, p. 255, for a discussion of renal ammonia production).

Increased aldosterone levels

Increased urinary losses also may develop due to increased aldosterone levels. Remember that aldosterone causes increased retention of sodium and water and increased secretion and urinary excretion of potassium. Aldosterone levels may increase due to primary or secondary causes. In primary hyperaldosteronism there is an increased production of aldosterone due to an adrenal tumor or adrenal hyperplasia. Secondary hyperaldosteronism occurs in conditions associated with a decrease in effective circulating volume (ECV), such as congestive heart failure (CHF), cirrhosis, and nephrotic syndrome, in which there is a chronic stimulus to produce and release aldosterone. Administration of glucocorticoid medications, such as cortisone, may contribute to the development of hypokalemia due to their mineralocorticoid (aldosterone) effects.

Abnormal GI losses

Abnormal loss of GI fluids may result in hypokalemia due to a combination of factors. Both upper and lower GI fluids contain potassium, so their loss will contribute to the development of hypokalemia. Additionally, volume deficit induced by abnormal loss of GI contents stimulates the release of aldosterone, resulting in increased loss of potassium in the urine. The loss of upper GI contents secondary to vomiting or NG suction will potentiate hypokalemia due to the development of metabolic alkalosis. Upper GI fluids are rich in hydrogen ions (acid) and chloride. The loss of hydrogen causes a relative increase in the plasma bicarbonate levels, and thus metabolic alkalosis. The combination of hypovolemia and chloride loss contributes to the development and perpetuation of metabolic alkalosis. Metabolic alkalosis, in turn, decreases plasma potassium due to a pH-mediated intracellular shift and increased urinary excretion. The complex relationship between potassium and metabolic alkalosis will be reviewed again in Chapter 18.

Foods High in Potassium

Apricots	Nuts
Artichokes	Oranges, orange juice
Avocado	Peanuts
Banana	Potatoes
Cantaloupe	Prune juice
Carrots	Pumpkin
Cauliflower	Spinach
Chocolate	Sweet potatoes
Dried beans, peas	Swiss chard
Dried fruit	Tomatoes, tomato juice, tomato sauce
Mushrooms	Watermelon

Increased sweat losses

Increased sweating may result in hypokalemia due to direct loss of potassium in the sweat and increased urinary losses secondary to hypovolemia. As with abnormal loss of GI fluids, increased production of sweat may result in volume depletion with increased release of aldosterone. Hypokalemia is most likely to develop in hot climates where sweat losses are chronically increased.

Decreased intake

A poorly balanced or inadequate diet may contribute to hypokalemia but rarely will cause it since significant amounts of potassium are contained in a variety of foods (see the accompanying box). Hypokalemia may develop, though, in the individual who is maintained on parenteral fluids with inadequate replacement of potassium or when reduced intake is combined with increased losses.

Intracellular shift of potassium into the cells

A temporary shift of potassium into the cells is another important cause of plasma hypokalemia. In this case there has been no loss of potassium from the body; overall body stores are unchanged. However, the individual still experiences the signs and symptoms of hypokalemia. Causes of intracellular shifts include increased insulin production, alkalosis, and tissue repair after burns, trauma, or starvation (usually this is accompanied by inadequate intake/replacement of potassium). The mechanisms involved in the development of hypokalemia secondary to alkalosis and tissue repair already have been discussed. Insulin facilitates the movement of potassium into skeletal muscle and hepatic cells. Thus any condition that is associated with an increase in insulin may result in hypokalemia. This may occur with increased exogenous administration of insulin (e.g., previously uncontrolled diabetes mellitus) or increased endogenous production (e.g., after the administration of a carbohydrate load in a nondiabetic person). Total parenteral nutrition containing large glucose loads may result in hypokalemia, especially if the individual is undergoing a period of tissue repair.

Stress

Any condition causing physical or emotional stress will result in a loss of potassium from the ECF. Stress is associated with an increase in the release of hormones from the adrenal cortex (e.g., aldosterone) and adrenal medulla (e.g., epinephrine). As discussed above, aldosterone will reduce ECF potassium by increasing urinary excretion. Epinephrine promotes the entry of potassium into the cells. Stress-induced hypokalemia may be quite significant when combined with preexisting hypokalemia. The risk of sudden cardiac death secondary to ventricular dysrhythmias is increased, for example, in individuals with coronary ischemia who are taking potassium-wasting diuretics (Rose, 1989). This observation has important implications for nurses working in coronary care units. Potassium levels should be closely monitored in all individuals at risk for hypokalemia, and hypokalemia should be ruled out as a possible cause of ventricular dysrhythmias.

Assessment

The signs and symptoms of hypokalemia occur as the result of an altered ratio between ECF and ICF potassium levels (see Table 9-1 for quick assessment guidelines for individuals with hypokalemia). This ratio determines the magnitude of the resting membrane potential (the electrical charge of the nerve and muscle cell membranes). Small changes in the ECF potassium level have a significant affect on the ECF to ICF potassium ratio, and thus a significant affect on the resting membrane potential. Before discussing the significance of an altered resting membrane potential, it is first necessary to review the basics of impulse conduction.

Impulse conduction

An *action potential* is the electrical impulse necessary for nerve and muscle fibers to transmit their messages (Figure 9-1). Without proper generation of an action potential, nerves cannot relay their messages and muscles cannot contract. The resting state between messages is termed the *resting membrane potential* (RMP). The RMP is expressed in negative millivolts (mV), the normal being approximately −85 to −90 mV. Generation of an action potential requires that the electrical charge of the membrane be "raised" to a critical point, approximately −65 mV, termed the *threshold potential* (TP). Once the membrane has reached this new value, the action potential is generated and the *message* is sent.

Table 9-1 Quick assessment guidelines for individuals with hypokalemia

Potassium < 3.5 mEq/L

Signs and Symptoms

Fatigue, muscle weakness, leg cramps, nausea, vomiting, ileus, paresthesias, enhanced digitalis effect, dysrhythmias

Physical Assessment

Decreased bowel sounds owing to smooth muscle weakness, weak and irregular pulse, decreased reflexes, and decreased muscle tone (soft and flabby muscles)

History and Risk Factors

- **Reduction in total body potassium:**

  Hyperaldosteronism
  Diuretics or abnormal urinary losses
  Increased GI losses
  Increased loss through diaphoresis

NOTE: Inadequate dietary intake may contribute to hypokalemia but rarely causes it. Hypokalemia, may develop in any individual who is maintained on parenteral fluids with inadequate replacement of potassium.

- **Intracellular fluid shift**

  Increased insulin (e.g., from total parenteral nutrition)
  Alkalosis
  During periods of tissue repair after burns, trauma, or starvation. Usually, this is accompanied by inadequate intake or replacement of potassium

- **Stress**

The difference between the RMP and TP determines how easily nerves and muscles can relay their messages. If the gap between the RMP and TP is narrowed, nerve and muscle cells become more excitable and will more readily transmit their messages. If the gap increases, nerves and muscles become less excitable and are less capable of conducting an impulse. In hypokalemia, the RMP becomes increasingly negative, widening the gap between the RMP and TP. An easy way to remember the effects of ECF (plasma) potassium on the RMP, is to remember that a "low" plasma potassium level "lowers" the RMP, and an "elevated" plasma potassium level "elevates" it (Figure 9-1). The level of the TP also may be "lowered" and "elevated". This occurs with changes in the plasma calcium level and will be discussed in Chapter 10.

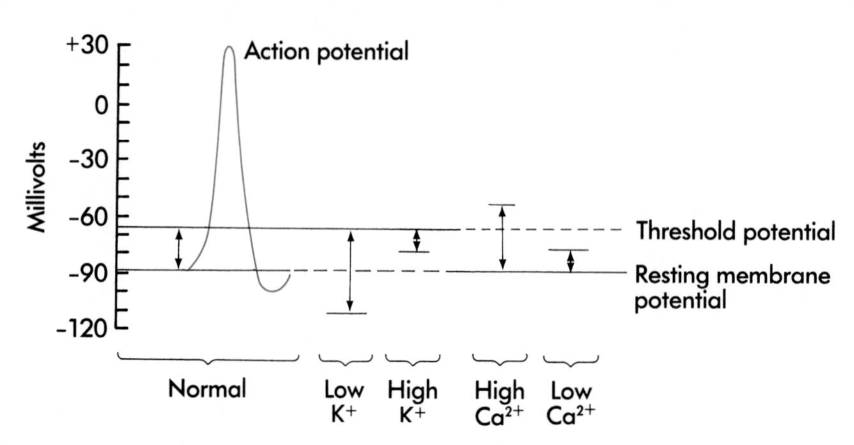

Figure 9-1 **Relationships of ECF concentrations of potassium and calcium to impulse conduction.**

(Modified from Rose BD: Clinical physiology of acid-base and electrolyte disorders, ed 3, New York, 1989, McGraw-Hill.)

The symptoms of hypokalemia are usually not apparent until the serum potassium level drops to 2.5 to 3.0 mEq/L but are variable, depending on the cause of the hypokalemia. Symptoms are more likely to develop at a given serum potassium level when hypokalemia occurs because of an intracellular shift of potassium rather than external loss of potassium. When potassium is lost from the ECF because of increased urinary losses, for example, there is an increased gradient for potassium to flux out of the cells. Both the ECF and ICF potassium levels decrease, thereby minimizing the change in the ECF-ICF potassium ratio. In contrast, an intracellular shift of potassium decreases the ECF potassium while increasing the ICF potassium level, creating a far greater change in the ECF-ICF potassium ratio.

Neuromuscular effects

Since the gap between the RMP and TP widens in hypokalemia, the individual experiences fatigue, muscle weakness, leg cramps, and soft, flabby muscles. Physical assessment may reveal decreased reflexes and muscle tone. Extreme hypokalemia may result in lethal muscle paralysis of the respiratory muscles.

Cardiac effects

An irregular pulse may be present secondary to irregularities of cardiac rhythm. Because of the affects of hypokalemia on the shape and slope of the action potential, the heart muscle is slow to return to its fully resting or *repolarized* state. During this slow return to resting, weak impulses from an irritable myocardium may initiate abnormal beats. See the section on Diagnostic studies, which follows, for a description of electrocardiographic changes with hypokalemia. The presence of extra or early (ectopic) beats may provide the first clue to the presence of hypokalemia in individuals in whom other symptoms might not be obvious. Dangerous ventricular tachycardia may develop in severe hypokalemia.

In addition, hypokalemia potentiates the actions of digitalis. Digitalis works by inhibiting the myocardial cell sodium-potassium pump, thereby decreasing intracellular potassium. Digitalis-induced dysrhythmias may develop when the individual is hypokalemic despite a therapeutic digoxin level.

Gastrointestinal (GI) effects

A common complication of hypokalemia is altered GI function. Decreased contraction of GI smooth muscle and response to parasympathetic stimulation leads to the development of a paralytic ileus with anorexia, nausea, and vomiting. Physical assessment will reveal abdominal distention and decreased bowel sounds. The presence of GI symptoms may necessitate potassium replacement therapy via IV fluids.

Renal effects

Hypokalemia reduces the kidneys' ability to produce a concentrated urine. The exact mechanism for this is unclear but may be due to decreased renal responsiveness to ADH. The individual with hypokalemia may complain of polyuria with nocturia and increased thirst (secondary to polyuria-induced volume depletion). Hypokalemia also increases the production of ammonia by the renal tubular cell, which may precipitate hepatic coma in individuals with severe hepatic dysfunction. For this reason, potassium-wasting diuretics are used with extreme caution in individuals with advanced liver disease. See Chapter 13, p. 255, for a discussion of renal ammonia production.

Diagnostic Studies

Serum potassium level

Hypokalemia is defined as a serum potassium level of <3.5 mEq/L. Keep in mind though, that plasma potassium levels reflect the potassium level of the ECF only, and do not necessarily reflect the level within the ICF. Even if overall body stores of potassium are normal, a decreased potassium in the plasma may result in symptoms due to alteration in the ECF to ICF ratio. Symptoms of muscle weakness usually are not evident until the plasma potassium level drops below 2.5 mEq/L.

Arterial blood gas (ABG) values

ABGs may show metabolic alkalosis (increased pH and HCO_3^-) since hypokalemia is often associated with this condition. Metabolic alkalosis may cause hypokalemia, as discussed above, but hypokalemia also contributes to the development of metabolic alkalosis (see Chapter 18). Severe hypokalemia (potassium level <2.0 mEq/L) may cause respiratory acidosis secondary to respiratory muscle weakness.

Urine potassium levels

Urine potassium levels may be obtained to help identify the cause of hypokalemia. When hypokalemia develops due to nonrenal causes, the kidneys attempt to compensate by decreasing the amount of potassium lost in the urine. Even with an extreme deficit, however, the kidneys are unable to excrete a potassium-free urine. Urine potassium levels of less than 20 mEq/L indicate maximal urinary conservation of potassium and potassium loss from some extrarenal cause.

Electrocardiogram (EKG)

Since the signs and symptoms of hypokalemia may be vague and quite variable, changes noted on the EKG may provide the first clue to the presence of hypokalemia. Bedside rhythm strips also may show changes although not truly diagnostic ones. The progression of changes noted on the EKG are as follows: flattened T wave, presence of U wave, ST-segment depression, and ventricular dysrhythmias (Figure 9-2). Since hypokalemia potentiates the effects of digitalis, the EKG may reveal signs of digitalis toxicity in spite of a normal serum digitalis level.

Medical Management

Potassium replacement

Although removal of the underlying cause is important, treatment usually begins with potassium replacement, either PO (via increased dietary intake or medication) or IV. The usual dose is 40 to 80 mEq/day in divided doses (for adults). IV potassium is necessary if hypokalemia is severe or the patient is unable to take potassium orally. IV potassium is *never* given IV push. It must be diluted appropriately and administered to adults at rates not greater than 10 to 20 mEq/hr nor in concentrations greater than 30 to 40 mEq/L (when added to IV solutions), unless hypokalemia is *severe.* Too rapid an administration can result in life-threatening hyperkalemia. If potassium is administered via a peripheral line, the rate of administration may need to be reduced to prevent irritation of vessels. Patients receiving 10 to 20 mEq/hr or greater should receive the medication through a central line and be on a continuous cardiac monitor. The development of peaked T waves suggests the presence of hyperkalemia and requires immediate MD notification. IV potassium is most commonly administered as potassium chloride (KCl) since most potassium loss is associated with chloride loss, but it also may be given as potassium phosphate. Potassium must be administered with extreme caution to persons with decreased renal function due to the risk of hyperkalemia.

Oral potassium supplements

Oral potassium supplements are available in a variety of forms. Oral preparations may contain potassium combined with chloride, citrate, bicarbonate, or gluconate. As with IV preparations, oral potassium is most commonly combined

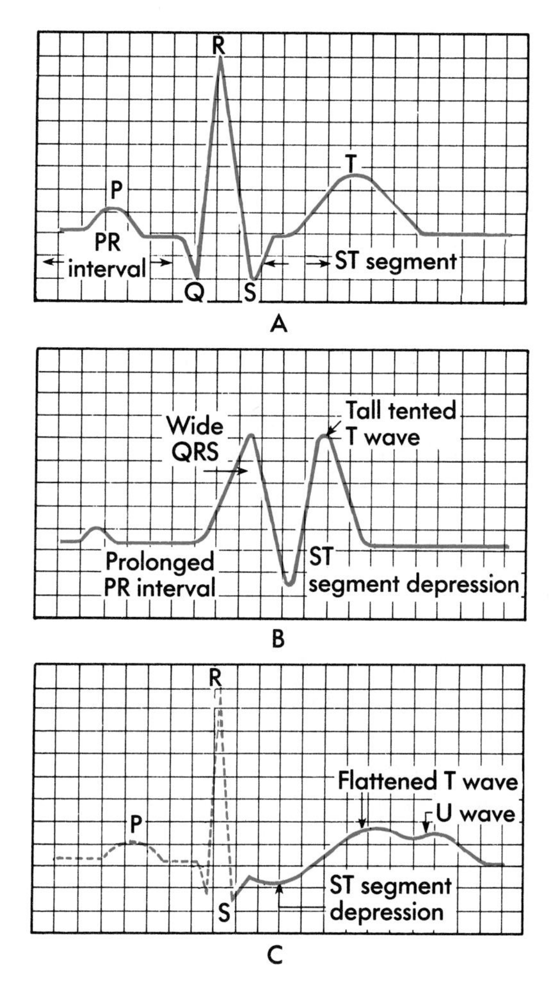

Figure 9-2 The effects of potassium on EKG tracings. **A,** Normal electrocardiographic tracing. **B,** Serum potassium level above normal. **C,** Serum potassium level below normal.

(From Horne M and Swearingen P: Pocket guide to fluids and electrolytes, St Louis, 1989, The CV Mosby Co.)

with chloride. Oral supplements should be administered with food or after meals to avoid gastric irritation. Enteric-coated potassium tablets are available but must be used with caution due to the risk of bowel irritation and ulceration. Enteric-coated medications are not absorbed in the stomach, thus protecting the stomach from ulceration. Instead, they are absorbed in the intestine, exposing the bowel to high and potentially irritating concentrations of potassium.

Table 9-2 Potassium content of commonly used salt substitutes

Potassium Content	Salt Substitute
12.2 mEq/g (61 mEq/tsp)	Adoph's
12.2 mEq/g (61 mEq/tsp)	Co-Salt
10.8 mEq/g (54 mEq/tsp)	Diasal
11.6 mEq/g (58 mEq/tsp)	Featherweight *K*
10.5 mEq/g (52.5 mEq/tsp)	Featherweight seasoned salt
12.8 mEq/g (64 mEq/tsp)	Morton's
11.8 mEq/g (59 mEq/tsp)	Nu-salt (Sweet & Low)

Modified from Progress in drug research, Birkhäuser Verlag, 30:29, 1986.

Potassium-sparing diuretics, which reduce the excretion of potassium in the urine, are sometimes used in place of oral potassium supplements. Potassium chloride salt substitutes (Table 9-2) also may be used to supplement potassium intake (one teaspoon equals approximately 60 mEq KCl).

Nursing Diagnoses and Interventions

Decreased cardiac output related to electrical factors secondary to hypokalemia or too rapid correction of hypokalemia with resulting hyperkalemia

Desired outcomes: EKG shows normal configuration and absence of ventricular dysrhythmias. Pulse rate and regularity are normal for the patient. Serum potassium levels are within normal range (3.5 to 5.0 mEq/L).

NOTE: Disorders of potassium balance are potentially life-threatening due to the effects of altered potassium levels on neuromuscular and cardiac function. Suspected alterations in potassium balance require prompt MD notification.

1. Administer potassium supplement as prescribed. Avoid giving IV potassium chloride at a rate faster than recommended, as this can lead to life-threatening hyperkalemia. For this reason, concentrated potassium solutions should be administered via a pump. Do not add potassium to IV solution containers in the hanging position because this can cause layering of the medication. Instead, invert the solution container before adding the medication and mix well.

   IV potassium can cause local irritation of veins and chemical phlebitis. Assess IV insertion site for erythema, heat, or pain. Alert MD to symptoms. Irritation may be relieved by applying an ice bag, giving mild sedation, or numbing the insertion site with a small amount of local anesthetic. Phlebitis may necessitate changing of IV site.

2. Administer oral and IV potassium supplements as prescribed. Oral supplements may cause GI irritation. Administer with a full glass of water or fruit juice; encourage patient to sip slowly. Alert patients to the possibility of GI irritation and instruct them to notify RN or MD of the following symptoms: abdominal pain, distention, nausea, or vomiting. Do not switch potassium supplements without a doctor's prescription.
3. Encourage intake of foods high in potassium (see the box on p. 169). Since many foods containing potassium are expensive, review examples of low-cost foods that are high in potassium for individuals with limited finances. Salt substitutes may be used as an inexpensive potassium supplement.
4. Monitor I&O. Alert MD to urine output <30 ml/hour. Unless severe, symptomatic hypokalemia is present, potassium supplements should not be given if the patient has an inadequate urine output because hyperkalemia can develop rapidly in patients with oliguria (<15 to 20 ml/hour). Alert MD to elevated BUN or creatinine levels.
5. Monitor for the presence of an irregular pulse or pulse deficit (a discrepancy between the apical and radial pulse rates). Alert MD to changes.
6. Physical indicators of abnormal potassium levels are difficult to identify in the patient who is critically ill. Monitor EKG for signs of continuing hypokalemia (ST-segment depression, flattened T wave, presence of U wave, ventricular dysrhythmias) or hyperkalemia (tall, thin T waves; prolonged PR interval; ST depression; widened QRS; loss of P wave), which may develop during potassium replacement (see Figure 9-2).
7. Monitor serum potassium levels carefully, especially in individuals at risk for developing hypokalemia, such as patients taking diuretics or receiving NG suction. Alert MD to abnormal levels. Review signs and symptoms of hypokalemia with individuals at risk.
8. Administer potassium cautiously in patients receiving potassium-sparing diuretics (e.g., spironolactone or triamterene) because of the potential for the development of hyperkalemia.
9. Because hypokalemia can potentiate the effects of digitalis, monitor patients receiving digitalis for signs of increased digitalis effect: development of an irregular pulse; and if the individual is on a cardiac monitor, multifocal or bigeminal PVCs; paroxysmal atrial tachycardia with varying AV block; and Wenckebach (type I AV) heart block. Review with the patient and significant others the importance of taking prescribed potassium supplements if taking diuretics or digitalis. Review indicators of digitalis toxicity.

Potential ineffective breathing pattern related to weakness of respiratory muscles secondary to *severe* hypokalemia (potassium <2 to 2.5 mEq/L)

Desired outcome: Patient has effective breathing pattern as evidenced by normal respiratory depth and pattern (eupnea) and rate of 12 to 20 breaths/min.

1. If patient is exhibiting signs of worsening hypokalemia, be aware that severe hypokalemia can lead to weakness of respiratory muscles, resulting in shallow respirations and eventually, apnea and respiratory arrest. Assess character, rate, and depth of respirations. Alert MD promptly if respirations become rapid and shallow.
2. Keep manual resuscitator at patient's bedside if severe hypokalemia is suspected.
3. Reposition patient q2h to prevent stasis of secretions; suction airway as needed.

HYPERKALEMIA

Hyperkalemia (serum potassium level >5.0 mEq/L) occurs because of an increased intake of potassium, decreased urinary excretion of potassium, or movement of potassium out of the cells. Remember that changes in serum potassium levels reflect changes in ECF potassium, not necessarily changes in total body levels. In diabetic ketoacidosis, for example, a large quantity of potassium may be lost in the urine due to the glucose-induced osmotic diuresis. Although this causes a significant reduction in total body potassium, the patient initially may present with a normal or elevated potassium. This misleadingly high potassium level occurs because of a shift of potassium out of the cells secondary to acidosis, lack of insulin, and increased tissue catabolism. Hypokalemia does not develop until treatment is initiated and potassium returns to the cells. Hyperkalemia from any cause is most likely to develop when combined with a decreased ability to excrete potassium.

History and Risk Factors

Inappropriately high intake of potassium

Hyperkalemia solely due to increased intake is almost always the result of inappropriate administration of IV potassium. Individuals with hyperkalemia are most likely to be symptomatic when there is a sudden increase in the potassium level, as might occur with too rapid an administration of IV potassium preparations. This risk can be minimized by the use of monitored pumps for potassium administration. IV potassium preparations must be given cautiously, in amounts and at rates no greater than recommended, due to the risk of dangerous hyperkalemia.

Care also must be taken to avoid inadvertent boluses of IV potassium. Potassium additives never should be added to IV containers in the hanging position due to the risk of layering of potassium at the bottom of the container. The needle used to inject the potassium into a bag of IV solution should be long enough to extend past the injection port tubing to prevent pooling of potassium within the port. After potassium is added to an IV bag or bottle, the container should be inverted several times to ensure adequate mixing. The container must be clearly labeled so that all personnel are aware of its contents.

Decreased excretion of potassium

Decreased excretion of potassium occurs with renal disease, decreased ECV, hypoaldosteronism, or use of potassium-sparing diuretics (e.g., triamterene and spironolactone). Hyperkalemia is a common complication of both acute and chronic renal failure due to decreased urine output. Decreased ECV also may cause hyperkalemia due to decreased production of urine. Hypoaldosteronism (decreased production of aldosterone), a condition that sometimes develops in individuals with diabetes mellitus, may cause hyperkalemia because of decreased secretion and excretion of potassium by the renal tubule secondary to decreased production of aldosterone. Use of potassium-sparing diuretics may cause hyperkalemia, especially when combined with increased intake or decreased renal function.

Decreased urinary excretion and increased intake of potassium are an especially dangerous combination. The individual with acute or chronic renal failure must be monitored closely for hidden sources of potassium, such as those found in salt substitutes (see Table 9-2), potassium penicillin, or banked blood. The amount of potassium contained in a unit of banked blood increases over time due to the release of intracellular potassium as red blood cells die. Thus the longer the unit has been stored, the greater the potential potassium load. A unit of blood that is about to expire may have a concentration of 30 mEq/L of potassium. This is most likely to be of concern for the individual requiring multiple units of blood.

Movement of potassium out of the cells

Movement of potassium out of the cells occurs with metabolic acidosis, insulin deficiency/hyperglycemia, or tissue catabolism (e.g., occurring with fever, sepsis, trauma, or surgery).

Metabolic acidosis

In metabolic acidosis there is an increase in the number of H^+ (acids). These excess H^+ move into the cells to be buffered. Because hydrogen is a positively charged ion, another positively charged ion (i.e., potassium—K^+) must move out of the cells to maintain electroneutrality.

Insulin deficiency/hyperglycemia

Acute hyperglycemia due to decreased production of insulin (diabetes mellitus) is often associated with hyperkalemia. As water moves out of the cells in response to hyperglycemia (i.e., hypertonicity), potassium is pulled as well. This occurs due to solvent drag (see the section on osmosis, Chapter 1). In addition, the loss of intracellular water causes an increase in the intracellular concentration of potassium, favoring simple diffusion of potassium out of the cells. Lack of insulin prevents potassium from returning to the cells.

Increased tissue catabolism

As discussed above, increased tissue catabolism and metabolic acidosis are factors in the development of hyperkalemia in diabetic ketoacidosis. Any time there is increased tissue breakdown, the risk of hyperkalemia exists due to the release of intracellular potassium. Examples of conditions associated with increased tissue catabolism include trauma, fever, sepsis, and administration of cytotoxic agents. Remember that the concentration of potassium within the cells is approximately 150 mEq/L as compared with the approximately 5 mEq/L in the ECF.

Assessment

The signs and symptoms of hyperkalemia occur as a result of altered neuromuscular and cardiac function. See Table 9-3 for quick assessment guidelines for individuals with hyperkalemia.

Neuromuscular effects

As discussed in the hypokalemia section, changes in the ratio of ECF to ICF potassium alter the resting membrane potential of nerve and muscle cells. Hyperkalemia (increased ECF potassium) causes the RMP to become less negative, thus narrowing the gap between the RMP and the TP (see Figure 9-1). This increases neuromuscular excitability. The initial symptoms of hyperkalemia are increased neuromuscular activity: irritability, anxiety, abdominal cramping, diarrhea, and paresthesias (abnormal sensations). As the potassium level increases, though, symptoms change from irritability to weakness (especially of the lower extremities) and are similar to those seen in hypokalemia. This change occurs when the RMP is at the level of or *above* the TP. The cell is unable to generate a new action potential.

Table 9-3 **Quick assessment guidelines for individuals with hyperkalemia**

Potassium > 5.0 mEq/L

Signs and Symptoms

Irritability, anxiety, abdominal cramping, diarrhea, weakness (especially of the lower extremities), paresthesias

Physical Assessment

Irregular pulse rate; cardiac standstill may occur at K^+ levels >8.5 mEq/L

History and Risk Factors

- **Inappropriately high intake of potassium:** Usually, with IV potassium delivery
- **Decreased excretion of potassium:** For example, with renal disease or use of potassium-sparing diuretics
- **Extracellular shift:** acidosis, hyperglycemia, increased tissue catabolism

Cardiac effects

Cardiac dysrhythmias occur due to the effects of hyperkalemia on atrial and ventricular depolarization (the upward curve of the action potential) and repolarization (the downward curve of the action potential). Ventricular fibrillation or cardiac standstill may occur at levels >8.5 mEq/L.

Symptoms are more likely to be present at a given potassium level when hyperkalemia has developed acutely. Symptoms are also more likely to be present when hyperkalemia is complicated by acidosis or hypocalcemia (a decreased plasma calcium level).

Diagnostic Studies

Serum potassium level

The serum potassium will be >5.0 mEq/L. Keep in mind, however, that several factors may cause a falsely high serum potassium due to increased release of intracellular potassium in the laboratory specimen (e.g., a high platelet count, prolonged use of a tourniquet at the time of venipuncture, **hemolysis** of the blood specimen, or delayed separation of serum and cells). Care must be taken in obtaining and handling blood specimens to prevent erroneous results. A repeat specimen should be obtained if reported levels are inconsistent with the individual's history and condition.

Arterial blood gas (ABG) values

Arterial blood gasses may show metabolic acidosis (decreased pH and HCO_3^-) since hyperkalemia often accompanies acidosis.

Electrocardiogram (EKG)

The EKG will demonstrate changes in configuration, indicating altered atrial and ventricular depolarization and repolarization. Progressive changes include tall, thin T waves; prolonged PR interval; ST depression; widened QRS; and loss of P wave. Eventually the QRS widens further and cardiac arrest occurs. (See Figure 9-2.)

Medical Management

The goal of medical management is to treat the underlying cause (e.g., reduce potassium intake or tissue catabolism in the individual with renal failure) and return the serum potassium level to normal. Treatment will depend on the level of serum potassium and the severity of symptoms.

Cation exchange resins

Mild or subacute hyperkalemia may be treated with cation exchange resins (e.g., sodium polystyrene sulfonate—Kayexalate), which exchange sodium ions for potassium ions in the gut. Kayexalate may be given either orally, through a nasogastric tube, or via retention enema. It is usually combined with sorbitol to prevent constipation and induce diarrhea, thus increasing potassium loss in the bowels. Unfortunately, Kayexalate may bind with other cations in the GI tract and contribute to the development of hypomagnesemia or hypocalcemia. Due to the exchange of sodium for potassium, it provides a potentially unwanted sodium load.

Glucose and insulin

Moderate to severe hyperkalemia is treated with a combination of medications. IV glucose and insulin are usually administered to decrease serum potassium levels on a temporary basis. Insulin facilitates the movement of potassium into the cells, and glucose is necessary for preventing the development of dangerous hypoglycemia. Typically hypertonic glucose, either an ampule of 50% dextrose in water or 250 to 500 ml of 10% dextrose in water, is given with 10 to 15 units of regular insulin (adults). This decreases serum potassium for approximately 6 hours.

Sodium bicarbonate

Administration of IV sodium bicarbonate also will temporarily shift potassium into the cells, reducing serum potassium for approximately 1 to 2 hours. Usually 1 to 2 ampules are given IV over 5 to 10 minutes.

Since the therapeutic effects of glucose and insulin and sodium bicarbonate are temporary, it is necessary to follow these medications with a therapy that actually removes potassium from the body, for example, dialysis or administration of cation exchange resins. Continued monitoring of serum potassium levels is essential.

Calcium gluconate

Severe hyperkalemia that may result in fatal cardiotoxicity requires the additional administration of IV calcium gluconate. Calcium counteracts the effects of an elevated potassium on the nerves and muscles and will protect the heart, allowing time for other therapies to lower the serum potassium level. Calcium causes the TP to be less negative, thus restoring the normal gap between the RMP and the TP (see Figure 9-1). Calcium does not lower the serum potassium, and thus, serum potassium levels will remain elevated. Calcium chloride also may be used, however the two forms of calcium are *not* interchangeable. Although both come in 10 ml ampules, calcium gluconate contains only 4.5 mEq of calcium, while calcium chloride contains 13.6 mEq of calcium.

Dialysis

Dialysis is the other form of therapy besides cation exchange resins that actually removes excess potassium from the body. Dialysis is a life-saving procedure used to treat persons with severely decreased renal function. It also may be used in the

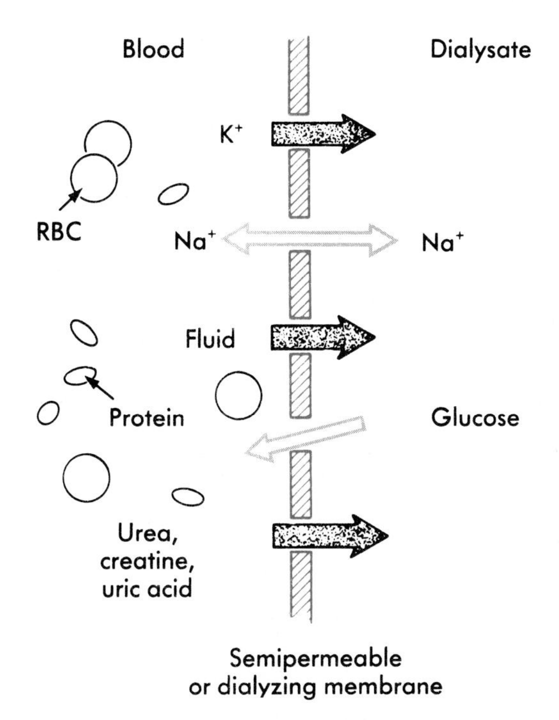

Figure 9-3 **Osmosis and diffusion in dialysis. Net movement of major particles and fluid is illustrated.**

(From Phipps W: Medical-surgical nursing, ed 3, St Louis, 1987, The CV Mosby Co.)

presence of normal renal function to treat dangerous fluid overload, severe electrolyte imbalances, or certain drug ingestions. Dialysis uses the principles of osmosis, diffusion, and filtration to remove excess water, uremic toxins, and certain medications, in addition to correcting electrolyte imbalances and restoring body buffers (Figure 9-3). Dialysis may be defined as the selective movement of water and solutes from one fluid compartment to another across a semipermeable membrane. The two fluid compartments are the individual's blood and the dialysate (electrolyte and glucose solution).

Peritoneal dialysis

Peritoneal dialysis uses the peritoneum as the semipermeable dialysis membrane. Dialysate is instilled into the peritoneal cavity via a specialized catheter, and the movement of solutes and fluid occurs between the individual's capillary blood and the dialysate. At set intervals the peritoneal cavity is drained and new dialysate is instilled.

Hemodialysis

During hemodialysis, the individual's blood is continuously pumped through an artificial kidney (dialyzer) that exposes the blood to dialysate across an artificial membrane. Hemodialysis is the most effective means of removing excess potassium from the body but requires specialized equipment, trained personnel, and an adequate vascular access (e.g., surgically created fistula or insertion of a catheter into the subclavian vein).

Nursing Diagnoses and Interventions

Potential for decreased cardiac output related to electrical factors secondary to severe hyperkalemia or too rapid correction of hyperkalemia with resulting hypokalemia

Desired outcomes: EKG shows no evidence of ventricular dysrhythmias related to hypokalemia (e.g., U wave, PVCs) or hyperkalemia (e.g., peaked T wave). Serum potassium levels are within normal range (3.5 to 5.0 mEq/L).

NOTE: Disorders of potassium balance are potentially life-threatening due to the effects of altered potassium levels on neuromuscular and cardiac function. Suspected alterations in potassium balance require prompt MD notification.

1. Monitor I&O. Alert MD to urine output <30 ml/hr. Oliguria increases the risk for developing hyperkalemia.
2. Monitor for indicators of hyperkalemia (e.g., irritability, anxiety, abdominal cramping, diarrhea, weakness of lower extremities, paresthesias, irregular pulse). Also be alert to indicators of hypokalemia (e.g., fatigue, muscle weakness, leg cramps, nausea, vomiting, decreased bowel sounds, paresthesias, weak and irregular pulse) following treatment. Assess for hidden sources of potassium: medications (e.g., potassium penicillin G); banked blood (the older the blood, the greater the amount of potassium owing to the release of potassium as RBCs die and breakdown); salt substitute; GI bleeding (due to breakdown of blood in the gut); or conditions causing increased catabolism, such as infection, burns, or trauma.
3. Monitor serum potassium levels, especially in patients at risk of developing hyperkalemia, such as individuals with renal failure. Notify MD of levels above or below normal range.
4. Physical indicators of abnormal potassium levels are difficult to identify in the patient who is critically ill. Monitor EKG for signs of hypokalemia (ST-segment depression, flattened T waves, presence of U wave, ventricular dysrhythmias), which may develop secondary to therapy, or continuing hyperkalemia (tall, thin T waves; prolonged PR interval; ST-segment depression; widened QRS; loss of P wave). Notify MD stat if EKG changes occur. EKG changes at a given potassium level may be less dramatic in the chronic renal patient who develops hyperkalemia more slowly. See Figure 9-2 for EKG changes with hypokalemia and hyperkalemia.
5. Administer calcium gluconate as prescribed, giving it cautiously in patients receiving digitalis, because digitalis toxicity can occur.

   CAUTION: Do not add calcium gluconate to solutions containing sodium bicarbonate because precipitates may form. However, IV glucose and $NaHCO_3$ may be combined without harmful precipitate. Insulin should be given separately. For more information about calcium administration, see Chapter 10.

6. Monitor individuals receiving sodium bicarbonate for signs of fluid volume excess due to the osmotic shift of fluid into the intravascular space, which may occur after the administration of hypertonic sodium bicarbonate.
7. If administering cation exchange resins by enema, encourage patient to retain the solution for at least 30 to 60 minutes to ensure therapeutic effects. To facilitate retention, the enema may be instilled into the rectum and sigmoid colon via an indwelling urinary catheter with the balloon inflated and clamped. The onset of action is approximately 30 to 90 minutes when administered via retention enema. In contrast, oral or NG Kayexalate may take several hours to work.
8. Monitor individuals receiving glucose and insulin for signs of hypoglycemia: diaphoresis, increased heart rate, altered sensorium, and confusion. Suspected hypoglycemic reactions may be confirmed by fingerstick glucose monitoring. Alert MD to hypoglycemic episodes.

Knowledge deficit: Causes and treatment of hyperkalemia

Desired outcome: Patient and significant others identify the causes of hyperkalemia and the role of diet and medications in treatment.

1. Give patient and significant others verbal and written instructions for prescribed medications to be taken at home, including name, purpose, dosage, frequency, precautions, and potential side effects.
2. Explain the indicators of both hypokalemia and hyperkalemia. Alert patient to the following signs and symptoms that necessitate immediate medical attention: weakness, pulse irregularities, and fever or other indicator of infection. Teach patient and significant others how to measure pulse rate and detect irregularities.
3. Discuss the importance of preventing recurrent hyperkalemia; review potential causes.
4. For the individual on a potassium-restricted diet (e.g., chronic renal failure patient) provide a list of foods high in potassium that should be avoided (see the box on p. 169). Remind patient that salt substitute and *lite* salt also should be avoided. Fruits that are relatively low in potassium include apples, grapes, and cranberries.

Case Study 9-1

L.S., a 42-year-old man who has diabetes and end stage renal disease (ESRD—chronic renal failure that has reached the stage at which dialysis or kidney transplantation becomes necessary), arrives in the emergency room with extreme muscle weakness, especially in his lower extremities. Although he does not have shortness of breath (SOB) or chest pain, he does exhibit signs of fluid volume excess (e.g., 2+ pitting pretibial edema, increased neck vein distention, and increased blood pressure). His wife points to an open draining area on the heel of his right foot. The open area began as a small blister 3 weeks ago and has progressed to the size of a half dollar. The surrounding tissue is markedly red and warm with red streaks extending half way to the calf.

L.S.'s wife states that he is dialyzed at the nearby outpatient dialysis unit three times a week (M-W-F) and is due for dialysis tonight at 5 PM. It is now noon on Monday. L.S.'s last treatment was Friday evening. He and his wife have just returned from 2 days out of town visiting friends, who have a fruit orchard. L.S.'s wife reports that her husband ate many fresh fruits and vegetables, apricots in particular.

The nurse on duty completes her assessment and, suspecting hyperkalemia, notifies the emergency room physician stat of L.S.'s condition.

The physician requests stat laboratory work (electrolytes, BUN, creatinine, and blood sugar), a wound culture of the right heel, and a stat EKG.

1. Question: After a quick nursing assessment of L.S.'s condition, the nurse immediately notified the MD. Why was this an important nursing action?

Answer: Hyperkalemia is potentially a medical emergency.

2. Question: What factors may have led the nurse to suspect hyperkalemia?

Answer: L.S.'s renal failure, combined with intake of high-potassium foods (fresh fruits and vegetables, specifically apricots), place him at great risk for hyperkalemia. For the individual on chronic hemodialysis, hyperkalemia is controlled through a combination of dietary restrictions and hemodialysis. Serum potassium levels increase steadily between hemodialysis treatments and will be at their highest just prior to the next dialysis treatment. L.S. is at his longest interdialytic space (i.e., Friday to Monday), increasing his potential for hyperkalemia. The infected area on his heel also increases the risk of hyperkalemia due to increased tissue breakdown with the release of intracellular potassium. His complaint of muscle weakness in the lower extremities is strongly suggestive of hyperkalemia.

3. Question: Why has the MD requested a stat EKG?

Answer: Potassium affects the electrical activity of the heart and the stat EKG may identify the presence and severity of hyperkalemia before the results of the laboratory work, thus enabling quicker initiation of treatment.

The results of L.S.'s stat EKG show narrow, peaked T waves; absent P waves; and widened QRS complex. These changes strongly suggest hyperkalemia. Based on L.S.'s history, symptoms, and EKG findings, the physician prescribes stat treatment for hyperkalemia.

The goal of medical treatment is to lower L.S.'s potassium level as quickly as possible. One method of removing excess potassium from the body is hemodialysis. L.S. is to be hemodialyzed as soon as possible. It will take approximately 1 hour for the hemodialysis nurse to set up the necessary equipment and begin treatment.

4. Question: Which two therapeutic measures temporarily could reduce L.S.'s serum potassium level and help prevent fatal cardiotoxicity while he awaits hemodialysis?

Answer:

a. IV glucose and insulin.
b. IV sodium bicarbonate.

5. Question: The emergency room physician prescribes 1 ampule of D_{50} (25 grams of dextrose) and 10 units of regular insulin to be given slowly IV and 1 ampule of sodium bicarbonate (44 mEq $NaHCO_3$) to be given IV over a 5-minute period. L.S.'s wife states that her husband took his AM insulin but did not eat breakfast or lunch. The emergency room nurse will monitor L.S. closely for complications of therapy. What might these include?

Answer:

a. Hypoglycemia secondary to additional insulin.
b. Intravascular volume overload secondary to the IV sodium load from the ampule of $NaHCO_3$. Remember that L.S. is already experiencing fluid volume excess, as evidenced by the 2+ pretibial edema.

6. Question: Glucose and insulin, along with $NaHCO_3$, will cause a temporary drop in serum potassium. Why does the serum potassium level drop with these therapies, and how long will the decrease last?

Answer: Both therapies cause a shift of potassium into the cell (review p. 182). $NaHCO_3$ therapy works within 15 to 30 minutes and is effective for approximately 2 hours. Glucose and insulin therapy works within 30 to 60 minutes, and its effect lasts approximately 6 hours. These therapies are only temporary measures and should be followed by treatments that remove potassium from the body.

The laboratory results obtained on admission are as follows:

L.S.'s Lab Values	Normal Values
Na^+ 132 mEq/L	137-147 mEq/L
Cl^- 98 mEq/L	95-108 mEq/L
K^+ 9.6 mEq/L	3.5-5.0 mEq/L
HCO_3^- 17 mEq/L	22-26 mEq/L
BUN 120 mg/dl	6-20 mg/dl
with ESRD, predialysis:	70-90 mg/dl
Creatinine 14.3 mg/dl	0.6-1.5 mg/dl
Blood sugar 393 mg/dl	<145 mg/dl

The potassium results confirm that L.S. is dangerously hyperkalemic. The low HCO_3^- suggests the presence of metabolic acidosis, which occurs with ESRD, especially in an individual who has been noncompliant with dietary restrictions or is catabolic. In addition to regulating potassium, the kidneys are responsible for excreting a portion of the daily acid load. Remember that acidosis will contribute to the development of hyperkalemia due to a shift of potassium out of the cells. The BUN results also suggest poor dietary control, catabolism, or underdialysis. The predialysis BUN for an individual with ESRD should be 70 to 90 mg/dl. L.S.'s blood sugar is elevated, most likely due to the stress of the infected heel.

7. **Question:** Besides hemodialysis, what other therapy will remove potassium from the body?

   Answer: Cation exchange resins (e.g., Kayexalate). Hemodialysis, however, is the treatment of choice for L.S. since it is also necessary to treat his marked azotemia (retention of metabolic wastes). In addition to glucose and insulin and $NaHCO_3$ therapy, the physician prescribes 1 ampule of calcium gluconate (10 ml of a 10% solution) to be administered IV over a 2 to 3 minute period. L.S. is to be on continuous EKG monitoring.

8. **Question:** How does calcium administration affect serum potassium levels?

   Answer: It does not change serum potassium levels, but rather, it protects the heart against the cardiotoxic effects of hyperkalemia. Calcium gluconate works within minutes to make the threshold potential of the cardiac membrane less negative. (Recall that potassium reduces the negativity of the resting potential; administration of calcium helps to restore the normal gap between the RMP and TP.)

9. **Question:** Why is it important that L.S.'s heart be monitored continuously during calcium administration?

   Answer: Like potassium, sudden increases in the calcium level may be cardiotoxic. Calcium levels affect the threshold potential of nerve and muscle cells. This will be discussed in greater detail in Chapter 10.

10. **Question:** Identify *potential* nursing diagnoses for this patient.

    Answer:

a. **Decreased cardiac output** related to risk of hyperkalemia-induced dysrhythmias
b. **Impaired tissue integrity** related to open wound on the heel
c. **Fluid volume excess** related to decreased urinary output secondary to chronic renal failure

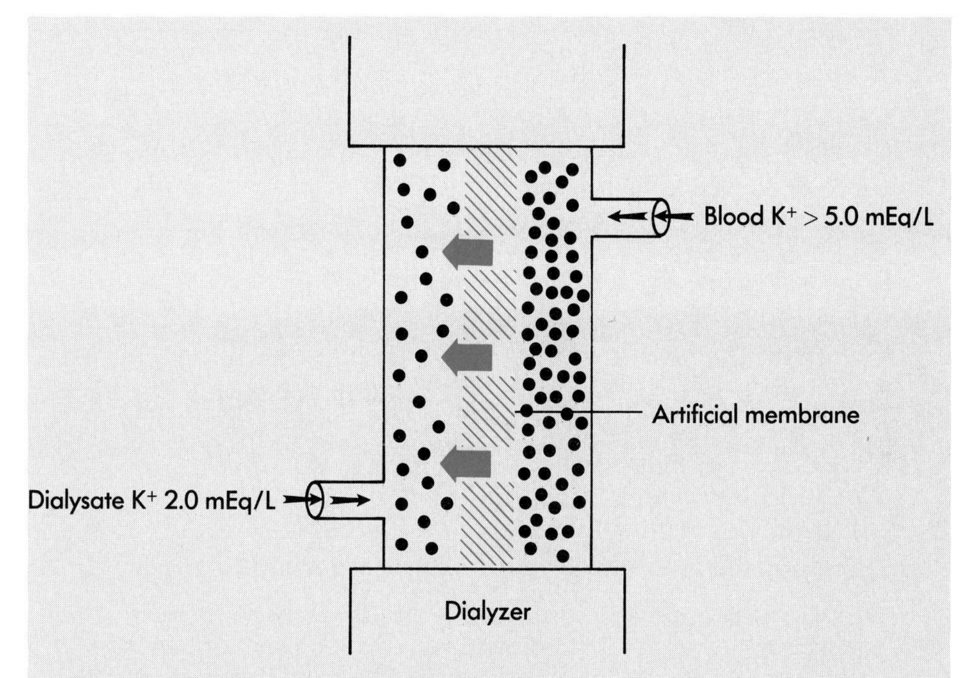

Figure 9-4 Potassium loss during hemodialysis.

d. **Potential for infection** related to open sore on heel
e. **Potential for trauma** related to lower extremity muscle weakness
f. **Impaired physical mobility** related to lower extremity muscle weakness
g. **Noncompliance** with treatment related to risk of dietary indiscretion
h. **Knowledge deficit:** Importance of dietary restriction of potassium-containing foods
i. **Ineffective individual coping:** Denial

11. **Question:** L.S. is to be hemodialysed as soon as possible. If L.S.'s blood is dialyzed using a dialysate with a potassium concentration of 2 mEq/L, in which direction will net diffusion of potassium occur? From the dialysate to the blood, or from the blood to the dialysate? Assume that despite other therapy L.S. is still hyperkalemic predialysis.

Answer: Because L.S. remains hyperkalemic, the net diffusion of potassium will occur from the blood to the dialysate due to the concentration gradient between the two fluid compartments (Figure 9-4).

Case Study 9-2

R.H., is a 68-year-old man who has a long history of chronic congestive heart failure (CHF). He is seen in his internist's office today for a routine office visit. Medications he takes at home include:

digoxin 0.25 mg/day
furosemide (Lasix) 40 mg bid
20% KCl liquid 15 ml/day (40 mEq/15 ml).

Lab work drawn earlier today shows the following:

R.H.'s Lab Values (mEq/L)	Normal Values (mEq/L)
Sodium 140	137-147
Potassium 3.0	3.5-5.0
Chloride 96	95-108
Bicarbonate 33	22-26

1. Question: What is a possible cause of R.H.'s hypokalemia?

Answer: Lasix (a potent diuretic) will cause increased excretion of potassium in the urine. This increased loss of potassium, however, should be corrected by the potassium supplement. The most likely explanation for R.H.'s hypokalemia is that he is not taking his potassium supplement. The nurse reviews R.H.'s medications with him and determines that he is, in fact, not taking his potassium supplement. R.H. states that he hates his KCl and refuses to take it.

2. Question: After assessing his level of knowledge, the RN has identified **Knowledge deficit:** *Rationale for and importance of potassium therapy* as a nursing diagnosis for R.H. What are some potential interventions related to this diagnosis?

Answer:

a. Review the importance of taking his potassium supplement in terms that are appropriate to his level of comprehension.

b. Identify his specific objections to the potassium supplement and discuss possible alternate forms of supplementation with the MD, if appropriate. In addition to hyperkalemia, the most common complication of potassium supplements is GI irritation, resulting in nausea, vomiting, and abdominal discomfort. GI irritation may be relieved or avoided by diluting the liquid or effervescent preparations and by administering all supplements with meals or food.

c. Review foods that are high in potassium content. Help R.H. identify foods high in potassium content that he enjoys, are affordable, and can easily be incorporated into his diet.

d. Discuss with the MD the possibility of using salt substitute as an economical alternative. This is a logical approach, since R.H. is on a sodium-restricted diet for his CHF. Salt substitutes contain approximately 10 to 12 mEq potassium per gram, or approximately 50 to 60 mEq/tsp (see Table 9-1).

e. Review the signs and symptoms of hypokalemia and the importance of notifying the MD promptly should they occur.

f. Discuss with the MD the possibility of prescribing a potassium-sparing diuretic in lieu of potassium supplementation. If R.H. is started on a potassium-sparing diuretic, he must be cautioned not to take his potassium supplement or use salt substitute unless prescribed by MD.

3. Question: Why is hypokalemia of particular concern for R.H.?

Answer: Hypokalemia potentiates the actions of digitalis and may contribute to the development of digitalis toxicity.

References

Arieff AI and De Fronzo RA: Fluid, electrolyte, and acid-base disorders, New York, 1985, Churchill Livingstone, Inc.

Barta MA: Correcting electrolyte imbalances, RN 50:30-34, 1987.

Brater DC: Serum electrolyte abnormalities caused by drugs, Prog Drug Res 30:9-69, 1986.

Goldberger E: A primer of water, electrolyte, and acid-base syndromes, ed 7, Philadelphia, 1986, Lea & Febiger.

Guyton AC: Textbook of medical physiology, ed 7, Philadelphia, 1986, WB Saunders Co.

Horne M and Swearingen PL: Pocket guide to fluids and electrolytes, St Louis, 1989, The CV Mosby Co.

Kokko JP and others: Fluids and electrolytes, Philadelphia, 1986, WB Saunders Co.

Lowery SH and Ash SR: Diminishing the risks of IV potassium chloride, Nursing 88 18(6):64, 1988.

Maxwell MH and others: Clinical disorders of fluid and electrolyte metabolism, ed 4, New York, 1987, McGraw-Hill Book Co.

Metheny NM: Fluid and electrolyte balance: nursing considerations, Philadelphia, 1987, JB Lippincott Co.

Rice V: The role of potassium in health and disease, Crit Care Nurs 50:54-73, May/June 1982.

Rose BD: Clinical physiology of acid-base and electrolyte disorders, ed 3, New York, 1989, McGraw-Hill, Inc.

Schwartz MW: Potassium imbalances, Am J Nurs 87:1292-1299, 1987.

Toto KH: When the patient was hyperkalemic, RN 50:34-38, April 1987.

Chapter

10

Disorders of Calcium Balance

Calcium, the body's fifth most abundant ion, is primarily combined with phosphorus to form the mineral salts of the bones and teeth. In addition to providing the structure for bones and teeth, calcium is involved in the contraction of muscles, is necessary for normal clotting, and alters normal cell membrane permeability (thus affecting the development of the action potential of nerve and muscle cells). Only 1% to 2% of the body's calcium is contained within extracellular fluid (ECF), yet this concentration is regulated carefully by parathyroid hormone (PTH) and calcitonin. PTH is released by the parathyroid gland in response to a low serum calcium level. It increases resorption of bone (movement of calcium and phosphorus out of the bone), activates vitamin D, which increases the absorption of calcium from the GI tract (review the discussion of vitamin D in Chapter 2), and stimulates the kidneys to conserve calcium and excrete phosphorus. Calcitonin is produced by the thyroid gland when serum calcium levels are elevated. It inhibits bone resorption thereby lowering serum calcium levels.

The ECF gains calcium from intestinal absorption of dietary calcium and resorption from bones. Bone is living tissue that is constantly being deposited and reabsorbed. This continuous turnover of bone provides a ready source of calcium for the ECF. Although the bones act as a great reservoir for calcium, day-to-day calcium balance depends largely on changes in intestinal absorption. The amount of calcium absorbed from the gut is variable but is usually only about one third of the total amount ingested. Absorption does increase with growth and pregnancy, and it decreases with advancing age. In chronic conditions when increased intestinal absorption or increased renal conservation of calcium are not enough to maintain calcium levels, the ECF calcium may remain within normal limits due to increased bone resorption. Unfortunately, balance is then maintained at the expense of bone integrity. Calcium is lost from the ECF via secretion into the GI tract, urinary excretion, deposition in the bone, and a small amount in sweat.

Calcium is present in three different forms in the plasma: ionized, bound, and complexed. Approximately half of plasma calcium is free, ionized calcium. Slightly less than half is bound to protein, primarily to albumin. The remaining small percentage is complexed, that is, it is combined with nonprotein anions such as phosphate, citrate, and carbonate. Most laboratories report serum calcium as total calcium (bound, complexed, and ionized). However, it is only the ionized calcium that is physiologically important. On request, the laboratory may measure and report ionized calcium. This is usually not necessary since the percentage of total calcium that is ionized is normally fairly constant. However, such factors as plasma pH, phosphorus level, and albumin level will alter the normal relationship between ionized, bound, and complexed calcium. Therefore these factors must be considered when evaluating serum calcium levels.

The relationship between ionized calcium and plasma pH is reciprocal: an increase in pH decreases the percentage of calcium that is ionized. Patients with alkalosis (an elevated pH), for example, may show signs of hypocalcemia despite a normal total calcium level. Changes in plasma albumin levels will affect total serum calcium level without changing the level of ionized calcium. In hypoalbuminemia less protein is available to bind with calcium, and the total calcium level drops; yet the level of ionized calcium is unchanged. When there is an increase or decrease in the plasma phosphorus, the ionized calcium level is altered in the opposite direction. For example, if the plasma phosphorus level increases, the ionized calcium level decreases due to increased binding with phosphorus.

HYPOCALCEMIA

Symptomatic hypocalcemia may occur because of a reduction of total body calcium or a reduction of the percentage of ionized calcium. Total calcium levels may be decreased due to increased calcium loss, reduced intake secondary to altered intestinal absorption, or altered regulation (e.g., hypoparathyroidism). Elevated phosphorus levels and decreased magnesium levels also may precipitate hypocalcemia.

History and Risk Factors

Reduction in the percentage of ionized calcium

A decrease in ionized calcium may occur with alkalosis (increased bicarbonate levels) since calcium forms a complex with bicarbonate. Increased calcium binding also may occur with the administration of large quantities of citrated blood. Citrate, which is added to banked blood to prevent clotting, binds with a portion of the ionized calcium, causing symptomatic hypocalcemia. This is a potential problem only after massive transfusions or with exchange transfusions in the neonate. Volume replacement with large quantities of normal saline after hemorrhage reduces the level of ionized calcium due to hemodilution.

Increased calcium loss

Increased calcium loss from the body can occur with the use of the loop diuretics (e.g., furosemide) or renal disease that results in increased urinary excretion of calcium. Massive subcutaneous infections or burns also may lead to significant losses via the damaged skin.

Decreased intestinal absorption

Decreased intake, impaired vitamin D metabolism (e.g., with liver disease and renal failure), chronic diarrhea, and small bowel disease may result in decreased intestinal absorption of calcium. Gastrectomy and intestinal resection also will affect calcium absorption due to decreased absorption of dietary vitamin D.

Hypoparathyroidism

Any condition that causes a decrease in the production of PTH may result in the development of hypocalcemia. Hypoparathyroidism can occur with surgical removal of a portion of the parathyroid glands or injury to the parathyroid glands during thyroid or neck surgery. Hypocalcemia is most likely to develop within the first 24 to 48 hours after surgery.

Hyperphosphatemia and hypomagnesemia

Calcium and phosphorus have a reciprocal relationship, that is, as one goes up, the other tends to go down. This is an important protective physiologic mechanism since simultaneously increased calcium and phosphorus levels would lead to excessive binding of calcium and phosphorus with precipitation in the tissues (this concept will be discussed in greater detail later in this chapter in the section on hypercalcemia). An elevated phosphorus level decreases the calcium level due to decreased production of the active form of vitamin D and decreased movement of calcium and phosphorus out of the bone. Rapid administration of IV phosphate, which is used to treat hypophosphatemia, may cause a sudden drop in the level of ionized calcium due to increased calcium-phosphorus binding. Hypomagnesemia may cause hypocalcemia because of decreased secretion and action of PTH. Magnesium deficiency also increases the exchange of magnesium and calcium along the bone surface. Bone uptake of calcium increases as magnesium is released from the bone.

Acute pancreatitis

Hypocalcemia often occurs with acute pancreatitis. The exact reason for this is unclear, but it may be caused by a combination of factors, including the deposition of calcium in areas of fat necrosis, decreased secretion of PTH, hypoalbuminemia, and increased production of calcitonin.

Medications and chronic alcohol abuse

Certain medications (e.g., phenytoin, phenobarbital) alter hepatic metabolism of vitamin D, thereby decreasing the serum calcium level. Chronic alcohol abuse also may alter hepatic metabolism of vitamin D, thus contributing to the development of hypocalcemia.

Assessment

Neuromuscular effects

As with alterations in potassium balance, many of the signs and symptoms of hypocalcemia (see Table 10-1 for quick assessment guidelines) develop because of altered generation of the neuromuscular action potential (see Figure 9-1). The nerves and muscles become increasingly irritable due to the reduction in the threshold potential, that is, less electrical stimulus is necessary to initiate the action potential. Other neurologic signs and symptoms of hypocalcemia include personality changes, emotional irritability, depression, numbness with tingling in the fingers and circumoral region, hyperactive reflexes, muscle cramps, tetany (intermittent tonic muscle

Table 10-1 Quick assessment guidelines for individuals with hypocalcemia

Calcium < 8.5 mg/dl or 4.3 mEq/L

Signs and Symptoms

Numbness with tingling in the fingers and circumoral region, hyperactive reflexes, muscle cramps, tetany, convulsions. In chronic hypocalcemia, fractures may be present due to increased bone porosity

EKG changes: Prolonged QT interval caused by elongation of ST segment

Physical Assessment

- *Positive Trousseau's sign:* Ischemia-induced carpal spasm. It is elicited by applying a BP cuff to the upper arm and inflating it past systolic BP for 2 minutes
- *Positive Chvostek's sign:* Unilateral contraction of facial and eyelid muscles. It is elicited by irritating the facial nerve by percussing the face just in front of the ear

History and Risk Factors

- Decreased ionized calcium: Alkalosis, administration of large quantities of citrated blood and hemodilution
- Increased calcium loss in body fluids: Use of loop diuretics
- Decreased intestinal absorption: Decreased intake, impaired vitamin D metabolism (e.g., in renal failure), chronic diarrhea, post-gastrectomy
- Hypoparathyroidism
- Hyperphosphatemia: Renal failure
- Hypomagnesemia
- Acute pancreatitis
- Chronic alcoholism

spasms), and convulsions. In the presence of impending tetany, the following signs may be elicited:

Positive Chvostek's sign

A positive Chvostek's sign is a unilateral contraction of the facial and eyelid muscles. It is elicited by irritating the facial nerve when percussing the face just in front of the ear. (See Figure 10-1, *A*.)

Positive Trousseau's sign

A positive Trousseau's sign is ischemia-induced carpal spasm. This sign is elicited by applying a blood pressure cuff to the upper arm and inflating it past systolic blood pressure for 2 minutes. (See Figure 10-1, *B* and *C*.)

Respiratory effects

In severe hypocalcemia, stridor, dyspnea, and crowing may develop due to laryngeal muscle spasms. Spasms of bronchial smooth muscles may contribute to the respiratory distress.

Musculoskeletal effects

Skeletal abnormalities occur with chronic hypocalcemia but are more directly the result of the conditions that cause the hypocalcemia, e.g., altered vitamin D or phosphate metabolism. Osteomalacia, which is impaired mineralization of bone, is associated with vitamin D deficiency.

Additional effects

Additional signs and symptoms of hypocalcemia include dry skin, brittle hair, and decreased cardiac contractility, which sometimes results in congestive heart failure. Both hyperkalemia and hypomagnesemia will potentiate the cardiac and neuromuscular irritability caused by hypocalcemia.

Diagnostic Studies

Serum calcium levels

The total serum calcium level will be <8.5 mg/dl unless symptomatic hypocalcemia is the result of an increase in the percentage of calcium that is complexed. Serum calcium levels must be evaluated in combination with the serum albumin level, since decreases in the serum albumin level will result in a decrease in the total calcium level without affecting the amount of ionized calcium present. Thus decreases in total calcium that occur with decreases in serum albumin levels may not be clinically significant. For every 1.0 g/dl drop in the serum albumin level (below 4.0 g/dl), there is a 0.8 to 1.0 mg/dl drop in total calcium level. The individual with a reported total calcium level of 7 mg/dl and a serum albumin of 2.0 g/dl, for example, has a corrected calcium level of 8.6 to 9.0 mg/dl. Thus this individual should have a normal ionized

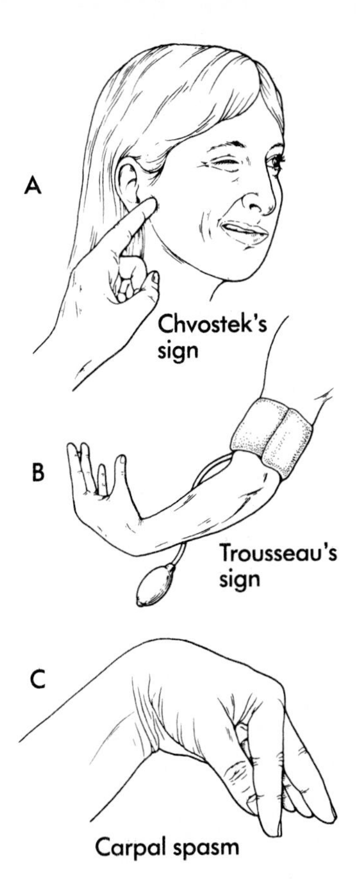

Figure 10-1 A, Chvostek's sign is a contraction of facial muscles in response to a light tap over the facial nerve in front of the ear. **B,** Trousseau's sign is a carpal spasm **(C)** induced by inflating a blood pressure cuff above the systolic pressure.

(From Lewis S and Collier I: Medical-Surgical nursing: assessment and management of clinical problems, ed 2, New York, 1987, McGraw-Hill, Inc.)

calcium level and will not be symptomatic. In true symptomatic hypocalcemia the ionized serum calcium will be less than 4.5 mg/dl.

Parathyroid hormone (PTH) levels

Decreased PTH levels occur in hypoparathyroidism; increased levels may occur with other causes of hypocalcemia. The normal range for PTH varies among laboratories and will depend on the exact form of PTH that is being measured.

Magnesium and phosphorus levels

Magnesium and phosphorus levels may be abnormal, depending on the cause of hypocalcemia. The phosphorus level is usually elevated (>2.5 to 4.5 mg/dl) due to the reciprocal relationship of calcium and phosphorus. Hypocalcemia is a frequent manifestation of severe or chronic hypomagnesemia. A magnesium level of 1 mg/dl (normal range 1.5 to 2.5 mEq/L) may induce hypocalcemia.

Electrocardiogram (EKG)

The primary EKG finding associated with hypocalcemia is a prolonged QT interval caused by elongation of the ST segment.

Medical Management

As with most electrolyte disturbances, the best treatment approach is to correct the underlying cause. Symptomatic hypocalcemia, however, is potentially lethal and necessitates immediate calcium replacement.

Calcium replacement

Hypocalcemia may be treated with either PO or IV calcium, depending on its severity. Tetany is treated with 10 to 20 ml of 10% calcium gluconate IV or a continuous drip of 100 ml of 10% calcium gluconate in 1,000 ml D_5W, infused over at least 4 hours (Goldberger, 1986). Calcium gluconate is considered preferable to calcium chloride due to the risk of local vessel and tissue damage with the latter treatment. Although both calcium gluconate and calcium chloride are available in 10 ml ampules, they do not contain equal quantities of calcium and are *not* interchangeable. One ampule of calcium gluconate contains 4.5 mEq of calcium, while calcium chloride contains 13.6 mEq of elemental calcium. Magnesium replacement should be initiated in the individual who is also magnesium depleted since hypomagnesemia-induced hypocalcemia is often refractory to calcium therapy alone.

Oral calcium is available as calcium carbonate, calcium lactate, or calcium gluconate. Calcium supplements are sometimes combined with vitamin D (e.g., Os-Cal + D) to enhance GI absorption. Vitamin D therapy alone (e.g., dihydrotachysterol, calcitriol) may be used to increase calcium absorption from the GI tract. See Table 10-2, for a list of vitamin D preparations.

Table 10-2 Vitamin D preparations

Generic Name	Trade Name	Chemical Abbreviation
ergocalciferol	Calciferol	D_2
dihydrotachysterol	Hytakerol	DHT
calcifediol	Calderol	$25(OH)D_3$
calcitriol	Rocaltrol	$1,25(OH)_2D_3$

Phosphorus-binding antacids

The combination of hypocalcemia and hyperphosphatemia is common in chronic renal failure (CRF). Providing calcium supplementation in the presence of an elevated phosphorus level runs the risk of excess calcium-phosphorus binding with precipitation of calcium phosphate in soft tissues. For this reason, treatment with phosphorus-binding antacids to reduce elevated phosphorus levels is usually necessary for the individual with CRF before treatment of hypocalcemia is initiated.

Nursing Diagnoses and Interventions

Potential for trauma related to risk of tetany and seizures secondary to severe hypocalcemia

Desired outcomes: Patient does not exhibit evidence of trauma caused by complications of severe hypocalcemia. Serum calcium levels are within normal range (8.5 to 10.5 mg/dl).

1. Monitor for the signs and symptoms of hypocalcemia in persons at increased risk (e.g., the postthyroidectomy patient). Monitor patients with hypocalcemia for evidence of worsening hypocalcemia: numbness and tingling in the fingers and circumoral region, hyperactive reflexes, and muscle cramps. Notify MD promptly if these symptoms develop because they occur prior to overt tetany. In addition, notify MD if patient has positive Trousseau's or Chvostek's signs, because they also signal latent tetany.
2. Administer IV calcium with caution. IV calcium should not be given faster than 0.5 to 1 ml/minute because rapid administration can cause hypotension. Use of an IV pump is recommended. Observe IV insertion site for evidence of infiltration because calcium will slough tissue. Concentrated calcium solutions should be administered through a central line. Do not add calcium to solutions containing bicarbonate or phosphate because dangerous precipitates will form. NOTE: Always clarify the type of IV calcium to be given. Both calcium chloride and calcium gluconate come in 10 ml ampules. One amp of calcium chloride contains approximately 13.6 mEq of calcium, whereas 1 amp of calcium gluconate contains 4.5 mEq of calcium.

   Monitor patient receiving IV calcium therapy for signs and symptoms of hypercalcemia: lethargy, confusion, irritability, nausea, and vomiting. Ideally, individuals receiving IV calcium should be monitored for EKG changes. The development of ventricular dysrhythmias or heart block requires immediate MD notification. **CAUTION:** Digitalis toxicity may develop in patients taking digitalis because calcium potentiates its effects.
3. For individuals with chronic hypocalcemia, administer oral calcium supplements and vitamin D preparations (see Table 10-2) as prescribed. Administer oral calcium 30 minutes before meals and at bedtime for maximal absorption. Administer

Foods High in Calcium

Cottage cheese	Molasses
Cheese	Seafood, especially sardines with bones
Milk and cream	Rhubarb
Eggnog	Brazil nuts
Yogurt	Sesame seeds
Soy flour	Broccoli
Oat flakes	Collard, mustard, and turnip greens
Milk chocolate	Spinach
Ice cream	

aluminum hydroxide antacids immediately after meals. Give patient and significant others verbal and written instructions for medications, including drug name, purpose, dosage, frequency, precautions, and potential side effects. Review with the patient and significant others the indicators of hypercalcemia and hypocalcemia. Identify symptoms that necessitate immediate medical attention: numbness and tingling in the fingers and circumoral region and muscle cramps.

4. Encourage intake of foods high in calcium: milk products, meats, leafy green vegetables. (See the accompanying box.) Many foods that are high in calcium, such as milk products, are also high in phosphorus and may need to be limited in persons with renal failure. A program of phosphorus control and calcium supplementation may be necessary for the individual with CRF.
5. Notify MD if response to calcium therapy is ineffective. Tetany that does not respond to IV calcium may be caused by hypomagnesemia.
6. Keep symptomatic patients on seizure precautions; decrease environmental stimuli.
7. Avoid hyperventilation in patients in whom hypocalcemia is suspected. Respiratory alkalosis may precipitate tetany due to increased calcium-bicarbonate binding.

Decreased cardiac output related to decreased cardiac contractility (negative inotropy) and/or altered conduction secondary to hypocalcemia or digitalis toxicity occurring with calcium replacement therapy

Desired outcome: Patient's cardiac output is adequate as evidenced by CVP $\leq$ 6 mm Hg (< 12 cm H_20), HR $\leq$ 100, BP within patient's normal range, and absence of the clinical signs of heart failure or pulmonary edema. Critical care patients exhibit a PAP of 20 to 30/8 to 15 mm Hg.

1. Monitor EKG for signs of worsening hypocalcemia (prolonged QT interval) or digitalis toxicity with calcium replacement: multifocal or bigeminal PVCs, paroxysmal atrial tachycardia with varying AV block, Wenckebach (Type I AV) heart block.
2. Hypocalcemia may decrease cardiac contractility. Monitor patient for signs of heart failure or pulmonary edema: crackles (rales), rhonchi, SOB, decreased BP, increased HR, increased PAP, or increased CVP.

Potential for impaired gas exchange related to altered oxygen supply secondary to laryngeal spasm occurring with severe hypocalcemia

Desired outcome: Patient exhibits respiratory depth, pattern, and rate (12 to 20 breaths/min) within normal range and is asymptomatic of laryngeal spasm: laryngeal stridor, dyspnea, or crowing.

1. Assess patient's respiratory rate, character, and rhythm. Be alert to laryngeal stridor, dyspnea, and crowing, which occur with laryngeal spasm, a life-threatening complication of hypocalcemia.
2. Keep an emergency tracheostomy tray and manual resuscitator (e.g., Ambu or Laerdal) at the bedside of symptomatic patients.

HYPERCALCEMIA

Symptomatic hypercalcemia can occur because of an increase in total serum calcium or in the percentage of free, ionized calcium. If hypercalcemia is accompanied by a normal or elevated serum phosphorus level, calcium phosphate crystals may precipitate in the serum and deposit throughout the body. Soft tissue calcifications usually occur when the product of the serum calcium and serum phosphorus (i.e., calcium × phosphorus) exceeds 70 mg/dl.

History and Risk Factors

Increased calcium intake

Excessive IV administration of calcium during cardiopulmonary arrest may result in hypercalcemia, as may excessive intake of oral calcium supplements. **Milk-alkali syndrome** is a condition that occurs with chronic ingestion of calcium carbonate antacids and milk and is associated with hypercalcemia and metabolic alkalosis. In addition to increased calcium intake, the pathogenesis involves increased intestinal absorption and decreased urinary excretion of calcium (see the section on metabolic alkalosis in Chapter 18 for additional information.)

Increased gastrointestinal (GI) absorption

Both vitamin D overdose and hyperparathyroidism will cause increased GI absorption of calcium. Increased levels of PTH stimulate the conversion of vitamin D to its metabolically active form, which, in turn, facilitates the absorption of calcium from the GI tract and acts with PTH to increase bone resorption.

Increased bone resorption

Increased movement of calcium out of the bone may occur with hyperparathyroidism, malignancies, prolonged immobilization, or Paget's disease, a disorder associated with impaired action of the osteoclasts (bone-resorbing cells) and osteoblasts (bone-building cells). The development of hypercalcemia in individuals with malignancies occurs due to the effects of certain medications (such as androgens used in the treatment of carcinoma of the breast), release of PTH and PTH-like substances from malignant tissue, and metastasis to the bone.

Prolonged immobilization causes an increased loss of calcium from the bones due to the lack of bone stress. The stresses of weightbearing and muscular activity are essential for maintaining the normal balance between bone resorption and deposition. Hypercalcemia secondary to immobilization is most likely to occur in children because of the rapid rate at which their bones grow. Paget's disease may cause hypercalcemia due to increased turnover of bone.

Decreased urinary excretion

Factors causing decreased calcium excretion include renal failure, medications (e.g., thiazide diuretics and lithium), milk-alkali syndrome, and hyperparathyroidism. The thiazide diuretics decrease urinary excretion of calcium via direct action on the renal tubule and diuretic-induced volume depletion.

Increased ionized calcium

Acidosis can increase the level of ionized calcium without affecting the total calcium level by decreasing the percentage of calcium that is complexed.

Assessment

Neuromuscular effects

The signs and symptoms of hypercalcemia (see Table 10-3 for quick assessment guidelines) are largely the result of calcium's effects on cell membrane permeability and neuromuscular irritability. Increased calcium levels decrease neuromuscular irritability due to a reduction in the negativity of the threshold potential (see Figure 9-1). The gap between the resting membrane potential and the threshold potential is increased. Thus a greater electrical impulse is necessary to initiate the neuromuscular action potential. Symptoms of neuromuscular depression include weakness, fatigue, paresthesias, decreased deep tendon reflexes, and incoordination. The central nervous system effects of hypercalcemia are manifested by depression, confusion, personality changes, psychotic behavior, and, if the calcium level is severely elevated, stupor and coma.

Gastrointestinal (GI) effects

Decreased contractility of GI smooth muscle results in decreased peristalsis, with resultant anorexia, nausea, vomiting, and constipation.

Table 10-3 Quick assessment guidelines for individuals with hypercalcemia

Calcium > 10.5 mg/dl or 5.3 mEq/L

Signs and Symptoms

Lethargy, weakness, anorexia, nausea, vomiting, polyuria, fractures, flank pain (secondaryto renal calculi), depression, confusion, paresthesias, personality changes, stupor, coma

EKG findings: Shortening of ST segment and QT interval. PR interval is sometimes prolonged. Ventricular dysrhythmias can occur with severe hypercalcemia

History and Risk Factors

- Increased intake of calcium: Excessive administration during cardiopulmonary arrest
- Increased intestinal absorption: Vitamin D overdose or hyperparathyroidism
- Increased release of calcium from bone: Occurs with hyperparathyroidism, malignancies, prolonged immobilization, Paget's disease
- Decreased urinary excretion: Renal failure, medications (e.g., thiazide diuretics)
- Increased ionized calcium: Acidosis

Renal effects

Hypercalcemia adversely affects renal function to the point that acute (reversible) or chronic (permanent) renal failure may develop. The severity of hypercalcemic nephropathy will depend on the severity and duration of hypercalcemia. Prolonged, severe hypercalcemia is most likely to result in permanent renal disease. Increased urinary calcium concentrations decrease the kidneys' ability to concentrate the urine, leading to polyuria and potentially to volume depletion. This is a type of nephrogenic diabetes insipidus treated by correcting the hypercalcemia. Increased urine calcium also may contribute to the development of kidney stones (nephrolithiasis). The signs and symptoms of nephrolithiasis include flank or groin pain (depending on where the stone lodges in the urinary tract) and hematuria.

Diagnostic Studies

With hypercalcemia total serum calcium level will be >10.5 mg/dl, while the ionized calcium level will be >5.3 mg/dl. Parathyroid hormone levels may be increased in primary or secondary hyperparathyroidism. X-ray findings may reveal the presence of osteoporosis, bone cavitation, or urinary calculi.

Electrocardiogram (EKG)

EKG findings in hypercalcemia are the result of alterations in the configuration of the cardiac action potential. The repolarization period is decreased, as evidenced by shortening of the ST segment and QT interval. As hypercalcemia progresses, the PR interval is prolonged. Ventricular dysrhythmias can occur with severe hypercalcemia.

Medical Management

The medical management of asymptomatic hypercalcemia may be limited to treatment of the underlying cause, for example, antitumor chemotherapy for malignancy or partial parathyroidectomy for hyperparathyroidism. Symptomatic hypercalcemia requires more immediate treatment to reduce the plasma calcium level. Calcium levels greater than 15 mg/dl require emergency treatment.

Saline and furosemide treatment

One approach to treatment is increasing urinary excretion of calcium via administration of several liters of IV saline, either 0.45% or 0.9%, combined with IV furosemide (Lasix). This therapeutic approach requires normal or near normal renal function. Calcium reabsorption in the proximal portion of the renal tubule is tied to sodium reabsorption. In volume expanded states when sodium reabsorption is decreased, calcium reabsorption also will be decreased. Thus administration of several liters of saline solution will expand the vascular volume and increase the excretion of calcium. The increased plasma water will also dilute the existing plasma calcium level. Furosemide, a potent loop diuretic, is administered to prevent fluid overload and increase calcium excretion. The thiazide diuretics are not used since they decrease urinary calcium excretion.

IV phosphate administration

A rapid, although potentially dangerous, approach to the treatment of hypercalcemia is the administration of IV phosphates (sodium phosphate or potassium phosphate). Phosphates cause a drop in the level of ionized calcium due to increased calcium-phosphorus binding. Unfortunately, this mode of therapy poses the risk of soft tissue calcifications if the calcium-phosphorus product exceeds 70 mg/dl.

Low-calcium diet/steroid treatment

A low-calcium diet may be combined with cortisone to limit intake and intestinal absorption of calcium. Steroids compete with vitamin D, thereby reducing intestinal absorption of calcium. For a list of foods high in calcium, see the box on p. 200.

Decreased bone resorption

Decreased bone resorption may be achieved via increased activity level, or mithramycin. Mithramycin, a cytotoxic antibiotic, acts directly on bone to reduce decalcification, and it is used primarily to treat hypercalcemia associated with neoplastic disease. Compressional loads (e.g., weight-bearing) stimulate bone deposition; thus, increased activity decreases bone resorption. Calcitonin also reduces bone resorption in addition to increasing bone deposition and urinary excretion of calcium and phosphorus.

Nursing Diagnoses and Interventions

Potential for injury related to neuromuscular, sensorium, or cardiac changes secondary to hypercalcemia

Desired outcomes: Patient does not exhibit evidence of injury due to neuromuscular or sensorium changes. Patient verbalizes orientation to person, place, and time. Serum calcium levels are within normal range (8.5 to 10.5 mg/dl).

1. Monitor patient for worsening hypercalcemia. Assess and document level of consciousness; patient's orientation to person, place, and time; and neurologic status with each VS check.
2. Personality changes, hallucinations, paranoia, and memory loss may occur with hypercalcemia. Inform patient and significant others that altered sensorium is temporary and will improve with treatment. Utilize reality therapy: clocks, calendars, and familiar objects; keep them at the bedside within patient's visual field. Have patient wear glasses or hearing aid if needed.
3. Hypercalcemia causes neuromuscular depression with poor coordination, weakness, and altered gait. Provide a safe environment. Keep side rails up and bed in lowest position with wheels locked. Assist patient with ambulation if it is allowed.
4. Because hypercalcemia potentiates the effects of digitalis, monitor patient taking digitalis for signs and symptoms of digitalis toxicity: anorexia, nausea, vomiting, and irregular pulse. EKG changes may include multifocal or bigeminal PVCs, paroxysmal atrial tachycardia with varying AV block, Wenckebach (Type I AV) heart block. Monitor for pulse changes in the nonEKG-monitored setting.
5. Monitor serum electrolyte values for changes in serum calcium (normal range is 8.5 to 10.5 mg/dl); potassium (normal range is 3.5 to 5.0 mEq/L); and phosphorus (normal range is 2.5 to 4.5 mg/dl) secondary to therapy. Notify MD of abnormal values.
6. Encourage increased mobility to reduce bone resorption. Ideally, patient should be out of bed and sitting up in a chair at least 6 hr/day. Instruct patient and significant others regarding the need for weightbearing activities, especially in the individual who is housebound.
7. Avoid vitamin D preparations (see Table 10-2) since they increase intestinal absorption of calcium.

Potential alterations in fluid volume: *Excess* secondary to administration of IV saline solutions; and *deficit* secondary to concomitant administration of furosemide for treating hypercalcemia

Desired outcomes: Patient is normovolemic as evidenced by RR 12 to 20 breaths/min with normal depth and pattern (eupnea), balanced I&O, stable weights, HR 60 to 100 bpm, normal jugular vein presentation, edema <2+ on a 1-4+ scale, good skin turgor, and absence of adventitious breath sounds. Blood pressure and CVP are within patient's normal range.

1. Monitor for indicators of fluid volume excess during administration of IV saline: increased BP, heart rate, and CVP; increased jugular venous distention; SOB; and crackles/rhonchi.
2. Observe for and document presence of edema: pretibial, sacral, and periorbital. If present, rate pitting on a 1+ (barely detectable) through 4+ (deep, persistent) scale and alert MD.
3. Monitor and record I&O and daily weights, which are indicative of overall fluid loss or gain.
4. Monitor for indicators of fluid volume deficit after administration of furosemide: decreased BP, increased HR, decreased CVP, and poor skin turgor (except in the older adult, whose skin elasticity may be decreased due to aging).

Altered pattern of urinary elimination related to dysuria, urgency, frequency, and polyuria secondary to administration of diuretics, calcium stone formation, or changes in renal function occurring with hypercalcemia

Desired outcome: Patient exhibits voiding pattern and urine characteristics that are normal for patient.

1. Monitor I&O hourly. Alert MD to unusual changes in urine volume, for example, oliguria alternating with polyuria, which may signal urinary tract obstruction, or continuous polyuria, which may be indicative of nephrogenic diabetes insipidus.
2. Because hypercalcemia can impair renal function, monitor individual's renal function carefully: urine output, BUN, creatinine.
3. Provide patient with a low-calcium diet (see the box on p. 200 for foods to avoid) and avoid use of calcium-containing medications (e.g., antacids such as Tums)
4. If stone formation is a concern, encourage intake of fruits, such as cranberries, prunes, or plums, which leave an acid ash in the urine that reduces the risk of calcium stone formation; give patient and significant others verbal and written instructions for foods that leave an acid ash; review foods (see the box on p. 200) and over-the-counter medications (e.g., antacids) that are high in calcium; and explain the importance of increased fluid intake (3 to 4 liters in nonrestricted patients). Encourage regular voiding to prevent urinary stasis. Assess patient for indicators of kidney stone formation: intermittent pain, nausea, vomiting, hematuria. Review signs and symptoms of nephrolithiasis with the patient and significant other.

Case Study 10-1

A.W., a 56-year-old librarian on chronic hemodialysis for the last 12 years, has been admitted to the surgical unit. He is scheduled for a subtotal (partial) parathyroidectomy tomorrow. The surgery is necessary to reduce A.W.'s markedly elevated PTH level. Hyperparathyroidism develops in chronic renal failure due to a combination of factors, including elevated phosphorus levels (secondary to decreased excretion of phosphorus) and decreased calcium levels (secondary to decreased activation of vitamin D and end organ resistance to PTH). For the individual with chronic renal failure, excessive production of PTH contributes to the development of renal bone disease (osteodystrophy), metastatic calcifications (PTH increases the entry of calcium into soft tissues), and pruritus. In addition to showing radiographic evidence of bone disease, A.W. suffers from severe pruritus that has been unresponsive to other forms of therapy. The itching is so constant that A.W. has been unable to sleep at night or function effectively at work. The nurse caring for A.W. notes multiple scabs and scratch marks on both of his arms.

1. Question: What are the actions of PTH?

Answer:

a. Increased renal conservation of calcium.
b. Increased urinary excretion of phosphorus.
c. Activation of vitamin D, which increases GI absorption of calcium.
d. Increased bone resorption (movement of calcium and phosphorus out of the bone).

2. Question: How might excessive production of PTH contribute to the development of bone disease in the individual with chronic renal failure?

Answer: Chronic excessive production of PTH increases the movement of calcium and phosphorus out of the bone. Demineralization weakens bone structure and may lead to spontaneous fractures.

3. Question: What are some potential nursing diagnoses for A.W.?

Answer:

a. **Impaired skin integrity** related to continuous scratching
b. **Potential for infection** related to broken skin
c. **Sleep pattern disturbance** related to pruritus
d. **Potential for injury** related to risk of fractures secondary to renal osteodystrophy
e. **Knowledge deficit:** Preoperative and postoperative care

Preoperatively, A.W.'s calcium level is in the upper limits of normal. Postoperatively, he will be monitored closely for evidence of hypocalcemia, since it is the most common complication of a subtotal parathyroidectomy.

4. **Question:** Identify nursing considerations for the individual who is or who may be hypocalcemic.

Answer:

a. Assess for the signs and symptoms of hypocalcemia, e.g., muscle cramps, numbness and tingling of fingers and circumoral region, hyperactive reflexes, positive Trousseau's sign, positive Chvostek's sign, and tetany; instruct the patient to report immediately muscle cramps, numbness, or tingling.
b. If the calcium level is markedly decreased, implement seizure precautions (i.e., padded siderails, oral airway at the bedside) and keep the room quiet and dimly lit to minimize stimuli, which might excite the patient's nervous system and thus precipitate tetany or seizures.
c. Ensure that IV calcium is readily available if needed.
d. If the individual is symptomatic with hypocalcemia, keep emergency tracheostomy tray, suction equipment, and manual resuscitator at the bedside in case of laryngeal stridor.
e. Provide clear, simple explanations for nursing and medical interventions and provide emotional support to patient and family. Remember that anxiety-induced hyperventilation may result in alkalosis, which can potentiate the hypocalcemia.

5. **Question:** What are the nursing precautions for IV calcium administration?

Answer:

a. Do not administer faster than 0.5 to 1.0 mg/min due to the risk of hypotension.
b. Guard against infiltration, since calcium may cause tissue to slough.
c. Do not add to solutions containing phosphate or carbonate, since precipitates will form.
d. Administer cautiously to patients receiving digitalis due to increased risk of digitalis toxicity. Monitor for evidence of dysrhythmias.

Case Study 10-2

M.M. is a 67-year-old retired postal worker who has been admitted to the oncology floor for treatment of multiple myeloma. She has mild hypertension currently being treated with a thiazide diuretic and a 6-week history of gastric upset, which she has been self-medicating with frequent small glasses of milk and Tums (calcium carbonate). Her serum calcium level is 14.5 mg/dl (normal value is 8.5 to 10.5 mg/dl).

1. **Question:** What are the potential causes of M.M.'s hypercalcemia?

Answer:

a. Increased movement of calcium out of the bone secondary to multiple myeloma.
b. Decreased urinary excretion of calcium secondary to thiazide diuretic therapy.
c. Increased calcium intake secondary to milk and calcium carbonate consumption, with possible milk-alkali syndrome.

M.M. exhibits angry, paranoid behavior because of her hypercalcemia. Her husband is confused and upset about her personality changes, stating, "I don't understand this. Please don't be angry with her. She really is a pleasant and cooperative person."

2. **Question:** What are some potential nursing diagnoses for M.M?

Answer:

a. **Sensory/perceptual alterations** related to the central nervous system effects of hypercalcemia
b. **Knowledge deficit** (Family): Temporary effects of hypercalcemia
c. **Potential for trauma:** Self-inflicted, related to delusional thinking and other central nervous system complications of hypercalcemia
d. **Potential altered patterns of urinary elimination:** Dysuria and polyuria, secondary to the renal effects of hypercalcemia

M.M. is to be treated with 2 L of IV isotonic sodium chloride solution to be infused at a rate of 500 ml/hr and 20 mg IV furosemide (Lasix) after each liter of fluid.

3. **Question:** The nurse caring for M.M. will need to monitor for which potential complications of this medical therapy?

Answer:

a. Potential fluid volume excess secondary to rapid administration of isotonic sodium chloride solution.
b. Potential fluid volume deficit secondary to excessive diuresis.
c. Hypokalemia secondary to diuretic therapy.

References

Arieff AI and De Fronzo RA: Fluid, electrolyte, and acid-base disorders, New York, 1985, Churchill Livingstone, Inc.

Calloway C: When the problem involves magnesium, calcium, or phosphate, RN 50:30-35, May 1987.

Corbett JV: Laboratory tests and diagnostic procedures with nursing diagnoses, ed 2, Norwalk, Conn and San Mateo, Calif, 1987, Appleton & Lange.

Goldberger E: A primer of water, electrolyte, and acid-base syndromes, ed 7, Philadelphia, 1986, Lea & Febiger.

Guyton AC: Textbook of medical physiology, ed 7, Philadelphia, 1986, WB Saunders Co.

Holick M and others: Calcium, phosphorus, and bone metabolism: calcium regulating hormones. In Braunwald E and others: Harrison's principles of internal medicine, ed 11, New York, 1987, McGraw-Hill, Inc.

Horne M and Swearingen PL: Pocket guide to fluids and electrolytes, St Louis, 1989, The CV Mosby Co.

Kokko JP and others: Fluids and electrolytes, Philadelphia, 1986, WB Saunders Co.

Lockhart JS and Griffin CW: Action stat: tetany, Nursing 88 18(8):33, 1988.

Malluche HH and Faugere M: Renal osteodystrophy, Mediguide to Nephrology 1(1):1-8, 1986.

Maxwell MH and others: Clinical disorders of fluid and electrolyte metabolism, ed 4, New York, 1987, McGraw-Hill, Inc.

Metheny NM: Fluid and electrolyte balance: nursing considerations, Philadelphia, 1987, JB Lippincott Co.

Rice V: Magnesium, calcium, and phosphate imbalances: their clinical significance, Crit Care Nurs 3:90-112, May/June 1983.

Sutton R: Diuretics and calcium metabolism, Am J Kidney Dis 5(1):4-9, 1985.

Vander AJ: Renal physiology, ed 3, New York, 1985, McGraw-Hill, Inc.

Chapter

11

Disorders of Phosphorus Balance

Phosphorus is the primary anion of the intracellular fluid (ICF), and it is vital to normal cellular function. Approximately 85% of the body's phosphorus is in the bones and teeth, 14% is in the soft tissue, and less than 1% is within the extracellular fluid (ECF). Because of the large intracellular store, under certain acute conditions, phosphorus may move into or out of the cells, causing dramatic changes in plasma phosphorus. Chronically substantial increases or decreases can occur in intracellular phosphorus levels without significantly altering plasma levels. Thus plasma phosphorus levels do not necessarily reflect intracellular levels. Although most laboratories measure and report elemental phosphorus, nearly all the phosphorus in the body exists in the form of phosphate (PO_4), and the terms phosphorus and phosphate are often used interchangeably. Normal range for serum phosphorus is 2.5 to 4.5 mg/dl (1.7 to 2.6 mEq/L).

Phosphorus is an important constituent of all body tissues and has a wide variety of important functions, including formation of energy-storing substances (e.g., adenosine triphosphate—ATP); formation of red blood cell 2,3-DPG, which facilitates oxygen delivery to the tissues; metabolism of carbohydrates, protein, and fat; and maintenance of acid-base balance. In addition, phosphorus is critical to normal nerve and muscle function and combines with calcium to provide structural support to bones and teeth. Plasma phosphorus levels vary with age, sex, and diet. Levels decrease with increasing age, with the exception of a slight rise in phosphorus in women following menopause. Glucose, insulin, or sugar-containing foods cause a temporary drop in phosphorus due to a shift of serum phosphorus into the cells.

Acid-base status also affects phosphorus balance. Alkalosis, particularly respiratory alkalosis, may cause hypophosphatemia secondary to an intracellular shift of phosphorus. The exact mechanism for this shift is not fully understood but may be related to an alkalosis-induced cellular glycolysis (breakdown of glucose and other sugars) with increased formation of phosphorus-containing metabolic intermediates. Respiratory acidosis may cause phosphorus to shift out of the cells and contribute to hyperphosphatemia.

The level of ECF phosphate is regulated by a combination of factors, including dietary intake, intestinal absorption, and hormonally-regulated bone resorption and deposition. The ECF gains phosphorus through dietary intake and the movement of phosphorus out of the cells or bones. A variety of foods are high in phosphorus, including red meat, poultry, fish, dairy products, and nuts. The primary regulators of phosphorus balance, however, are the kidneys. Elevated plasma phosphorus levels cause increased excretion of phosphorus, while decreased levels result in decreased excretion of phosphorus.

As discussed in Chapter 10, phosphorus balance is closely tied to that of calcium. Parathyroid hormone (PTH), the primary regulator of the plasma calcium level, increases the movement of both calcium and phosphorus out of the bone, and through the actions of vitamin D, increases GI absorption of both calcium and phosphorus. At the same time, however, PTH increases the renal excretion of phosphorus. This is necessary to facilitate an increase in the ionized calcium level. Recall that a simultaneous increase in calcium and phosphorus will result in increased calcium-phosphorus binding and potential soft tissue calcifications.

HYPOPHOSPHATEMIA

Hypophosphatemia (serum phosphorus <2.5 mg/dl) may occur due to transient intracellular shifts, increased urinary losses, decreased intestinal absorption, or increased utilization. Severe phosphorus deficiency also may result from a combination of factors in conditions such as alcoholism and diabetic ketoacidosis (DKA). As discussed at the beginning of this chapter, changes in plasma phosphorus levels do not necessarily correlate with changes in intracellular or total body levels. Hypophosphatemia associated with a shift of phosphorus into the cell may be clinically significant even though the total body phosphorus level is unchanged (see the next section).

History and Risk Factors

Intracellular shifts

Glucose loads

The movement of phosphorus into the cells is the most common cause of hypophosphatemia. It has long been observed that the administration of a glucose load (e.g., IV solutions containing large concentrations of glucose) will precipitate hypophosphatemia. The exact mechanism is unclear, although increased insulin levels have been implicated. Since insulin facilitates the movement of both glucose and phosphorus into muscle cells, the administration of concentrated glucose solutions, which dramatically increase the release of insulin, would be expected to cause an intracellular shift of phosphorus. Additionally, any condition associated with accelerated tissue regeneration and repair will contribute to the development of hypophosphatemia due to increased utilization as new tissue is formed. Total parenteral nutrition (TPN) solutions with inadequate phosphorus content have caused dramatic decreases in serum phosphorus when used to treat individuals with malnutrition, increased tissue catabolism, or burns.

Respiratory alkalosis

Respiratory alkalosis (i.e., hyperventilation) also may cause a clinically significant intracellular shift of phosphorus. This is believed to be caused by an increase in pH within the cells, which increases the production of phosphorus-containing metabolites. Hypophosphatemia induced by respiratory alkalosis can occur with salicylate intoxication and sepsis. Although hypophosphatemia also may occur with metabolic alkalosis, it is less likely to develop because of a slower increase in intracellular pH.

An intracellular shift of phosphorus is most likely to result in symptoms when there is a preexisting or concomitant phosphorus deficit. A transient drop in phosphorus due to an intracellular shift that is not accompanied by a true phosphorus deficit usually does not require treatment. Thus intracellular shifts are most significant when they occur in the individual who is experiencing decreased intake, increased loss, or increased need (e.g., tissue repair).

Increased urinary losses

Hypomagnesemia, hypokalemia, glycosuria, hyperparathyroidism, and the use of thiazide diuretics increase the excretion of phosphorus by the kidneys and may contribute to the development of hypophosphatemia.

Reduced intestinal absorption or intestinal loss

Prolonged use or administration of large quantities of antacids (e.g., aluminum hydroxide antacids, such as Amphojel or ALternaGEL, or magnesium hydroxide antacids, such as Mylanta) can cause hypophosphatemia due to the binding of phosphorus in the GI tract. This is most commonly encountered in individuals being treated for peptic ulcer disease or in individuals with chronic renal failure. Phosphorus-binding antacids are the primary means of controlling hyperphosphatemia in renal failure; however, when they are administered in the presence of decreased intake, hypophosphatemia may develop. Decreased absorption of phosphorus from the GI tract also may occur with vomiting, diarrhea, malabsorption disorders, and vitamin D deficiency.

Mixed causes

Hypophosphatemia is associated with alcoholism (especially during acute withdrawal) secondary to poor intake, vomiting, diarrhea, hyperventilation, use of phosphorus-binding antacids, and increased urinary losses. A combination of factors are also responsible for the development of hypophosphatemia in DKA. With DKA there is a significant loss of phosphorus in the urine due to the glucose-induced osmotic diuresis. This developing hypophosphatemia is masked, however, by the movement of phosphorus out of the cells because of increased tissue catabolism (cellular breakdown). When ketoacidosis is treated with glucose, insulin, and fluids, there is a dramatic shift of phosphorus back into the cells, and the existing phosphorus depletion then becomes apparent.

Assessment

Individuals may show acute symptoms secondary to a sudden decrease in serum phosphorus, or their symptoms may develop gradually with chronic phosphorus deficiency. Signs and symptoms of chronic hypophosphatemia are often subtle or subclinical and include memory loss, lethargy, weakness, bone pain, joint stiffness, arthralgia, osteomalacia, and pseudofractures. In contrast, the signs and symptoms occurring with an acute drop in plasma phosphorus may be dramatic and potentially lethal. The clinical manifestations of acute hypophosphatemia are largely the result of decreases in ATP and 2,3-DPG. Decreased ATP impairs cellular function due to decreased available energy supplies, while decreased levels of 2,3-DPG cause oxygen to be more tightly bound to hemoglobin, reducing oxygen delivery to all tissues. Decreased availability of oxygen further decreases the production of ATP. The combination of decreased energy and oxygen supply adversely affects all organ systems. Thus, the acute manifestations of hypophosphatemia typically involve multiple organ systems. See Table 11-1 for quick assessment guidelines for individuals with hypophosphatemia.

Neuromuscular effects

Neurologic symptoms of hypophosphatemia include memory loss, confusion, obtundation, seizures, and coma. In the chronic alcoholic hypophosphatemia may be mistaken for hepatic encephalopathy or acute withdrawal (delirium tremens). The presence of hallucinations suggests delirium tremens or hypomagnesemia rather than hypophosphatemia. Additional neuromuscular symptoms include muscle pain, numbness and tingling of the fingers and circumoral region, incoordination, and decreased strength, as evidenced by difficulty speaking, weakness of the respiratory muscles, and a weak hand grasp. Rhabdomyolysis (destruction of striated muscle) may occur with severe hypophosphatemia, probably due to accumulation of sodium and water within the cells secondary to the lack of ATP. Reddish-brown discoloration of the urine due to the urinary excretion of myoglobin may be a signal of acute muscle damage.

Table 11-1 **Quick assessment guidelines for individuals with hypophosphatemia**

Phosphorus < 2.5 mg/dl (1.7 mEq/L)

Signs and Symptoms

Patients may present with acute symptoms due to sudden decreases in serum phosphorus, or symptoms may develop gradually owing to chronic PO_4 deficiency. The majority of symptoms are secondary to decreases in ATP and 2,3-DPG

- *Acute:* Confusion, seizures, coma, chest pain due to poor oxygenation of the myocardium, muscle pain, increased susceptibility to infection, numbness and tingling of the fingers and circumoral region, and incoordination
- *Chronic:* Memory loss, bone pain, arthralgia, and osteomalacia

Physical Assessment

- *Acute:* Decreased strength as evidenced by difficulty speaking, weakness of respiratory muscles, and weak hand grasp. Hypoxia may cause an increased respiratory rate and metabolic alkalosis (secondary to hyperventilation). Metabolic alkalosis causes phosphorus to move intracellularly, aggrevating the existing hypophosphatemia
- *Chronic:* Bruising and bleeding may occur due to platelet dysfunction. Lethargy, weakness, joint stiffness, cyanosis, and pseudofractures may occur

Hemodynamic Measurements

Severely depleted individuals may show signs of decreased myocardial function, including increased **PAWP**, decreased CO, and decreased BP with decreased response to pressor agents

History and Risk Factors

- *Intracellular shifts:* Carbohydrate load; respiratory alkalosis (see Chapter 16)
- *Increased utilization due to increased tissue repair:* Total parenteral nutrition with inadequate phosphorus content; recovery from protein-calorie malnutrition
- *Increased urinary losses:* Hypomagnesemia (see Chapter 12); hypokalemia; hyperparathyroidism; glucosuria
- *Reduced intestinal absorption or increased intestinal loss:* Use of phosphorus-binding antacids; vomiting and diarrhea; malabsorption disorders such as vitamin D deficiency
- *Mixed causes:* Alcoholism or DKA

Cardiovascular effects

Myocardial dysfunction and chest pain may develop secondary to poor myocardial oxygenation. Severe myocardial dysfunction may lead to the development of heart failure and shock. Additionally, inadequate ATP, combined with ischemia, affects the electrical activity of the myocardium, resulting in dysrhythmias. Hemodynamic measurements in severely depleted individuals will show signs of decreased myocardial function, including increased pulmonary artery pressures (PAP), decreased cardiac output (CO), and decreased BP with decreased response to pressor agents.

Respiratory effects

Hypoxia increases the respiratory rate, causing metabolic alkalosis secondary to hyperventilation. Metabolic alkalosis, in turn, will cause phosphorus to move intracellularly, aggravating the existing hypophosphatemia. If severe hypophosphatemia is not corrected, weakness of the respiratory muscles may result in shallow, ineffective respirations and eventual respiratory failure.

Hematologic effects

Hypophosphatemia alters the structure and function of the red blood cells (RBCs), contributing to the development of hemolytic anemia. The function of white blood cells is also affected, resulting in increased susceptibility to infection. Bruising and bleeding may be present secondary to platelet dysfunction and destruction. Treatment of GI bleeding with banked blood (deficient in 2,3-DPG) and antacids (bind phosphorus in the gut) may aggravate existing hypophosphatemia.

Gastrointestinal (GI) effects

Reduced GI motility occurs in hypophosphatemia, causing dysphagia, gastric atony, and intestinal ileus. Anorexia, nausea, and vomiting also may be present. Hepatic dysfunction occurs because of decreased hepatic oxygen extraction. The clinical significance of this increases for chronic alcohol abusers with associated liver disease, a population especially prone to the development of hypophosphatemia.

Diagnostic Studies

The serum phosphorus will be <2.5 mg/dl (1.7 mEq/L). A phosphorus level between 1.0 and 2.5 mg/dl is considered mild to moderate, while a level of <1.0 mg/dl is considered severe hypophosphatemia. PTH level will be elevated when hypophosphatemia is the result of hyperparathyroidism. The serum magnesium or potassium levels may be decreased since these imbalances increase urinary excretion of phosphorus. Hypomagnesemia, as well as hypocalcemia, may develop secondary to hypophosphatemia, since hypophosphatemia increases urinary excretion of calcium and magnesium. Creatine phosphokinase (CPK) is released when muscle cells are destroyed, thus the CPK level will be increased in the presence of rhabdomyolysis. In chronic hypophosphatemia, x-ray studies may reveal the skeletal changes of osteomalacia.

Foods High in Phosphorus

Meats, especially organ meats (e.g., brain, liver, kidney)
Fish
Poultry
Milk and milk products (e.g., cheese, ice cream, cottage cheese)
Whole grains (e.g., oatmeal, bran, barley)
Seeds (e.g., pumpkin, sesame, sunflower)
Nuts (e.g., Brazil nuts, peanuts)
Eggs and egg products (e.g., eggnog, souffles)
Dried beans and peas

Medical Management

Prevention

Prevention is considered the best treatment for hypophosphatemia. Prevention includes avoiding use of phosphorus-binding antacids (aluminum, magnesium, or calcium gels) in individuals at risk for developing hypophosphatemia (e.g., the chronic alcoholic) and ensuring that TPN solutions contain adequate phosphorus to meet the needs of the individual who is anabolic. Hypophosphatemia that develops secondary to a temporary intracellular shift of phosphorus and that is not accompanied by increased loss or preexisting deficit, usually does not require treatment.

Increasing phosphorus intake

Mild hypophosphatemia may be treated by increasing intake of high-phosphorus foods (see accompanying box). Low-fat or skim milk are most often recommended. Eight ounces of skim milk provide approximately 250 mg of phosphorus, the equivalent of 1 Neutra-Phos-K capsule. Mild to moderate hypophosphatemia usually can be treated with oral phosphate supplements, such as sodium and potassium phosphate (Neutra-Phos) or sodium phosphate (Phospho-Soda). A phosphorus level below 1.0 mg/dl is considered serious and requires treatment with IV sodium phosphate or potassium phosphate. Use of IV supplements is also indicated when the GI tract is nonfunctional.

The need for phosphorus supplementation in treating DKA remains controversial (Fisher and Kitabchi, 1983). Remember that initially phosphorus is lost in the urine secondary to glycosuria, but this deficit may be masked by the movement of phosphorus out of the cells with increased tissue catabolism. As the diabetes is treated with insulin and fluids, there is a dramatic intracellular shift. Authorities disagree about whether there is a significant clinical benefit in routinely treating DKA with IV potassium phosphate. In any case, careful nursing assessment for the clinical indicators of hypophosphatemia is indicated in all individuals with DKA, especially those already at risk for hypoxia.

Nursing Diagnoses and Interventions

Potential for injury related to sensory or neuromuscular dysfunction secondary to hypophosphatemia-induced CNS disturbances

Desired outcome: Patient does not exhibit evidence of injury due to altered sensorium or neuromuscular dysfunction.

1. Monitor serum phosphorus levels in patients at increased risk. Notify MD of decreased levels.
2. Apprehension, confusion, and paresthesias suggest hypophosphatemia. Assess and document LOC, orientation, and neurologic status with each VS check. Reorient patient as necessary. Alert MD to significant changes.
3. Inform patient and significant others that altered sensorium is temporary and will improve with treatment.
4. Do not administer IV phosphate at a rate greater than that recommended by the manufacturer. Potential complications of IV phosphorus administration include *tetany,* owing to hypocalcemia (serum calcium levels may drop suddenly if serum phosphorus levels increase suddenly (see Chapter 10 for additional information); *soft tissue calcification* (if the patient develops hyperphosphatemia, the calcium and phosphorus in the ECF may combine and form deposits in tissue—see the discussion on hypercalcemia in Chapter 10); and *hypotension,* caused by a too rapid a delivery. When IV phosphorus is administered as potassium phosphate, the infusion rate should not exceed 10 mEq/hr. Monitor the IV site for signs of infiltration, as potassium phosphate can cause necrosis and sloughing of tissue (see Chapter 9 for precautions when administering IV potassium).
5. Monitor urine output of individuals receiving phosphorus supplements, especially IV supplements. Oliguria or anuria greatly increases the risk of hyperphosphatemia, volume overload (secondary to high sodium content of sodium phosphate), hyperkalemia (secondary to potassium load with potassium phosphate administration), and calcium imbalances.
6. Give patient and significant others verbal and, if medication is to be taken at home, written instructions for oral phosphate supplements, including drug name, purpose, dosage, frequency, precautions, and potential side effects. Complications of oral phosphate therapy include diarrhea and metabolic acidosis. Gradually increasing the dosage reduces the risk of diarrhea. Although Neutra-Phos involves less risk of acidosis than Phospho-Soda, both preparations have an acidic pH and should be given with caution to individuals at risk for metabolic acidosis.
7. Keep the side rails up and the bed in its lowest position, with wheels locked.
8. Use reality therapy; use clocks, calendars, and familiar objects. Keep these articles at the bedside, within patient's visual field. Have individual wear glasses or hearing aid if needed.
9. If patient is at risk for seizures, pad the side rails and keep an airway at the bedside.

Impaired gas exchange related to altered oxygen supply to the tissues secondary to decreased strength of respiratory muscles, decreased cardiac function, and decreased 2,3-DPG.

NOTE: With decreased 2,3-DPG levels, the oxyhemoglobin dissociation curve will shift to the right. That is, at a given Pao_2 level, more oxygen will be bound to hemoglobin and less will be available to the tissues (see Figure 14-1).

Desired outcome: Patient exhibits adequate gas exchange as evidenced by RR 12 to 20 breaths/min with normal depth and pattern (eupnea); normal skin color; absence of chest pain; and orientation to person, place, and time.

1. Monitor rate and depth of respirations in patients who are severely hypophosphatemic. Alert MD to changes.
2. Assess patient for signs of hypoxia: restlessness, confusion, increased RR, complaints of chest pain, and cyanosis (a late sign).
3. Caution individuals at risk for hypophosphatemia to use phosphorus-binding antacids sparingly or only as prescribed by physician.
4. Also see interventions with **Potential for decreased cardiac output,** below.

Potential for disuse syndrome related to risk of osteomalacia with bone pain and fractures owing to movement of phosphorus out of the bone secondary to chronic hypophosphatemia; or muscle weakness and acute rhabdomyolysis (breakdown of striated muscle) secondary to severe hypophosphatemia

Desired outcome: Patient exhibits ability to move purposefully without evidence of weakness, pain, or fractures.

1. Monitor all patients with suspected hypophosphatemia for evidence of decreasing muscle strength. Perform serial assessments of hand grasp strength and clarity of speech. Alert MD to changes.
2. Monitor serum phosphorus levels for evidence of increasing hypophosphatemia. Alert MD to changes.
3. Assist the patient with ambulation and ADL. Keep personal items within easy reach.
4. Encourage the intake of foods high in phosphorus. See the box on p. 217. Give patient and significant others verbal and written instructions for foods that are high in phosphorus content.
5. Alert individuals at risk for chronic hypophosphatemia to the need for notifying MD of the presence of bone pain. Medicate for pain as prescribed.

Potential for decreased cardiac output related to negative inotropic changes associated with reduced myocardial functioning secondary to severe phosphorus depletion

Desired outcomes: Patient's cardiac output is adequate as evidenced by CVP <6 mm Hg (<12 cm H_20), HR $\leq$ 100 bpm, BP within patient's normal range, RR 12 to 20 breaths/min with normal depth and pattern (eupnea), and absence of adventitious breath sounds. Critical care patients exhibit PAP 20 to 30/8 to 15 mm Hg.

1. Monitor patient for signs of heart failure or pulmonary edema: crackles (rales), rhonchi, shortness of breath, decreased BP, increased HR, increased PAP, or increased CVP.
2. Prevent patient from hyperventilating because respiratory alkalosis will cause increased movement of phosphorus into the cells.

Potential for infection related to impaired function of WBCs secondary to reduced ATP

Desired outcome: Patient is infection free, as evidenced by afebrile state and absence of erythema, swelling, warmth, and purulent drainage at invasive sites.

1. Monitor temperature and secretions q4h for evidence of infection. Culture suspicious secretions.
2. Use meticulous, aseptic technique when changing dressings or manipulating indwelling lines (e.g., TPN catheters, IV needles).
3. Provide oral hygiene and skin care at regular intervals. Intact skin and membranes are the body's first line of defense against infection.

HYPERPHOSPHATEMIA

Hyperphosphatemia occurs most often in the presence of renal insufficiency or failure due to the kidneys' decreased ability to excrete excess phosphorus. In addition to renal failure, other causes of hyperphosphatemia include increased intake of phosphates, extracellular shifts (i.e., movement of phosphorus out of the cells and into the ECF), cellular destruction with concomitant release of intracellular phosphorus, and decreased urinary losses unrelated to decreased renal function.

When the serum phosphorus level increases, there is increased phosphorus and calcium binding, which may result in hypocalcemia (decreased ionized serum calcium level). Hypocalcemia is most likely to occur in sudden, severe hyperphosphatemia (e.g., after IV administration of phosphates) or when the patient is already prone to hypocalcemia (e.g., with chronic renal failure). As the ionized calcium level drops, the parathyroid gland responds by increasing the release of PTH. PTH in turn increases the release of calcium and phosphorus from the bone and increases the absorption of calcium and phosphorus from the GI tract. An increased calcium level combined with an increased phosphorus level leads to further calcium-phosphorus binding, with precipitation of calcium phosphate in the soft tissue, joints, and arteries (metastatic calcification). Thus the primary complications of hyperphosphatemia are hypocalcemia and metastatic calcification. **Metastatic calcifications** usually occur when the calcium and phosphorus product (calcium × phosphorus) exceeds 70 mg/dl.

History and Risk Factors

Renal failure

The primary risk factor for the development of hyperphosphatemia is renal failure, both acute and chronic. The phosphorus level begins to increase with advanced renal insufficiency (glomerular filtration rate less than 25 ml/min) and may be quite marked in renal failure. For the individual with chronic renal failure, chronic hyperphosphatemia plays a central role in the development of renal osteodystrophy.

Increased phosphorus intake

Excessive administration of phosphorus supplements, vitamin D excess with increased GI absorption, or excessive use of phosphorus-containing laxatives (e.g., Fleet Phospho-Soda) or enemas (e.g., Fleet enema) may result in hyperphosphatemia. This is most likely to occur in children, individuals with decreased renal function, or conditions that limit the kidneys' ability to excrete phosphorus. The healthy kidney is able to compensate for most phosphorus loads by increasing renal excretion of phosphorus. The presence of renal disease or factors such as hypoparathyroidism, milk-alkali syndrome, or volume depletion, limit the kidneys' ability to compensate.

Extracellular shift/cellular release

Phosphorus may be added to the ECF from cellular stores due to extracellular shifts or cellular destruction with release of intracellular phosphorus. Respiratory acidosis may cause phosphorus to shift out of the cell. Decreased utilization of phosphorus within the cells is the most likely cause of such a shift. Conditions that result in cellular destruction with the release of intracellular phosphorus include neoplastic disease (e.g., leukemia and lymphoma), especially when treated with cytotoxic agents, increased tissue catabolism (breakdown), rhabdomyolysis (breakdown of striated muscle), and acute hemolytic anemia. Hyperphosphatemia may occur in DKA prior to treatment due to increased tissue catabolism.

Hyperthyroidism

Hyperthyroidism, especially thyrotoxicosis, is often associated with an elevated phosphorus level. Although the exact mechanism is unclear, it is believed that excess levels of thyroid hormone reduce renal excretion of phosphorus (Kokko and Tannen, 1986).

Assessment

Typically, individuals experience few symptoms with hyperphosphatemia. The majority of symptoms that do occur relate to the development of hypocalcemia or soft tissue (metastatic) calcifications. The most common sites of calcium phosphate deposition in the individual with reduced renal function are the corneas, lungs, kidneys, gastric mucosa, and the small- to medium-sized blood vessels. Indicators of metastatic calcification include oliguria, corneal haziness, conjunctivitis, and papular eruptions. Deposition of calcium phosphate in the heart may lead to dysrhythmias and conduction disturbances. Signs and symptoms associated with hypocalcemia include

Table 11-2 Quick assessment guidelines for individuals with hyperphosphatemia

Phosphorus > 4.5 mg/dl (2.6 mEq/L)

Signs and Symptoms

Usually, patients experience few symptoms with hyperphosphatemia. The majority of symptoms that do occur relate to the development of hypocalcemia (muscle cramps, hyperreflexia, tetany) or soft tissue (metastatic) calcifications. Indicators of metastatic calcification include oliguria, corneal haziness, conjunctivitis, irregular heart rate, and papular eruptions

Physical Assessment

See the section on hypocalcemia in Chapter 10. In addition see the previous section, *Signs and Symptoms,* for indicators of metastatic calcifications

EKG Changes

See the section on hypocalcemia in Chapter 10. Deposition of calcium phosphate in the heart may lead to dysrhythmias and conduction disturbances

History and Risk Factors

- *Renal failure:* Acute and chronic
- *Increased intake:* Excessive administration of phosphorus supplements; vitamin D excess with increased GI absorption; excessive use of phosphorus-containing laxatives or enemas
- *Extracellular shift:* Respiratory acidosis; diabetic ketoacidosis (prior to treatment)
- *Cellular destruction:* Neoplastic disease (e.g., leukemia and lymphoma) treated with cytotoxic agents; increased tissue catabolism (breakdown); rhabdomyolysis (breakdown of striated muscle)
- *Decreased urinary losses:* Hypoparathyroidism; volume depletion
- *Hyperthyroidism*

numbness and tingling of fingers and circumoral region, muscle cramps, hyperreflexia, and tetany. See Table 11-2 for quick assessment guidelines for individuals with hyperphosphatemia.

Diagnostic Studies

The serum phosphorus level will be >4.5 mg/dl (2.6 mEq/L). Levels of 8 to 9 mg/dl are not uncommon in untreated hyperphosphatemia of chronic renal failure. The creatinine level may be measured when reduced renal function is suspected. Hyperphosphatemia in the presence of normal renal function suggests the presence of hypoparathyroidism, hyperthyroidism, lymphoma, or administration of phosphorus-containing medications. The PTH level will be decreased if hyperphosphatemia is the result of hypoparathyroidism. X-ray studies may reveal skeletal changes of osteodystrophy in the individual with chronic phosphorus excess.

Medical Management

Although identification and elimination of the cause is obviously the best approach to hyperphosphatemia, it is not always feasible. Elimination of the cause is not practical, for example, for the hyperphosphatemia associated with chronic renal failure. However, contributing factors, such as increased phosphorus intake, may be eliminated.

The most common treatment of hyperphosphatemia is the use of aluminum, magnesium, or calcium gels or antacids, which bind phosphorus in the gut, thereby increasing GI elimination of phosphorus. Magnesium antacids are avoided in renal failure due to the risk of hypermagnesemia. Reduced dietary intake of phosphorus may be combined with administration of phosphorus-binding antacids in the treatment of hyperphosphatemia that is associated with renal insufficiency or renal failure. Dialysis therapy required for the management of end-stage renal disease also will help to reduce serum phosphorus levels. For adequate control, however, the individual on chronic hemodialysis usually requires a combined therapy, including reduced phosphorus intake and phosphorus-binding antacids. (See the box on p. 217 for a list of foods that are high in phosphorus.) Serum phosphorus levels may be allowed to remain slightly elevated (4.5 to 6.0 mg/dl) in chronic renal failure to ensure adequate levels of 2,3-DPG. This helps to minimize the effects of chronic anemia on oxygen delivery to the tissues. Dialytic therapy may be necessary for acute, severe hyperphosphatemia that is accompanied by symptomatic hypocalcemia, regardless of cause.

The management of hyperphosphatemia associated with neoplastic disease includes treatment with allopurinol and maintenance of an adequate fluid volume prior to the administration of chemotherapy. The combination of rapid cell turnover and chemotherapy-induced cell lysis with chemotherapy releases phosphorus, potassium, and uric acid. Allopurinol and fluids help to prevent the development of uric acid nephropathy that may cause reduced excretion of phosphorus and potassium.

Nursing Diagnoses and Interventions

Knowledge deficit: Purpose of phosphate binders and the importance of reducing gastrointestinal absorption of phosphorus to control hyperphosphatemia and prevent long-term complications

Desired outcome: Patient describes the potential complications of uncontrolled hyperphosphatemia and ways in which they can be prevented.

NOTE: Because symptoms of hyperphosphatemia may be minimal, the prevention of long-term complications relies primarily on adequate patient and family education.

1. Teach patients and significant others the purpose of phosphate binders. Stress the need to take binders as prescribed with or after meals to maximize effectiveness.
2. Prepare patients for the possibility of constipation secondary to binder use. Encourage use of bulk-building supplements or stool softeners if constipation occurs. Phosphate-containing laxatives and enemas must be avoided.
3. Phosphate binders are available in liquid or capsule form. Confer with MD regarding an alternate form or brand for individuals who find a particular medication unpalatable or difficult to take. Phosphate binders vary in their aluminum, magnesium, or calcium content, however, and one may not be exchanged for another without first ensuring that the individual is receiving the same amount of elemental aluminum, magnesium, or calcium. Review with patient and significant others the medication name, purpose, dosage, frequency, precautions, and potential side effects.
4. Encourage patient to avoid or limit foods high in phosphorus (see the box on p. 217). Provide a list of foods that are high in phosphorus.
5. Review the importance of avoiding over-the-counter medications that contain phosphorus: certain laxatives, enemas, multivitamins, and mixed vitamin-mineral supplements. Instruct individual and significant others to read the label for the words *phosphorus* and *phosphate.*

Potential for injury related to internal factors associated with precipitation of calcium phosphate in the soft tissue (e.g., corneas, lungs, kidneys, gastric mucosa, heart, blood vessels) and periarticular region of the large joints (e.g., hips, shoulders, and elbows) or development of hypocalcemic tetany

Desired outcomes: Patient exhibits no evidence of metastatic calcification or hypocalcemia as evidenced by urinary output $\geq$ 30 ml/hr, clear corneas and conjunctiva, baseline vision, regular HR, clear and nonerupted skin, normal deep tendon reflexes, and absence of numbness, tingling, and muscle cramps. The calcium-phosphorus product (calcium $\times$ phosphorus) remains $<$70 mg/dl.

1. Monitor serum phosphorus and calcium levels. Alert MD to abnormal values. Remember that phosphorus values may be kept slightly higher (4 to 6 mg/dl) in chronic renal failure patients to ensure adequate levels of 2,3-DPG, thereby minimizing effects of chronic anemia on oxygen delivery to the tissues.
2. Avoid vitamin D products (see Table 10-2) and calcium supplements until the serum phosphorus level approaches normal.
3. Alert MD to indicators of metastatic calcification: oliguria, corneal haziness, conjunctivitis, irregular heart rate, and papular eruptions.

4. Monitor patient for evidence of increasing hypocalcemia: numbness and tingling of the fingers and circumoral region, hyperreactive reflexes, and muscle cramps. Notify MD promptly if these symptoms develop because they occur prior to overt tetany. In addition, alert MD if patient has a positive Trousseau's or Chovstek's signs (see Figure 10-1, *A*, *B*, and *C*) as they signal latent tetany (see Chapter 4 for a discussion of these signs and Chapter 10 for additional information regarding treatment and prevention of hypocalcemia). Provide verbal and written instructions for indicators of hyperphosphatemia and hypocalcemia. Review the symptoms that require immediate medical attention: muscle cramps and numbness and tingling of fingers and circumoral region. Alert patients with chronic hyperphosphatemia to the necessity of notifying MD if symptoms of metastatic calcification occur.
5. Because hyperphosphatemia can impair renal function due to precipitation of calcium phosphate in the renal tubule, monitor patient's renal function carefully: urine output, BUN, and creatinine.

Case Study 11-1

Case study 3-1 describes M.B., a 20-year-old female with diabetes mellitus. She had been admitted to the medical intensive care unit with a diagnosis of DKA. She exhibits the signs of fluid volume deficit: decreased blood pressure, increased heart rate, poor skin turgor, and decreasing urine output. Arterial blood gases drawn on admission reveal marked acidemia: pH 6.97 (normal pH is 7.35 to 7.45). An admission chemistry panel shows the following:

Test Values	Normal Values
Serum sodium 142 mEq/L	137-147 mEq/L
Serum chloride 99 mEq/L	95-108 mEq/L
Serum potassium 5.6 mEq/L	3.5-5.0 mEq/L
Serum CO_2 content 4.5 mEq/L	22-28 mEq/L
BUN 30 mg/dl	6-20 mg/dl
Creatinine 3.5 mg/dl	0.6-1.5 mg/dl
Serum phosphorus 8.6 mg/dl	2.5-4.5 mg/dl
Serum calcium 10.3 mg/dl	8.5-10.5 mg/dl

1. Question: What are the primary intracellular ions, and how do their intracellular concentrations compare with their extracellular concentrations? (If necessary review Table 1-4.)

Answer: The primary intracellular ions are potassium and phosphorus. The intracellular concentration of potassium (150 mEq/L) is approximately 33 times greater than that of the plasma; the intracellular concentration of phosphorus (40 mEq/L) is approximately 20 times greater than the plasma.

2. **Question:** The individual with DKA is catabolic, that is, experiencing increased tissue breakdown. What would you expect to happen to the serum levels for potassium and phosphorus?

   Answer: They would be expected to increase due to the release of intracellular stores secondary to increased tissue catabolism. The extracellular shift of potassium would be further accentuated by the presence of acidosis and the lack of insulin. Decreased renal perfusion or function, if present, also would tend to increase the ECF level of these two ions secondary to decreased excretion.

3. **Question:** Does M.B. have laboratory evidence of decreased renal perfusion or function?

   Answer: Yes. Both the BUN and creatinine are increased, indicating decreased renal perfusion or function.

As anticipated, both M.B.'s serum phosphorus and potassium levels are increased, yet it is likely that she has an overall potassium and phosphorus deficit. The glucose-induced diuresis that occurs with uncontrolled diabetes increases the urinary excretion of both potassium and phosphorus and may result in significant loss of these two important electrolytes. Currently, M.B.'s urine output is decreased, but initially she experienced a profound diuresis. As described in Chapter 3, M.B. will now be treated with insulin and fluids, and, due to the severity of her acidosis, IV sodium bicarbonate.

4. **Question:** What would be expected to happen to the ECF phosphorus and potassium levels with treatment and increased tissue anabolism (building and repair)?

   Answer: The ECF potassium and phosphorus levels will decrease. Insulin will facilitate the movement of both potassium and phosphorus back into the cells. Correction of M.B.'s acidosis will also result in an intracellular shift of potassium. Restoration of normal fluid volume should improve renal function and increase the urinary excretion of both electrolytes. If renal function is normal, potassium supplementation is usually initiated with insulin therapy. As discussed, routine phosphorus supplementation remains controversial, although it is recommended by many sources. As with potassium, phosphorus supplementation is usually not initiated until there is evidence of adequate renal function.

5. **Question:** Which medical therapy will correct both deficits?

   Answer: Administration of potassium phosphate (either IV or PO).

6. **Question:** What are some potential nursing diagnoses for M.B. related to her potential hypophosphatemia?

Answer:

a. **Potential for injury/trauma** related to altered LOC, incoordination, and altered sensations secondary to hypophosphatemia-induced CNS disturbances
b. **Impaired gas exchange** related to diminished oxygen delivery to the tissues secondary to decreased 2,3-DPG

NOTE: M.B. is already at risk for decreased tissue oxygenation due to increased incidence of vascular disease in individuals with diabetes.

References

Brater DC: Serum electrolyte abnormalities caused by drugs, Prog Drug Res 30:9-69, 1986.

Calloway C: When the problem involves magnesium, calcium, or phosphate, RN 50:30-35, May 1987.

Fisher JN and Kitabchi AE: A randomized study of phosphate therapy in the treatment of diabetic ketoacidosis, J Clin Endocrinol Metab 57(1):177-180, 1983.

Goldberger E: A primer of water, electrolyte, and acid-base syndromes, ed 7, Philadelphia, 1986, Lea & Febiger.

Horne M: Fluid and electrolyte disturbances. In Swearingen PL and others: Manual of critical care: applying nursing diagnoses to adult critical illness, St Louis, 1988, The CV Mosby Co.

Horne M and Swearingen PL: Pocket guide to fluids and electrolytes, St Louis, 1989, The CV Mosby Co.

Kokko JP and Tannen RL: Fluids and electrolytes, Philadelphia, 1986, WB Saunders Co.

Malluche HH and Faugere M: Renal osteodystrophy, Mediguide to Nephrology, 1(1), 1-8, 1986.

Maxwell MH and others: Clinical disorders of fluid and electrolyte metabolism, ed 4, New York, 1987, McGraw-Hill Book Co.

Metheny NM: Fluid and electrolyte balance: nursing considerations, Philadelphia, 1987, JB Lippincott Co.

Rice V: Magnesium, calcium, and phosphate imbalances: their clinical significance, Crit Care Nurs 3:90-112, May/June 1983.

Chapter

12

Disorders of Magnesium Balance

Magnesium is the body's fourth most abundant cation, yet its measurement and evaluation are often overlooked. Of the body's magnesium, approximately 50% to 60% is located in bone and approximately 1% is located in the extracellular fluid (ECF). The remaining magnesium is contained within the cells, thereby constituting the second most abundant intracellular cation after potassium. Magnesium is regulated by a combination of factors, including vitamin D-regulated gastrointestinal absorption and renal excretion. Normally only about 30% to 40% of dietary magnesium is absorbed. The major site of GI absorption of magnesium is the distal small bowel. Renal excretion of magnesium changes to maintain magnesium balance and is affected by sodium and calcium excretion, ECF volume, and the presence of parathyroid hormone (PTH). Renal excretion of magnesium is decreased with increased PTH, decreased excretion of sodium or calcium, and fluid volume deficit.

Because magnesium is a major intracellular ion, it is vital to normal cellular function. Specifically, magnesium activates enzymes involved in the metabolism of carbohydrates and protein, is essential to the storage and utilization of intracellular energy, and triggers the sodium-potassium pump, thus affecting intracellular potassium levels. Magnesium is also important in the transmission of neuromuscular activity, neural transmission within the central nervous system (CNS), intracellular calcium levels, and myocardial functioning. Magnesium exerts a sedative effect on nerves and muscles by reducing the release of the neurotransmitter acetylcholine.

Normal serum magnesium level is 1.5 to 2.5 mEq/L. As with the other important intracellular ions, plasma magnesium levels do not necessarily reflect total body stores. Intracellular shifts of magnesium may occur with administration of carbohydrate loads, with refeeding after starvation, and during periods of increased stress. Approximately one third of the plasma magnesium is bound to protein, a small portion is combined with other substances (complexed), and the remaining portion is free or ionized. It is the free, ionized magnesium that is physiologically important.

As with calcium levels, magnesium levels should be evaluated in combination with serum albumin levels. Low serum albumin levels will decrease the total magnesium level, while the amount of free, ionized magnesium may remain unchanged.

HYPOMAGNESEMIA

Hypomagnesemia (serum magnesium level < 1.5 mEq/L) usually occurs because of decreased gastrointestinal (GI) absorption or increased urinary loss. It also may occur with excessive GI loss (e.g., vomiting and diarrhea) or with prolonged administration of magnesium-free parenteral fluids. Individuals with chronic alcohol ingestion (see the next section, *History and Risk Factors*) and critical care patients are the two most common patient populations with this disorder. Typically hypomagnesemia is associated with hypocalcemia and hypokalemia (see *Diagnostic Studies,* p. 231, for additional information). One study has estimated that 42% of all cases of hypokalemia are also associated with hypomagnesemia (Nutrition Reviews, 1985).

History and Risk Factors

Decreased GI absorption

Decreased absorption of magnesium from the GI tract may occur with such conditions as colitis, malabsorption syndrome, pancreatic insufficiency with steatorrhea, surgical resection of the small bowel, and cancer involving the GI tract. Medications, such as cholestyramine and neomycin, may reduce the GI absorption of magnesium, as may laxatives and purgatives.

Increased urinary losses

Drugs that enhance urinary excretion of magnesium include diuretics, amphotericin, tobramycin, gentamicin, cisplatin, and digoxin. Renal magnesium wasting may be present for months after an individual has been treated with cisplatin. Although any medication or condition that induces diuresis may increase urinary losses of magnesium, the loop diuretics are especially likely to cause hypomagnesemia because of their site of action. Recall that the loop diuretics (furosemide and ethacrynic acid) inhibit tubular reabsorption in the ascending loop of Henle. This is the major site of magnesium reabsorption within the renal tubule.

Hyperaldosteronism, both primary and secondary, will increase magnesium excretion due to volume expansion. Diabetic ketoacidosis (DKA) may cause hypomagnesemia owing to catabolic release of intracellular magnesium, combined with increased loss in the urine secondary to osmotic diuresis. Renal tubular abnormalities associated with the diuretic phase of acute tubular necrosis (the most common cause of acute renal failure), postobstructive diuresis, and the diuresis that can follow renal transplantation may result in increased renal magnesium wasting and clinically significant hypomagnesemia.

Increased GI losses

Upper GI fluids contain relatively little magnesium (1 to 2 mEq/L) so that only prolonged, large volume losses will result in the development of hypomagnesemia. Vomiting or NG suction are most likely to result in hypomagnesemia when combined with intake of magnesium-free or magnesium-poor parenteral fluids. In contrast, the lower GI fluid has a relatively high magnesium concentration (10 to 14 mEq/L). Chronic diarrhea, high ileostomies, and intestinal fistulas may result in significant hypomagnesemia.

Decreased intake

Administration of low-magnesium or magnesium-free parenteral solutions may result in hypomagnesemia, especially in individuals maintained solely on parenteral fluids for long periods of time. Dietary magnesium deficiency occurs in children with chronic protein-calorie malnutrition or in any individual suffering from starvation. As with the other major intracellular electrolytes (potassium and phosphorus), magnesium depletion may develop with refeeding after starvation. During starvation tissue catabolism increases as body protein is metabolized to provide needed energy. As cells are broken down, intracellular ions are released and excreted in the urine. With refeeding there is an intracellular shift of these ions as new cells are formed. Thus individuals who have suffered a period of starvation will require supplementation of potassium, phosphorus, and magnesium during refeeding.

Intracellular shifts

Increased cellular uptake of glucose or amino acids will cause a shift of magnesium into the cells. Thus administration of concentrated glucose solutions, amino acid solutions, or insulin may contribute to the development of hypomagnesemia. This potentiates the drop in plasma magnesium that typically occurs with refeeding after starvation.

Chronic alcoholism

The most common cause of hypomagnesemia is chronic alcoholism. Hypomagnesemia develops due to a combination of factors, including poor dietary intake, decreased GI absorption, and increased urinary excretion secondary to ethanol effect. The presence of diarrhea or the use of diuretics also may contribute to the development of hypomagnesemia.

Assessment

Hypomagnesemia contributes to the development of hypokalemia, hypocalcemia, and hypophosphatemia; hence the signs and symptoms of hypomagnesemia also reflect these accompanying disturbances. The serum potassium level decreases due to the failure of the cellular sodium-potassium pump to move potassium into the cell and the accompanying loss of potassium in the urine. This hypokalemia may be resistant to potassium replacement until the magnesium deficit has been corrected. Hypomagnesemia may lead to hypocalcemia due to a reduction in the release and action of PTH. Recall that PTH is the primary regulator of serum calcium levels. As with hypokalemia, hypocalcemia associated with hypomagnesemia may be refractory to calcium therapy alone, yet may improve with magnesium therapy alone. Hypomagnesemia increases renal excretion of phosphorus. See Table 12-1 for quick assessment guidelines for individuals with hypomagnesemia.

Central nervous system (CNS)/neuromuscular effects

Since magnesium normally decreases the transmission of nerve impulses, hypomagnesemia will result in symptoms associated with increased nerve transmission, such as mood changes, hallucinations, confusion, agitation, and paresthesias. Changes noted on physical assessment may include hyperreactive reflexes, tremors, convulsions, tetany, and positive Chvostek's and Trousseau's signs (see Chapter 10, p. 196). Muscle weakness also may develop, including weakness of the respiratory muscles. Symptoms of hypomagnesemia usually do not develop until the serum magnesium level drops below 1 mEq/L.

Gastrointestinal (GI) effects

The GI effects of hypomagnesemia are most likely the result of accompanying hypokalemia. Symptoms include anorexia, nausea, vomiting, and abdominal distention.

Cardiovascular effects

Dysrhythmias may occur with hypomagnesemia, especially if the individual is receiving digoxin. Digoxin increases the excretion of magnesium, and hypomagnesemia increases the uptake of digoxin by the myocardial cells. Both hypomagnesemia and digoxin inhibit the myocardial cell sodium-potassium pump, thus decreasing intracellular potassium. Decreased intracellular potassium results in enhanced digoxin effect and possible digoxin toxicity. This situation is further compounded by the concomitant administration of diuretics, which enhance urinary losses of magnesium. See the section on diagnostic studies for a discussion of related electrocardiogram changes and dysrhythmias.

Diagnostic Studies

Serum magnesium

Serum magnesium level will be less than 1.5 mEq/L. A decreased serum albumin level may cause a decreased magnesium level due to a reduction in protein-bound magnesium, yet the amount of free, ionized magnesium may be unchanged. Like the

Table 12-1 Quick assessment guidelines for individuals with hypomagnesemia

Magnesium < 1.5 mEq/L

Signs and Symptoms

Agitation, mood changes, hallucinations, confusion, anorexia, nausea, vomiting, paresthesias

Physical Assessment

Increased reflexes, tremors, convulsions, tetany, and positive Chvostek's and Trousseau's signs (see Chapter 10) in part owing to accompanying hypocalcemia—the patient also may have tachycardia and hypotension

Hemodynamic Measurements

See the section on hypocalcemia in Chapter 10 and the section on hypokalemia in Chapter 9

History and Risk Factors

- *Chronic alcoholism:* The most common cause of hypomagnesemia due to a combination of poor dietary intake, decreased GI absorption, and increased urinary excretion secondary to ethanol effect
- *Malabsorption:* For example, due to cancer, colitis, pancreatic insufficiency, surgical resection of the GI tract
- *Increased GI losses:* Prolonged vomiting or NG suction, diarrhea, and laxatives
- *Administration of low-magnesium or magnesium-free parenteral solutions*
- *Hyperaldosteronism:* Due to volume expansion
- *Diabetic ketoacidosis:* Owing to movement of magnesium out of the cell and loss in the urine because of osmotic diuresis
- *Drugs that enhance urinary excretion:* For example, diuretics, amphotericin, tobramycin, gentamicin, cisplatin, digoxin
- *Intracellular shifts:* Administration of concentrated glucose solutions, amino acid solutions, and insulin

serum calcium levels, the serum magnesium levels must be assessed in relationship to the serum albumin level. As previously discussed, magnesium depletion contributes to the development of potassium, calcium, and phosphorus depletion and is often seen in combination with these imbalances.

Urinary magnesium

Measurement of the urinary magnesium level is often useful in identifying the etiology of hypomagnesemia. Increased urinary magnesium levels in the presence of decreased serum magnesium suggest increased renal loss as the cause of the deficiency, while decreased urinary levels indicate some extrarenal cause for the hypomagnesemia.

Electrocardiogram (EKG)

EKG evaluations may reflect magnesium as well as calcium and potassium deficiencies. Changes noted on the EKG may include tachyarrhythmias, prolonged PR and QT intervals, widening of the QRS segment, ST segment depression, and flattened T waves. Increased digitalis effect, as evidenced by multifocal or bigeminal PVCs, paroxysmal atrial tachycardia with varying AV block, and Wenckebach (Type I AV) heart block also may occur.

Medical Management

Prevention

Identifying and eliminating factors that predispose the development of hypomagnesemia is the ideal approach to treatment, for example, adequate replacement of magnesium in total parenteral nutrition (TPN) solutions or prophylactic administration of magnesium to individuals receiving cisplatin chemotherapy. Serum magnesium levels should be monitored routinely in individuals at risk for hypomagnesemia. Additionally, any individual with hypomagnesemia also must be evaluated for potassium and calcium deficits.

Magnesium replacement therapy

Therapy is dictated by the severity of the deficit. Increased dietary intake of magnesium may be all that is necessary to treat mild hypomagnesemia (see the accompanying box). Oral magnesium in the form of magnesium-containing antacids may be used to treat mild to moderate deficit (see box on p. 238). Severe or symptomatic hypomagnesemia, as evidenced by dysrhythmias or neurologic changes, requires immediate treatment with IV magnesium sulfate ($MgSO_4$). Once symptoms have resolved, subsequent replacement may be given IM, although this route may be painful (Berkelhammer and Bear, 1985). Since chronic hypomagnesemia is associated with intracellular depletion, adequate replacement therapy may require days.

Foods High in Magnesium

Green, leafy vegetables (e.g., beet greens, collard greens)
Meat
Seafood
Nuts and seeds
Wheat bran
Dairy products
Soy flour
Legumes
Chocolate
Molasses

Nursing Diagnoses and Interventions

Potential for injury related to sensory or neuromuscular dysfunction secondary to hypomagnesemia

Desired outcomes: Patient does not exhibit evidence of injury caused by complications of severe hypomagnesemia. Serum magnesium levels are within normal range (1.5 to 2.5 mEq/L).

1. Monitor serum magnesium levels in persons at risk for developing hypomagnesemia, for example those who are alcoholics or in critical care units. Alert MD to abnormal values.

   NOTE: Symptomatic hypomagnesemia may be mistakenly attributed to the delirium tremens of chronic alcoholism. Be especially alert to indicators of magnesium deficit in these patients.
2. Administer IV $MgSO_4$ with caution, preferably using a pump. Refer to manufacturer's guidelines. Too rapid an administration may lead to dangerous hypermagnesemia with cardiac or respiratory arrest. Patients receiving IV magnesium should be monitored for decreasing BP, labored respirations, and diminished patellar reflex (knee jerk). An absent patellar reflex is a signal of hyporeflexia due to dangerous hypermagnesemia. Should any of these changes occur, stop the infusion and notify MD immediately (see the next section on hypermagnesemia). Advise individuals receiving IV magnesium that flushing and a sensation of heat may develop secondary to peripheral vasodilatation. Keep calcium gluconate at the bedside in case the MD prescribes it for hypocalcemic tetany or sudden hypermagnesemia. Magnesium sulfate administration may potentiate existing hypocalcemia due to sulfate and calcium binding, which reduces the level of ionized calcium.
3. For patients with chronic hypomagnesemia, administer oral magnesium supplements as prescribed.
4. Administer all magnesium supplements with caution to patients with reduced renal function due to an increased risk of the development of hypermagnesemia. Monitor urine output and alert MD to urine output of less than 30 ml/hr.
5. Encourage the intake of foods high in magnesium in appropriate patients (see the box on p. 233). For most patients, a regular diet is usually adequate.
6. Keep symptomatic patients on seizure precautions. Decrease environmental stimuli (e.g., keep the room quiet, use subdued lighting).
7. For patients in whom hypocalcemia is suspected, caution against hyperventilation. Metabolic alkalosis may precipitate tetany due to increased calcium binding.
8. Dysphagia may occur in hypomagnesemia. Test the patient's ability to swallow water prior to giving food or medications.

9. Assess and document LOC, orientation, and neurologic status with each VS check. Reorient patient as necessary. Alert MD to significant changes. Inform patient and significant others that altered mood and sensorium are temporary and will improve with treatment.
10. Alert MD to patients receiving magnesium-free solutions (e.g., TPN) for prolonged periods of time.
11. Refer to Alcoholics Anonymous, Al-Anon, and Alateen as appropriate for the alcoholic and his or her significant others.
12. Provide verbal and written instructions for the indicators of hypomagnesemia and hypocalcemia. Emphasize the symptoms that necessitate immediate medical attention: numbness and tingling of fingers and circumoral region, muscle cramps, altered sensorium, and irregular or rapid pulse. Also review medications, including drug name, purpose, dosage, frequency, precautions, and potential side effects.
13. See the section on hypokalemia in Chapter 9 and the section on hypocalcemia in Chapter 10 for nursing care of these disorders.

    NOTE: Because magnesium is necessary for the movement of potassium into the cells, intracellular potassium deficits cannot be corrected until hypomagnesemia has been treated effectively.

Potential for decreased cardiac output related to electrical alterations associated with tachyarrhythmias or digitalis toxicity secondary to hypomagnesemia

Desired outcome: EKG shows normal configuration, and heart rate is within normal range for the patient.

1. Monitor heart rate and regularity with each VS check. Alert MD to changes.
2. Assess EKG in patients on continuous EKG monitoring.
3. Because hypomagnesemia and hypokalemia potentiate the cardiac effects of digitalis, monitor patients taking digitalis for digitalis-induced dysrhythmias. EKG changes may include multifocal or bigeminal PVCs, paroxysmal atrial tachycardia with varying AV block, and Wenckebach (Type IAV) heart block. Monitor for pulse changes in the nonEKG-monitored setting.

Alteration in nutrition: Less than body requirements of magnesium related to history of poor intake or anorexia, nausea, and vomiting secondary to hypomagnesemia

Desired outcome: Patient eats foods high in magnesium.

1. Encourage intake of small, frequent meals.
2. Teach patient about foods high in magnesium (see the box on p. 233), and encourage intake of these foods during meals.
3. Medicate with antiemetics as prescribed.
4. Include patient, significant others, and dietitian in meal planning as appropriate.
5. Provide oral hygiene before meals to enhance appetite.

HYPERMAGNESEMIA

Hypermagnesemia (serum magnesium levels >2.5 mEq/L) occurs almost exclusively in individuals with renal failure who have an increased intake of magnesium (e.g., use of magnesium-containing medications). It also may occur in obstetric patients whose preeclampsia is treated with parenteral magnesium. In rare cases, hypermagnesemia occurs because of excessive use of magnesium-containing medications (e.g., antacids, laxatives, enemas).

History and Risk Factors

Decreased magnesium excretion

Decreased excretion of magnesium may occur with renal failure or adrenocortical insufficiency. As stated, most cases of hypermagnesemia are associated with decreased renal function, typically chronic renal failure. But even in chronic renal failure, the development of clinically significant hypermagnesemia usually only occurs with use of magnesium-containing medications or parenteral solutions.

Increased magnesium intake

Excessive use of magnesium-containing antacids, enemas, and laxatives or overly rapid administration of magnesium sulfate in the treatment of hypomagnesemia or eclampsia may result in hypermagnesemia.

Assessment

The primary symptoms of hypermagnesemia occur as the result of depressed peripheral and central neuromuscular transmission. Excess magnesium decreases the release of neurotransmitters at the neuromuscular junction, leading to progressive muscle weakness and paralysis as the magnesium level increases. Symptoms usually do not occur until the magnesium level exceeds 4 mEq/L. See Table 12-2 for quick assessment guidelines for individuals with hypermagnesemia.

Central nervous system (CNS)/neuromuscular effects

CNS symptoms include altered mental functioning, drowsiness, and if the magnesium level reaches 12 to 15 mEq/L, coma. Neuromuscular effects include muscular weakness or paralysis and decreased deep tendon reflexes. The patellar reflex (knee jerk) is lost once the magnesium level exceeds 8 mEq/L. Muscle paralysis, including paralysis of the respiratory muscles, may occur with levels greater than 10 mEq/L.

Cardiovascular effects

Decreased arterial pressure caused by peripheral vasodilatation is an early sign of hypermagnesemia. Vasodilatation also may cause flushing, sweating, and the sensation of heat. These symptoms develop when the magnesium level exceeds 3 to 5 mEq/L. When the magnesium level exceeds 5 to 7 mEq/L, bradycardia may develop. Heart blocks occur with levels exceeding 12 mEq/L, and complete heart block leading to cardiac arrest is seen at levels of 15 to 20 mEq/L.

Gastrointestinal (GI) effects

GI symptoms of hypermagnesemia include nausea and vomiting, and, when present, occur fairly early in hypermagnesemia (3 to 5 mEq/L). The sensation of thirst also may be present.

Diagnostic Studies

The serum magnesium level will be >2.5 mEq/L. EKG findings associated with hypermagnesemia include prolonged PR interval, widened QRS segment, and atrioventricular blocks.

Table 12-2 Quick assessment guidelines for individuals with hypermagnesemia

Serum magnesium > 2.5 mEq/L

Signs and Symptoms
Nausea, vomiting, flushing, diaphoresis, sensation of heat, altered mental functioning, drowsiness, and coma

Physical Assessment
Hypotension, bradycardia, decreased deep tendon reflexes, and muscular weakness or paralysis. The patellar reflex (knee jerk) is lost once the magnesium level exceeds 8 mEq/L. Paralysis of the respiratory muscles may occur when the magnesium level exceeds 10 mEq/L

Hemodynamic Measurements
Decreased arterial pressure due to peripheral vasodilatation

History and Risk Factors
- *Decreased excretion of magnesium:* Renal failure or adrenocortical insufficiency
- *Increased intake of magnesium:* Excessive use of magnesium-containing antacids, enemas, and laxatives or excessive administration of magnesium sulfate (e.g., in the treatment of hypomagnesemia or preeclampsia)

Medical Management

Identification and elimination of the cause

The most common treatment approach to hypermagnesemia is eliminating the cause (e.g., discontinuing or avoiding use of magnesium-containing medications or supplements in individuals with decreased renal function). A complete nursing history is essential in identifying hidden sources of magnesium, such as in combined vitamin-mineral supplements or laxatives. See the accompanying boxes for a list of medications that contain magnesium.

Miscellaneous treatments

Moderate hypermagnesemia may be treated with diuretics and 0.45% sodium chloride solution to enhance magnesium excretion in persons with adequate renal function. Intravenous calcium gluconate is used to antagonize the neuromuscular effects of magnesium for patients with potentially lethal hypermagnesemia. Typically 10 ml of a 10% solution is administered. Dialysis using a magnesium-free dialysate may be necessary for the individual with decreased renal function who is symptomatic.

Magnesium-Containing Medications

Antacids

Aludrox
Camalox
Di-Gel
Gaviscon
Gelusil and Gelusil-II
Maalox and Maalox Plus
Mylanta and Mylanta-II
Riopan
Simeco
Tempo

Laxatives

Magnesium hydroxide (Phillips' Milk of Magnesia, Haley's M-O)
Magnesium citrate
Magnesium sulfate (Epsom salts)

Nursing Diagnoses and Interventions

Potential for injury related to sensory or neuromuscular dysfunction secondary to hypermagnesemia

Desired outcomes: Patient does not exhibit evidence of injury secondary to complications of hypermagnesemia. Serum magnesium levels are within normal range (1.5 to 2.5 mEq/L).

1. Monitor serum magnesium levels in patients at risk for developing hypermagnesemia, for example, those with chronic renal failure.
2. Assess and document LOC, orientation, and neurologic status (e.g., hand grasp) with each VS check. Assess patellar reflex (knee jerk) in patients with a moderately elevated magnesium level (>5 mEq/L). With patient lying flat, support the knee in a moderately fixed position and tap the patellar tendon firmly just below the patella. Normally the knee will extend. An absent reflex suggests a magnesium level of ≥ 7 to 8 mEq/L. Alert MD to changes. Monitor for evidence of respiratory muscle weakness or paralysis (e.g., alterations in respiratory pattern) in individuals with severe hypermagnesemia.
3. Reassure patient and significant others that altered mental functioning and muscle strength will improve with treatment.
4. Keep siderails up and the bed in its lowest position with the wheels locked.
5. Keep a manual resuscitator at the bedside in the event of respiratory depression or paralysis.

Knowledge deficit: Importance of avoiding excessive or inappropriate use of magnesium-containing medications, especially for individuals with chronic renal failure.

Desired outcome: Patient verbalizes the importance of avoiding unusual magnesium intake and identifies potential sources of unwanted magnesium.

1. Caution persons with chronic renal failure to review all over-the-counter medications with RN prior to use.
2. Provide a list of common magnesium-containing medications (see the box on p. 238).
3. Persons with renal failure are often on vitamin supplements. Caution these individuals to avoid combination vitamin-mineral supplements since they usually contain magnesium.

Case Study 12-1

H.K. is a 50-year-old disabled Viet Nam veteran who has a history of posttraumatic stress disorder complicated by alcohol abuse. After being admitted to the hospital for evaluation of gastrointestinal bleeding, H.K. confesses to the nurse that his drinking problem has progressed to the point that "I binge each month until my government check runs out, then I don't even have enough money to eat." Physical assessment confirms an inadequate nutritional intake. H.K. is emaciated, with a distended abdomen. He has loose, dry, scaly skin; thin, brittle hair and nails; and fissures at the corners of his mouth. Additional information gained through the nursing assessment includes a 3-day history of anorexia, nausea, epigastric pain, and generalized abdominal distress, for which H.K. has been taking Amphojel (aluminum hydroxide antacid) approximately every 4 hours; and 5 loose, burgundy-black stools over the last 24 hours.

1. **Question:** H.K.'s alcohol abuse and poor nutritional state place him at increased risk for which electrolyte deficits?

   Answer: Hypophosphatemia and hypomagnesemia. Recall that hypophosphatemia is associated with alcoholism, especially during acute withdrawal, secondary to a combination of factors, including poor oral intake, diarrhea, use of phosphorus-binding antacids, and increased urinary excretion. Poor oral intake and increased urinary excretion also play a role in the development of hypomagnesemia in chronic alcoholism. Ethanol significantly increases urinary excretion of magnesium, although the exact mechanism is unknown. Decreased GI absorption of magnesium, combined with diarrhea, contributes to magnesium depletion.

2. **Question:** Hypomagnesemia may contribute to the development of which additional electrolyte imbalances?

   Answer: Hypokalemia, hypocalcemia, and hypophosphatemia. Recall that an adequate magnesium level is necessary to maintain the cellular sodium-potassium pump. With malfunction of the sodium-potassium pump, potassium is lost from the cell and excreted in the urine. This deficit may be profound and largely unresponsive to potassium supplementation until the magnesium deficit has been corrected. Hypocalcemia develops due to decreased release of PTH combined with target organ response to PTH. Phosphorus depletion develops from increased urinary excretion of phosphorus in the presence of hypomagnesemia.

Alcohol increases the production of gastric acid, in addition to acting as a direct gastric irritant. The mucosal barrier that protects the stomach from the caustic effects of hydrochloric acid and pepsin may be damaged. As a result of these changes, hemorrhagic or bleeding gastritis or duodenal ulcers may develop. The physician suspects that either hemorrhagic gastritis or bleeding ulcer is the cause of H.K.'s symptoms. The nurse is to insert a double lumen nasogastric (NG) tube and attach it to continuous low suction in order to empty the stomach and determine the rate of bleeding, if present.

3. Question: Will gastric suction alter magnesium balance?

Answer: Yes, it will contribute to hypomagnesemia, although it will have less effect than the loss of lower GI fluids. Recall that upper GI fluids contain relatively little magnesium, while lower GI fluids have a relatively high magnesium content.

4. Question: What are some potential nursing diagnoses for H.K.?

Answer:

a. **Potential fluid volume deficit** related to abnormal fluid loss secondary to bleeding, diarrhea, or NG suction
b. **Alteration in nutrition:** Less than body requirements of magnesium
c. **Potential for injury** related to altered level of consciousness, incoordination, and altered sensations secondary to acute alcohol withdrawal with delirium tremens and neuromuscular symptoms of hypophosphatemia and hypomagnesemia
d. **Posttrauma response:** Viet Nam veteran
e. **Ineffective individual coping:** Alcohol abuse secondary to posttraumatic stress disorder

Although the current priority for H.K.'s care is assessment and replacement of blood, combined with prevention of further bleeding, ensuring adequate nutrition is also a consideration. Depending on the results of the endoscopy in the morning, the decision will be made whether to begin TPN.

5. Question: What may happen to the plasma magnesium and phosphorus levels with the initiation of TPN?

Answer: The administration of a glucose load, combined with increased tissue anabolism, will cause a shift of both phosphorus and magnesium into the cells. This may precipitate dangerous hypophosphatemia and hypomagnesemia if there are existing deficits and the TPN solution does not contain adequate quantities of these two important intracellular electrolytes.

6. Question: The nurse caring for H.K. will monitor him for the signs and symptoms of hypomagnesemia. What do these include?

Answer: Confusion, hallucinations, agitation, hyperreactive reflexes, tremors, tetany, nausea, and tachyarrhythmias.

Based on his hemoglobin and hematocrit results on admission, H.K. is to receive 2 units of whole blood. In addition to the blood, the following IV solution is to be infused at a rate of 125 ml/hr: 5% Dextrose in 0.45% NaCl with 2 g of magnesium sulfate/L of solution. The nurse also will monitor H.K. for signs of hypermagnesemia.

7. **Question:** What are the clinical indicators of hypermagnesemia?

   Answer: Nausea, vomiting, flushing, diaphoresis, sensation of heat, altered mental function, drowsiness, decreased deep tendon reflexes, coma, and muscular weakness or paralysis.

References

Arieff AI and De Fronzo RA: Fluid, electrolyte, and acid-base disorders, New York, 1985, Churchill Livingstone, Inc.

Berkelhammer C and Bear R: A clinical approach to common electrolyte problems: hypomagnesemia, Can Med Assoc J 132:360-368, Feb 1985.

Brater DC: Serum electrolyte abnormalities caused by drugs, Prog Drug Res 30:9-69, 1986.

Calloway C: When the problem involves magnesium, calcium, or phosphate, RN 50:30-35, May 1987.

Goldberger E: A primer of water, electrolyte, and acid-base syndromes, ed 7, Philadelphia, 1986, Lea & Febiger.

Horne M: Fluid and electrolyte disturbances. In Swearingen PL and others: Manual of critical care: applying nursing diagnoses to adult critical illness, St Louis, 1988, The CV Mosby Co.

Horne M and Swearingen PL: Pocket guide to fluids and electrolytes, St Louis, 1989, The CV Mosby Co.

Hypomagnesemia associated with other electrolyte imbalances, Nutr Rev 43(7):196-200, 1985.

Kokko JP and others: Fluids and electrolytes, Philadelphia, 1986, WB Saunders Co.

Maxwell MH and others: Clinical disorders of fluid and electrolyte metabolism, ed 4, New York, 1987, McGraw-Hill Book Co.

Metheny NM: Fluid and electrolyte balance: nursing considerations, Philadelphia, 1987, JB Lippincott Co.

Potter PA and Perry AG: Fundamentals of nursing, ed 2, St Louis, 1989, The CV Mosby Co.

Rice V: Magnesium, calcium, and phosphate imbalances: their clinical significance, Crit Care Nurs 3:90-112, May/June 1983.

Valle GA and others: Electrolyte imbalances in cardiovascular disease: the forgotten factor, Heart Lung 17(3):324-329, 1988.

Williams SR: Nutrition and diet therapy, ed 6, St Louis, 1989, Times Mirror/Mosby College Publishing.

Chapter

13

Overview of Acid-Base Balance

Alterations in acid-base balance can affect cellular metabolism and enzymatic processes. Consequently, acid-base balance is critical to the maintenance of homeostasis.

The study of acids and bases is often considered confusing. While it is a complex topic, it need not be overwhelming if taken one step at a time. Basic principles of acid-base balance, once understood, will aid the reader in assessing patients and determining their needs for specific interventions.

NORMAL ACID-BASE ENVIRONMENT

The normal acid-base environment is achieved and maintained by 3 primary mechanisms. These include the following: (1) chemical buffering (neutralizing) by the extracellular and intracellular buffers, (2) respiratory control of carbon dioxide (CO_2) via changes in rate and depth of respirations, and (3) renal regulation of bicarbonate (HCO_3^-) concentration and secretion of hydrogen ions (H^+). These mechanisms are complex and interrelated and will be discussed further after a brief review of fundamental acid-base terminology and principles.

FUNDAMENTAL ACID-BASE TERMINOLOGY AND PRINCIPLES

Acids and Bases

Acids are defined as *substances that can donate or give up a hydrogen ion* (H^+), while **bases** are defined as *substances that can accept or take on H^+*. A common acid is hydrochloric acid (HCl), which can dissociate (separate) into H^+ and Cl^-.

The ability of this substance to give up the H^+ makes it an acid. Bicarbonate (HCO_3^-) is a common base that can readily accept a H^+ to form H_2CO_3 ($HCO_3^- + H^+ = H_2CO_3$). This ability to accept an H^+ is one characteristic of a base. It is important to understand that while there are many components involved in acid-base balance and regulation, the two that are most important are the weak acid carbonic acid (H_2CO_3) and the base bicarbonate (HCO_3^-).

Acids

Strong acids such as hydrochloric acid readily give up H^+ and can greatly affect the pH of the solution. Weak acids, such as carbonic acid, do not donate H^+ as readily, and thus do not alter the pH as much.

While carbonic acid (H_2CO_3) plays a critical role in acid-base balance, it is difficult to measure directly because it dissolves in solution (i.e., plasma) and forms CO_2 (carbon dioxide) and H_2O (water).

$$\underset{\text{carbonic acid}}{H_2CO_3} \rightleftarrows \underset{\text{carbon dioxide}}{CO_2} + \underset{\text{water}}{H_2O}$$

Carbon dioxide is easier to measure directly, and since it is in balance (equilibrium) with carbonic acid in solution, the acid component of acid-base balance is expressed in terms of the amount of CO_2 present, not the amount of carbonic acid. By definition, CO_2 is not an acid since it has no H^+ to donate, but because of its relationship with H_2CO_3, it is referred to as an acid.

In acid-base terminology, CO_2 is expressed as P_{CO_2}. The *P* stands for partial pressure of CO_2, and it refers to the pressure or tension exerted by this dissolved gas in the blood. It is called *partial pressure* because CO_2 is only one of several gases contributing to the total pressure exerted (oxygen, nitrogen, and other gases are also present). In humans the P_{CO_2} in arterial blood is equal to the pressure of CO_2 in alveolar air, which is 40 mm Hg at sea level. The partial pressure of CO_2 in arterial blood is referred to as Pa_{CO_2}. The *a* that is added refers to *arterial blood*. Specific Pa_{CO_2} values will be discussed in subsequent sections.

Volatile acids

Because CO_2 can be vaporized and eliminated by the lungs, it is called a **volatile acid.** Along with H_2O, CO_2 is generated when the body metabolizes food for energy. Humans generate approximately 13,000 to 20,000 mmol of CO_2 every day. Because CO_2 is continually produced as a byproduct of cellular metabolism, the lungs must excrete an equal amount to maintain a constant acid-base ratio.

Nonvolatile acids

While excretion of the volatile acid CO_2 is critical in maintaining the acid-base environment, there are acids other than CO_2 that also must be excreted. The acids that cannot be eliminated by the lungs (i.e., acids other than CO_2) are called **nonvolatile** or **fixed acids** and they are excreted by the kidneys. These fixed acids come from protein, fat, and carbohydrate metabolism.

Protein metabolism contributes the greatest amount to the production of fixed acids, specifically, sulfuric and phosphoric acids. Depending on the amount of protein consumed, the body will generate 50 to 100 mmol of fixed acids daily, which must be excreted by the kidneys. Metabolism of carbohydrates and fat results in the production of organic acids, including lactic acid, acetoacetic acid, and β-hydroxybutrate (the last two are called *ketoacids*). Normally these acids are further metabolized by the body and do not affect pH. In some abnormal conditions, however, these organic acids can accumulate and the pH will decrease. For example, abnormal accumulation of acids occurs with diabetic ketoacidosis (DKA) and tissue hypoperfusion.

In addition to protein consumption, other factors that can increase endogenous acid production include starvation (fasting), fever, strenuous exercise, certain diseases, and drugs, such as tetracycline. These factors increase catabolism, which in turn generates additional fixed acids.

Both volatile and nonvolatile acids affect the acid-base environment. Altering the concentration of either type of acid can influence the regulation of the other. This will be discussed further in this chapter under respiratory and renal regulation.

Bases

Small amounts of base are added to the body through diet, mainly from fruits and vegetables. The daily dietary gain of base is offset by the normal loss of HCO_3^- through the stool. The base with the most significant role in acid-base balance is bicarbonate (HCO_3^-), and it is regulated by the kidneys. HCO_3^- combines or accepts H^+ to neutralize acids. In the process, CO_2 is generated and then excreted by the lungs. CO_2 also can combine with H_2O to form carbonic acid, which dissociates into HCO_3^- and H^+.

bicarbonate		hydrogen		carbonic acid		carbon dioxide		water
HCO_3^-	+	H^+	$\rightleftarrows$	H_2CO_3	$\rightleftarrows$	CO_2	+	H_2O
base		strong acid		weak acid		eliminated by lungs		

The role of bicarbonate as a buffer and its regulation by the kidneys will be discussed further in this chapter in the sections on the chemical buffer system and renal regulation.

pH

Acidity or alkalinity of body fluids is expressed in terms of the concentration of hydrogen ions (H^+). The body regulates H^+ concentration within a narrow range. Regulation of this narrow range is critical because minute changes in H^+ concentration have a dramatic influence on cellular function, such as enzyme activity, neuromuscular membrane-action potentials, and generation of adenosine triphosphate (ATP). The normal concentration of H^+ is 0.0004 mEq/L, a much smaller amount than any other ion found in body fluids. In the clinical setting this small number can be very cumbersome when analyzing the patient's acid-base status, so H^+

concentration is expressed in terms of pH. pH is defined as *the concentration of H^+ present in solution.* The pH value is obtained by taking the negative logarithm of the hydrogen ion concentration $[H^+]$. It can be expressed as follows:

$$pH = -\log[H^+] \quad \text{or} \quad pH = \frac{1}{\log[H^+]}$$

This expression means that the pH has an inverse relationship with the H^+ concentration, i.e., the fewer H^+ present, the higher the pH, and the more alkaline the solution; the more H^+ present, the lower the pH, and the more acidic the solution. Although pH is defined in terms of hydrogen ion (H^+) concentration in the body, it is determined by the *ratio* of acid to base in the body, not by their *absolute* concentration. To maintain a normal pH, *the ratio of 20 parts of base to 1 part acid must be maintained* (20 to 1 ratio).

The range of pH for all substances is 1 to 14, with 1 being the most acidic and 14 being the most basic. Water is neutral, with a pH of 7.0. The body maintains a slightly alkaline pH of 7.40, with a range of 7.35 to 7.45 considered normal for arterial blood. When the arterial pH is <7.20 or >7.55, aggressive medical intervention may be required to prevent serious consequences that could result from these alterations in the acid-base balance. An arterial pH outside the range of 6.8 to 7.8 for more than a brief period of time is considered incompatible with life. The pH of venous blood is approximately 7.35. It is more acidic than arterial blood because CO_2 that is generated by cellular metabolism is carried to the lungs for excretion. Since CO_2 combines with H_2O to produce carbonic acid (H_2CO_3), the increased concentration of the weak acid H_2CO_3 will cause the pH of venous blood to decrease slightly.

The pH of gastric fluid, which contains hydrochloric acid (HCl), a strong acid, is 1.0 to 2.0. The pH of fluid in the small intestine is alkaline, with a range of 7.0 to 8.0, and the pH of urine is acidic, ranging from 4.5 to 6.0. Although the pH of the various body fluids varies greatly, the optimal pH specific for a particular body fluid is carefully maintained and regulated by different feedback mechanisms.

Terms describing pH

Numerous terms are used to describe the ratio of acid to base within body fluids. **Acidosis** refers to *the process leading to the accumulation of acid or loss of base, with or without a change in pH.* **Acidemia** refers to *an actual decrease in the pH of arterial blood to less than 7.40.* **Alkalosis** refers to *the process leading to the accumulation of base or loss of acid, with or without a change in pH.* **Alkalemia** refers to *an increase in arterial pH to greater than 7.40.*

Acidosis that occurs because of the retention of nonvolatile acids or loss of base is termed *metabolic acidosis,* while the retention of volatile acid (CO_2) is termed *respiratory acidosis.* Similarly, a loss of nonvolatile acids or retention of base is termed *metabolic alkalosis,* and a reduction in the level of volatile acid (CO_2) is termed *respiratory alkalosis.* While a change in pH is usually present with acidosis or alkalosis, pathophysiologic processes can be present with a normal pH. For example, when both acidosis and alkalosis are present to the same degree, the normal base-to-acid ratio of 20:1 is maintained, and thus a normal pH can occur.

THE CHEMICAL BUFFER SYSTEM

Acids are being generated continuously as a result of cellular metabolism, but the serum pH remains constant. This constancy in pH is maintained by the numerous chemical buffers present in the body fluids and tissues. Chemical buffers are substances that act as sponges to soak up or release free hydrogen ions (H^+) so that the pH is not altered greatly. Chemical buffers do not change the absolute number of H^+ liberated by a strong acid or removed by a strong base. Buffers do, however, limit the number of free H^+ and thus lessen the effects a strong acid or base would have on the pH of a solution.

Buffers are present in all body fluids (i.e., plasma, intracellular fluid, interstitial fluid), tissue, and bone, and their response is instantaneous. Base stores, however, can be depleted if presented with large amounts of fixed acids that require buffering. For every 1 mEq of H^+ that must be buffered, 1 mEq of buffer is consumed. Strong bases are also buffered by the release of H^+ from tissue stores.

Buffers are described in terms of their buffering capacity or power, meaning the ability of the buffer to maintain the pH of a solution within a small variation when acids or bases are added. The primary chemical buffers include bicarbonate, organic phosphates, proteins, and bone. Bicarbonate is the most important buffer in the plasma and interstitial fluids, and it is considered an extracellular buffer. The organic phosphates and proteins play a major role in intracellular fluid (ICF). The bone does not have as large a role in normal buffering, but it is very important in certain chronic conditions, such as chronic renal failure.

Extracellular Buffers

The most important buffer pair in the extracellular fluid (ECF) is the bicarbonate-carbonic acid (HCO_3^-/H_2CO_3) pair. There are two other extracellular buffers that contribute to buffering, but they are less important than HCO_3^-. They are inorganic phosphates and plasma proteins.

Bicarbonate-carbonic acid buffer system

Bicarbonate is responsible for 80% of the buffering in ECF, and it buffers the fixed acid load (i.e., diet-generated acids, lactic acids, and ketoacids). The bicarbonate-carbonic acid buffer system consists of the base bicarbonate (HCO_3^-) and the weak acid carbonic acid (H_2CO_3).

This system reacts in the following way:

hydrogen		bicarbonate		carbonic acid
H^+	$+$	HCO_3^-	$\rightleftarrows$	H_2CO_3
acid		base		weak acid

Since carbonic acid also can dissociate into CO_2 and H_2O, there is a more complete equation for this buffering system:

hydrogen		bicarbonate		carbonic acid		carbon dioxide		water
H^+	$+$	HCO_3^-	$\rightleftarrows$	H_2CO_3	$\rightleftarrows$	CO_2	$+$	H_2O
acid		base		weak acid		weak acid		

The concentration of volatile and nonvolatile acids has a direct influence on each. When fixed acids are added to the body, they are buffered immediately by HCO_3^-. This buffering causes carbonic acid to increase, thus CO_2 also increases. This leads to increased respiratory rate and depth and elimination of the excess volatile acid (CO_2).

Here is an example of bicarbonate buffering a diet-generated nonvolatile acid:

strong acid		buffer salt		weak acid		salt		acid				
H_2SO_4	$+$	$2NaHCO_3$	$\rightarrow$	$2H_2CO_3$	$+$	$NaSO_4$	$\rightarrow$	$2CO_2$	$+$	$2H_2O$	$+$	$NaSO_4$
sulfuric acid (diet generated)		sodium bicarbonate		carbonic acid		sodium sulfate		carbon dioxide		water		sodium sulfate

The diet-generated acid has been buffered by the bicarbonate, with the end products being CO_2, which is excreted by the lungs, and the salt sodium sulfate, which is excreted by the kidneys. With the increase in CO_2 generated by the HCO_3^- buffering of the fixed acid load, respiratory rate and depth increase so that the excess CO_2 can be eliminated.

The bicarbonate usually appears as sodium bicarbonate ($NaHCO_3^-$) since sodium (Na^+) is the predominant cation in the ECF. Other cations, such as potassium (K^+), calcium (Ca^{2+}), and magnesium (Mg^{2+}), can be substituted for the Na^+. In ICF little bicarbonate is present, but that which is present exists as K^+ or Mg^{2+} bicarbonate.

HCO_3^- can be depleted quickly when buffering a large acid load, but in the short term it is very effective in maintaining a life-sustaining pH. The ultimate strength of the bicarbonate-carbonic acid buffer system is that bicarbonate is readily available in ECF and, even more importantly, both components can be regulated—CO_2 by the lungs and HCO_3^- by the kidneys. Another strength of HCO_3^- is that it is the only base that can be replenished by the kidneys as its concentration decreases from the buffering of fixed acids. Thus as body fluid stores of HCO_3^- are consumed during the buffering of fixed acids, they are restored by the regeneration of HCO_3^- by the kidneys. The regeneration of HCO_3^- is critical to the maintenance of a steady pH in light of the daily onslaught of acids. If HCO_3^- were not regenerated, alkali (base) stores would be depleted quickly by the normal dietary production of acids, and acidosis would ensue.

The components of the bicarbonate-carbonic acid buffer system are interdependent. The amount of HCO_3^- depends directly on the amount of CO_2 present, and vice versa. Remember that the pH of the body fluids is a function of the *ratio* of acids to base. Specifically, it is a function of the ratio of H_2CO_3 to HCO_3^-. In health, this ratio is 1 part H_2CO_3 (acid) to 20 parts HCO_3 (base), which maintains a pH of 7.40. Because of the relationship between HCO_3^- and H_2CO_3, it is clear that HCO_3^- cannot buffer H_2CO_3 (i.e., they are part of the same system). Rather, H_2CO_3 must be buffered by nonbicarbonate buffers, such as intracellular protein buffers. So when H_2CO_3 concentration increases, HCO_3^- is not depleted—only the ratio between them changes.

Henderson-Hasselbalch equation

The interdependency of H_2CO_3, HCO_3^-, and pH can be demonstrated by the Henderson-Hasselbalch equation. This equation has little direct application for nurses in the clinical setting, but it is the basis for understanding the relationship between the key components of the acid-base system. The Henderson-Hasselbalch equation states the following:

$$\mathrm{pH} = \mathrm{pK}^{*} + \log \frac{HCO_3^-}{H_2CO_3}$$

Because H_2CO_3 is not measured directly but exists in equilibrium in solution with CO_2, H_2CO_3 can be replaced in the equation by CO_2. The equation then is expressed as follows:

$$\mathrm{pH} = \mathrm{pK}(6.1) + \log \frac{HCO_3^-}{CO_2} \quad \begin{array}{l}\textbf{base regulated by the kidney}\\ \textbf{acid regulated by the lung}\end{array}$$

Stated more simply,

$$\mathrm{pH} \approx \frac{HCO_3^-}{CO_2}$$

To maintain a steady pH the ratio of the numerator (HCO_3^-) to the denominator (CO_2) must remain constant. If the numerator increases, the pH increases and the acid-base balance is shifted to the alkaline side. If the denominator increases, the pH decreases and the acid-base balance shifts to the acid side. If both the numerator and denominator increase or decrease proportionately within the normal ratio of 20:1, there will be no change in pH. Because of the relationship of these components one can determine the value for any one of the three components if the other two values are known. Table 13-1 shows the relationship between disturbances in acid-base and the effects of pH.

*pK = 6.1. This is the dissociation constant (i.e., the ability to release H^+ for carbonic acid).

Table 13-1 Acid-base regulation and effects on pH

Component	Regulation	Disturbance	pH
CO_2	Respiratory	Acidosis ↑ CO_2 $pH = \frac{HCO_3^- = 20}{CO_2 > 1}$	↓
CO_2	Respiratory	Alkalosis ↓ CO_2 $pH = \frac{HCO_3^- = 20}{CO_2 < 1}$	↑
HCO_3^-	Renal	Acidosis ↓ HCO_3^- $pH = \frac{HCO_3^- < 20}{CO_2 = 1}$	↓
HCO_3^-	Renal	Alkalosis ↑ HCO_3^- $pH = \frac{HCO_3^- > 20}{CO_2 = 1}$	↑

Intracellular Buffers

While the bicarbonate-carbonic acid buffer system is the most important chemical buffer in the ECF, intracellular buffers also play an important role in maintaining the acid-base balance. These intracellular buffers are critical in buffering the H^+ that results from excess CO_2 (H_2CO_3), which cannot be buffered by HCO_3^-. Intracellular buffers are responsible for buffering 95% of the carbonic acid and 50% of the fixed acids. They provide direct buffering via inorganic and organic phosphates and proteins, such as hemoglobin. A less direct buffering occurs as a result of the diffusion of CO_2, H^+, and HCO_3^- across the cell membrane. The diffusion of the two components of the bicarbonate-carbonic acid buffer system allows the intracellular pH to change in approximate proportion to the extracellular pH. Thus intracellular buffering helps buffer the extracellular space as well.

Protein buffers

Proteins provide the most powerful and plentiful buffer system in the body. Approximately 75% of all chemical buffering power is present inside the cells and is the result of intracellular proteins. Chemically, the protein buffer system acts in much the same manner as bicarbonate—it takes on hydrogen ions (H^+) to lessen their effect on pH. Because of the chemical composition of some proteins, they can operate as both acid and base buffering systems in that they either can accept or release H^+. This ability enhances their buffering power.

Hemoglobin

Hemoglobin is the most powerful protein buffer. Red blood cells (RBCs) are responsible for approximately 70% of all the buffering power of the blood. Hemoglobin is responsible for the transportation of most of the CO_2 to the lungs for elimination. This is accomplished in the following way: CO_2 from the tissue enters the plasma, diffuses into the RBCs, and combines with H_2O to form carbonic acid (H_2CO_3). This reaction occurs instantly because of the presence of large quantities of the enzyme carbonic anhydrase, which acts as a catalyst for the reaction. The carbonic acid quickly dissociates into HCO_3^- and H^+. The HCO_3^- diffuses back into the plasma and the H^+ is buffered by the hemoglobin. In the lungs, the H^+ is displaced from the hemoglobin by oxygen and the process is basically reversed—CO_2 escapes into the alveoli and is eliminated. The buffering power of hemoglobin allows large quantities of CO_2 to be transported, with only a small change in venous pH.

Phosphate buffers

Phosphate buffers have already been mentioned for their role as extracellular buffers. They also play an important role in intracellular fluids, where their concentration is much greater. The buffer salt in the phosphate buffer pair accepts a H^+, and a weak acid is formed that changes the pH minimally. The phosphate buffer also has an important buffering role in the tubular fluid of the kidneys. Phosphate buffers enable the kidneys to increase the excretion of H^+ in the urine. Without these buffers, the urine would become extremely acidic very quickly.

Bone buffers

Bone carbonate contributes to the buffering of acid and base loads. Up to 40% of the buffering of an acute acid load may occur in the bone. In the presence of a chronic acid load (e.g., chronic renal failure), the bone buffers play an even greater role. After an acid load, bone carbonate is released into the ECF. In individuals with chronic metabolic acidosis, such as those with **chronic renal failure (CRF)**, the persistent acidemia results in long-term leaching of alkaline salts (i.e., bone carbonate) for buffering. This chronic buffering by the bone contributes to the development of the numerous bone diseases associated with CRF.

Isohydric Principle

The buffers have been discussed as though they operate separately and independently, but they all work together. If the H^+ concentration is altered, the base-to-acid ratio (20:1) of all the buffer systems is affected. This is known as the **isohydric principle.** Because the buffer systems are so interdependent, the status of all the buffers can be determined, based on the values of a given buffer. Clinically, the buffer used to evaluate the total acid-base status of an individual is HCO_3^-/CO_2.

RESPIRATORY REGULATION

The lungs' role in acid-base balance involves the regulation of CO_2. The rate and depth of alveolar ventilation determine how much CO_2 is eliminated or retained.

Elimination of CO_2

Normally the amount of CO_2 eliminated equals the amount of CO_2 generated by metabolic processes. However, when the amount of CO_2 generated and the amount of CO_2 eliminated does not balance, then either respiratory acidosis (too little CO_2 eliminated for the amount produced) or respiratory alkalosis (too much CO_2 eliminated for the amount produced) occurs.

The respiratory center of the pons and medulla controls the initiation and regulation of ventilation. The medullary chemoreceptors are extremely sensitive to changes in the hydrogen ion (H^+) concentration of the cerebrospinal fluid (CSF). The pH of the CSF is primarily controlled by the blood CO_2 since H^+ and HCO_3^- cannot readily cross the blood-brain barrier. When changes occur in $Paco_2$ or in the pH of the CSF, the respiratory rate and depth are altered accordingly, almost instantaneously. If acidosis occurs, whether respiratory or metabolic, the respiratory rate and depth increase in an effort to decrease the volatile acid load (CO_2) by eliminating the excess CO_2, thereby protecting the body from a decrease in pH. In fact, the rate of alveolar ventilation can increase 6 to 7 fold. Although H^+ and HCO_3^- do not easily cross the blood-brain barrier, metabolic acidosis does cause an increase in respiratory rate and depth because of the alteration in CSF pH. If a metabolic or respiratory alkalosis occurs, with a decrease in $Paco_2$ or an increase in CSF pH, the rate and depth of respiration decrease in order to retain the acid CO_2 and reestablish the normal acid-base ratio. Arterial hypoxemia, a decrease in the oxygen level of arterial blood, also stimulates the respiratory center. Under normal conditions hypoxemia has a minimal role in the regulation of breathing but it may predominate in conditions of chronically elevated $Paco_2$ and at high altitudes.

The control of ventilation and the ability of the respiratory system to respond appropriately to abnormal conditions depend on the normal functioning of the respiratory center, appropriate innervations and functioning of the muscles of respiration, proper functioning of the lung components (i.e., airways and alveoli), and the normal and intact thoracic structures. When any of these are affected there is the potential for an alteration in the respiratory regulation of CO_2, thus potentially causing an acute or chronic alteration in acid-base balance.

Respiratory Compensation

A normally operating respiratory system has 1 to 2 times the overall buffering power of all the chemical buffers combined. Compensation, however, is limited with metabolic disturbances. The lungs eliminate only the volatile acid CO_2, so when faced with an acute increase in the fixed acid load, they are unable to excrete enough CO_2 to maintain a normal pH, but the compensation that occurs usually can keep the pH in a range compatible with life.

Respiratory compensation for alterations in hydrogen ion (H^+) concentration is approximately 50% to 75% effective. For example, if the pH suddenly decreases from 7.40 to 7.0 within a minute or so, the respiratory system will raise the pH to 7.20. Thus the lungs can compensate partially for metabolic disturbances but are unable to return the pH to normal. In cases of metabolic alkalosis (increased pH), the retention of CO_2 to prevent a further increase in pH is limited because too great a decrease in respiratory rate and depth will result in hypoxemia. Hypoxemia, in turn, will stimulate respirations.

RENAL REGULATION

The renal regulation of HCO_3^- and the secretion of H^+ are essential to the maintenance of a normal acid-base environment. Every day fixed acids are generated and buffered either by HCO_3^- or nonbicarbonate buffers. Without the kidneys' regulatory role, this process of continuous nonvolatile acid production and buffer consumption would lead to progressive metabolic acidosis, because for every 1 mEq of acid, 1 mEq of HCO_3^- is consumed or 1 mEq of nonbicarbonate buffer is acidified. Respiratory compensation is inadequate to deal with the steadily increasing fixed acids and decreasing buffers. Fortunately, normally functioning kidneys are able to prevent this acidosis through their role in reabsorption and regeneration of HCO_3^- and secretion of H^+. The mechanism for the regulation of HCO_3^- and H^+ are not as straightforward as the respiratory regulation of CO_2. Renal regulation involves many interrelated steps. While this makes the subject somewhat complex, it is important to understand that these interrelated steps allow the kidneys to make adjustments to maintain a normal acid-base environment, even in the presence of metabolic or respiratory abnormalities.

Reabsorption of HCO_3^-

When the arterial pH is normal and the kidneys are functioning normally, virtually all the bicarbonate that is present in the glomerular ultrafiltrate, about 4,000 to 4,300 mEq/L, will be reabsorbed. In states of alkalosis less of the HCO_3^- present in the ultrafiltrate will be absorbed. The reabsorption of HCO_3^-, which occurs in the proximal tubule, is not a passive process and is dependent on several variables. The most important variable is the secretion of hydrogen ions (H^+) by the kidneys. When H^+ are secreted in the tubular fluid, they combine with the filtered HCO_3^- and form H_2CO_3. Under the influence of the enzyme carbonic anhydrase, the H_2CO_3 rapidly dissociates into CO_2 and H_2O. The H_2O is excreted and the CO_2 is reabsorbed. The reabsorbed CO_2 is then converted back into HCO_3^- (again via the action of carbonic anhydrase) at the cellular level and is returned to the body fluids. The presence of carbonic anhydrase is critical in the reabsorption of HCO_3^-. Without it, the process of HCO_3^- reabsorption is slowed dramatically, and HCO_3^- is lost in the urine. This is the physiologic basis for the use of carbonic anhydrase inhibitors (e.g., Diamox) to treat some types of metabolic alkalosis (see Chapter 18).

Generation of HCO_3^-

Every day quantities of HCO_3^- are used up in the buffering of the acid load. Reabsorption does not add new HCO_3^- to body fluids to replace that which has been used up, but the kidneys have the ability to generate HCO_3^-, thus maintaining the HCO_3^- concentration in the presence of continuous buffering. The generation of HCO_3^- also involves the secretion of H^+, but the processes are distinct from those involved in HCO_3^- reabsorption. Generation of HCO_3^- occurs as carbonic acid dissociates into H^+ and HCO_3^-.

The liberated H^+ is excreted and the HCO_3^- moves back into the peritubular capillary. The bicarbonate generated by this process increases the plasma concentration. When alkalosis is present, less HCO_3^- is generated because fewer hydrogen ions (H^+) are secreted. Conversely, during acidosis an increased amount of HCO_3^- is generated because of the increase in H^+ secretion.

Secretion of Hydrogen Ions (H^+)

The critical component in both bicarbonate absorption and generation is the secretion of H^+ into the tubular lumen. The kidneys help maintain the 20:1 ratio of bases to acids by excreting excess H^+. The excess H^+ results from a daily net gain of 50 to 100 mmol of H^+ from metabolic and dietary sources (nonvolatile acids). Hydrogen ions (H^+) are secreted into the tubules and ultimately are excreted into the urine. All the epithelial cells of the proximal tubule, distal tubule, and proximal ducts secrete H^+. The majority (84%) of H^+ is secreted in the proximal tubule.

The amount of H^+ secreted is influenced by many factors (Table 13-2). The arterial pH is the most important physiologic regulator of H^+ secretion on a day-to-day basis, depending on the dietary acid load. A decrease in pH (acidosis) promotes a secretion of an increased amount of H^+, while an increase in pH (alkalosis) causes a decrease in H^+ secretion. Other factors that contribute to the regulation of H^+ are plasma K^+ level (hyperkalemia causes a decrease in H^+ secretion; hypokalemia increases H^+ secretion); effective circulating volume (ECV) (with a decreased circulating volume, secretion of H^+ is increased; conversely, volume expansion causes a decrease in H^+ secretion); aldosterone level (promotes H^+ secretion); and $Paco_2$ (increased $Paco_2$ causes an increase in H^+ secretion; decreased $Paco_2$ causes a decrease in H^+ secretion).

Table 13-2 Factors affecting H^+ secretion

Factor	Increases H^+ Secretion	Decreases H^+ Secretion
Decreased pH (acidosis)	↑	
Increased pH (alkalosis)		↓
Hyperkalemia		↓
Hypokalemia	↑	
Increased circulatory volume	↑	
Decreased circulatory volume		↓
Aldosterone	↑	
$Paco_2$ increased	↑	
$Paco_2$ decreased		↓

The kidneys must secrete H^+, but if free H^+ were secreted, the urinary pH would decrease rapidly, causing injury to the structural components of the kidney and urinary tract. Hydrogen ions (H^+) must be buffered in the urine just as they are in other body fluids. This urinary buffering occurs by the formation either of ammonium or phosphate buffers. Secretion of the usual acid load causes the normal urine pH to be approximately 6.0. When the acid load increases, the formation of greater quantities of ammonium and phosphate buffers allows excess H^+ to be secreted without decreasing the urinary pH below 4.5, which is the minimum urine pH for humans.

Ammonium

Ammonia (NH_3) is a base that is formed from the amino acid glutamine within the tubular cells. NH_3 can readily cross cell membranes. It crosses into the tubular lumen to combine with H^+ to form ammonium (NH_4). The NH_4 formed in the tubular lumen is trapped there because it is unable to cross cell membranes and is excreted in the urine. Thus by adding a H^+ to each NH_3, the formation of NH_4 enhances the secretion of H^+ without alterating urine pH (Figure 13-1). Normally, only 30 to 40 mmol of H^+ are secreted in this manner per day, but NH_4 excretion can dramatically increase H^+ secretion, up to 500 mmol/day when faced with an increase in fixed acids. Increased production of NH_3, and thus NH_4 is the kidneys' main adaptive response in the presence of an increased acid load. While this adaptation is very effective, it occurs slowly over 3 to 4 days, causing a delay in the kidneys' ability to respond to an acute rise in fixed acids.

Titratable acids

Another mechanism that aids in the secretion of H^+ is the formation of titratable acids. When HCO_3^- is reabsorbed, the tubular fluid is acidified and the H^+ present combine with buffers. The amount of strong base that would be required to titrate the

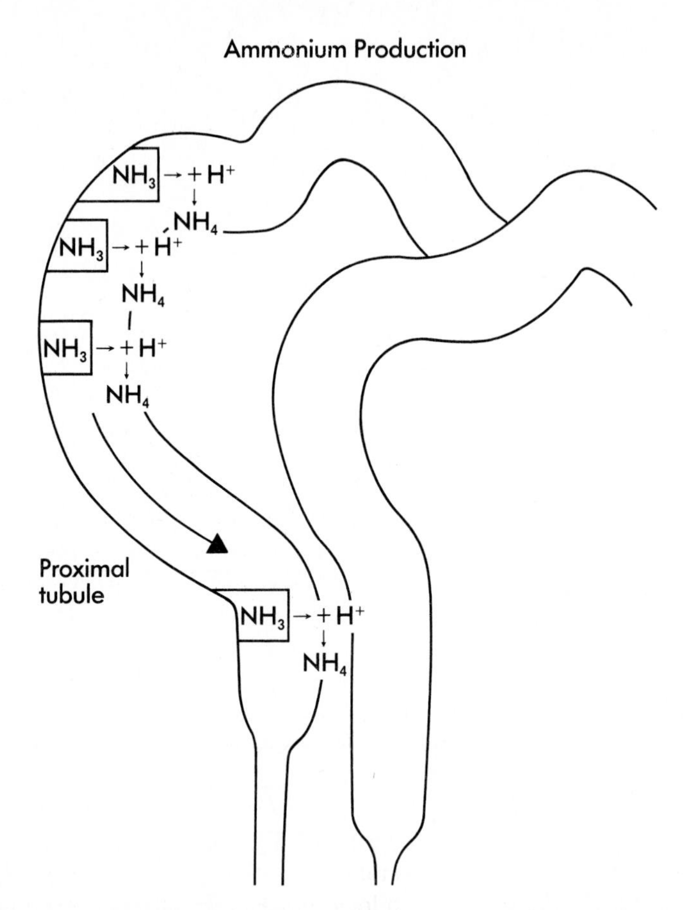

Figure 13-1 Ammonia (NH_3) is produced in the tubular cells. It moves into the tubular lumen and combines with H^+ to form ammonium (NH_4). NH_4 production occurs predominately in the proximal tubule. The ammonium is trapped in the tubular lumen and is excreted in the urine.

acidic urine back to a pH of 7.4 (the pH of the glomerular filtrate) is equal to the amount of acid excreted in the urine. This acid is known as the titratable acid. One of the buffers is Na_2HPO_4 (sodium monohydrogen phosphate), a neutral salt that combines with H^+ to become NaH_2PO_4 (sodium dihydrogen phosphate), an acid salt. This process occurs when the H_2CO_3 present in the tubular cells dissociates, the new HCO_3^- diffuses into the peritubular capillary, the free H^+ is buffered by the neutral phosphate salt in the tubular lumen, and a weak acid is formed that is then excreted. Normally about 10 to 40 mmol of H^+ are buffered in this manner. In the presence of an increased acid load, such as with diabetic ketoacidosis, 75 to 250 mmol of H^+ can be excreted in this way.

Table 13-3 Acid-base disturbances

Primary Disorder	pH	Primary Disturbance	Compensatory Response
Respiratory acidosis	↓	↑ Pa_{CO_2}	↑ HCO_3^-
Respiratory alkalosis	↑	↓ Pa_{CO_2}	↓ HCO_3^-
Metabolic acidosis	↓	↓ HCO_3^-	↓ Pa_{CO_2}
Metabolic alkalosis	↑	↑ HCO_3^-	↑ Pa_{CO_2}

CATEGORIES OF ACID-BASE DISORDERS

There are two categories of acid-base disturbances: (1) respiratory, which involves the derangement of CO_2 regulation by the lungs, (2) metabolic, which involves the derangement of HCO_3^- concentration by the kidneys (Table 13-3).

Simple Disorders

When there is one primary acid-base disturbance it is called a *simple disorder*. For example, when the renal regulation of HCO_3^- is altered and the HCO_3^- concentration decreases, the pH also decreases and a metabolic acidosis is present. Remember that a 20:1 ratio of bases to acids is necessary to maintain a pH of 7.40. If more base (HCO_3^-) is lost in proportion to acids (CO_2), then acidosis occurs. The decrease in pH stimulates the respiratory center and the rate and depth of respirations increases, causing a greater quantity of CO_2 to be eliminated by the lungs. This loss of excess CO_2 is an attempt to bring the base-to-acid ratio closer to the normal of 20:1. With this loss of *both* the base (the HCO_3^- by the kidneys) and the acid (the CO_2 eliminated by the lungs), the 20:1 ratio is closer to normal and the pH is not affected as greatly. This respiratory response to the metabolic acidosis is called a *compensatory response*.

Although during compensation the CO_2 levels may become abnormal, this is still a simple disorder because there is only one primary disorder—an alteration in HCO_3^- concentration. The change in CO_2 levels is the appropriate response to this alteration (Table 13-4). Renal compensatory responses also occur when there are alterations in the respiratory regulation of CO_2. Respiratory and metabolic disorders, along with the compensatory responses, will be discussed in greater depth in Chapters 15-18.

Mixed Disorders

Numerous clinical situations exist in which a combination of simple disorders leads to a mixed acid-base disturbance. For example, an individual may have respiratory acidosis (increased CO_2) from a drug overdose and metabolic acidosis (decreased HCO_3^-) from decreased tissue perfusion during shock (which leads to lactic acidosis).

Table 13-4 Quick assessment guidelines for simple acid-base disturbances

Acid-Base Disturbance	pH	Pa_{CO_2}	HCO_3^-	Clinical Signs and Symptoms	Common Causes
Acute respiratory acidosis	↓	↑	no change	Tachycardia, tachypnea, diaphoresis, headache, restlessness leading to lethargy and coma, dysrhythmias, hypotension, cyanosis (a late sign)	Acute respiratory failure, cardiopulmonary disease, drug overdose, chest wall trauma, asphyxiation, CNS trauma/lesions, impaired muscles of respiration
Chronic respiratory acidosis (compensated)	↓	↑	↑ *	Dyspnea & tachypnea, with increase in CO_2 retention that exceeds compensatory ability; progression to lethargy, confusion, and coma	COPD, extreme obesity (Pickwickian syndrome), superimposed infection on COPD
Acute respiratory alkalosis	↑	↓	no change (a decrease will occur if condition has been present for hours, providing that renal function is adequate)	Paresthesias, especially of the fingers; dizziness	Hyperventilation, salicylate poisoning, hypoxemia (e.g., with pneumonia, pulmonary edema, pulmonary thromboembolism), gram-negative sepsis, CNS lesion, decreased lung compliance, inappropriate mechanical ventilation
Chronic respiratory alkalosis	slightly ↑ or normal†	↓	↓ *	No symptoms	Hepatic failure; CNS lesion; pregnancy

Acute metabolic acidosis	↓	↓*	↓	Tachypnea leading to Kussmaul respirations, hypotension, cold and clammy skin, coma, and dysrhythmias	Shock, cardiopulmonary arrest (secondary to lactic acid production), ketoacidosis (e.g., diabetes, starvation, alcohol abuse), acute renal failure, ingestion of acids (e.g., salicylates), diarrhea
Chronic metabolic acidosis	↓	↓* (not as much as acute type)	↓	Fatigue, anorexia, malaise. Symptoms may be related to chronic disease process as well as acidosis	Chronic renal failure
Acute metabolic alkalosis	↑	↑* (can be as great as 60)	increased ↑	Muscular weakness and hyporeflexia (due to severe hypokalemia), dysrhythmias, apathy, confusion, and stupor	Volume depletion (Cl^- depletion) as a result of vomiting, gastric drainage, diuretic use. Posthypercapnea, hyperadrenocorticism (e.g., Cushing's syndrome), aldosteronism, severe potassium depletion, excessive alkali intake
Chronic metabolic alkalosis	↑	↑*	↑	Usually asymptomatic	Upper GI losses through continuous drainage; correction of hypercapnia if Na^+ and K^+ depletion remains uncorrected

*Compensatory response
†The only acid-base disturbance in which compensation can return the pH to normal

In this example, the ratio of bases to acids has been altered. The base (HCO_3^-) has decreased and the acid has increased, causing a great decrease in pH.

Many combinations of metabolic and respiratory disturbances can occur. Compensatory responses may or may not be present, depending on the primary disturbances involved in the mixed disorder. Mixed acid-base disturbances will be discussed in Chapter 19.

REVIEW OF ACID-BASE BALANCE

The chemical buffers are responsible for the immediate buffering of excess acids and bases. Chemical buffers are present in all body fluids and are vital in maintaining the acid-base environment on a moment-to-moment basis.

The respiratory system is involved in the maintenance of acid-base balance by its regulation of CO_2. The respiratory center in the brain responds to changes in pH by altering the breathing pattern (rate and depth) to maintain the appropriate level of CO_2 and return the pH to as near normal as possible. The respiratory response to alterations in pH caused by metabolic disturbances occurs almost instantaneously. Although respiratory compensation is 50% to 75% effective, the pH is not returned to normal with metabolic disturbances.

The kidneys' role in the maintenance of acid-base environment is dependent on their ability to secrete excess H^+ and to regulate the concentration of HCO_3^- through reabsorption and regeneration. The arterial pH and $Paco_2$ play dominant roles in the renal regulation of HCO_3^- and the secretion of H^+ enables the kidneys to compensate very effectively when faced with alterations in the acid-base balance, either metabolic or respiratory. While the kidneys compensate effectively, their response is slow, with an initial response occurring in approximately 24 hours and a maximal response requiring 3 to 4 days.

When alterations in the normal concentration of CO_2 or HCO_3^- or both occur, an acid-base disturbance is present. If the primary alteration is with CO_2, a respiratory acid-base disturbance is involved. If the HCO_3^- concentration is primarily affected, a metabolic disturbance is present. Disorders involving only one primary acid-base imbalance are called simple acid-base disturbances. When two or more primary acid-base disorders are present, a mixed acid-base disturbance is present.

Case Study 13-1

M.M., a 21-year-old college student, began reviewing his notes for an upcoming exam when he realized he had lost half of them. He became very anxious. His RR increased from 12 to 30 breaths/min, and after 10 minutes he began to feel tingling in his fingers and around his mouth.

1. **Question:** With the rapid, deep breathing would you expect his CO_2 to increase, decrease, or stay the same?

   Answer: Normally the CO_2 would decrease with rapid, deep breathing.

2. **Question:** If the CO_2 decreases quickly, what would you expect to happen to the pH?

   Answer: According to the interdependent relationship between pH, CO_2, and HCO_3^- (Henderson-Hasselbalch equation), if the CO_2 decreases with no change in the HCO_3^-, you would expect the pH to increase to >7.40.

3. **Question:** Knowing that the CO_2 has decreased, what do you anticipate has happened to the 20:1 ratio of bases to acids?

   Answer: With a decrease in CO_2 the ratio is altered, with a greater proportion of HCO_3^- to CO_2, even though there has not been a net gain of HCO_3^-. This results in an alkalosis that is respiratory in origin because the primary disturbance is the respiratory loss of the acid CO_2. Normally,

   $$\underset{(7.40)}{\text{pH}} = \frac{HCO_3^- : 20}{CO_2 : 1}$$

   In this instance, however,

   $$\underset{(>7.40)}{\text{pH}} = \frac{HCO_3^- : 20}{CO_2 < 1}$$

4. **Question:** Which acid-base system, respiratory or renal, would you expect to initiate a compensatory response?

   Answer: The renal system would compensate for the respiratory alkalosis by excreting more HCO_3^- and retaining H^+. Renal compensation takes from hours to days to occur, so no renal compensation would be expected in this situation. The respiratory system is responsible for the acid-base disturbance—a greater quantity of CO_2 is being excreted than is being produced—so it cannot compensate for the disturbance.

The student finds his notes; his anxiety decreases, and his RR returns to normal.

5. **Question:** What would you expect to happen to the $Paco_2$ to HCO_3^- ratio and the pH when the RR normalizes?

   Answer: The $Paco_2$ will increase to normal levels, causing the acid-base ratio to return to 1 part CO_2 to 20 parts HCO_3^-. The pH will then be normal at 7.40.

References

Easterday U: Acid-base balance. In Horne M and Swearingen PL: Pocket guide to fluids and electrolytes, St Louis, 1989, The CV Mosby Co.

Felver L: Acid-base balance and imbalances. In Patrick M and others: Medical-surgical nursing: pathophysiologic concepts, Philadelphia, 1986, JB Lippincott Co.

Fishbach T: A manual of laboratory diagnostic tests, Philadelphia, 1988, JB Lippincott Co.

Goldberger E: A primer of water, electrolyte, and acid-base syndromes, Philadelphia, 1986, Lea & Febiger.

Guyton AC: Textbook of medical physiology, ed 7, Philadelphia, 1986, WB Saunders Co.

Hricik D and Kassirer J: Understanding and using anion gap, Consultant 23(7):130-143, 1983.

Kee JL: Laboratory and diagnostic tests with nursing implications, Norwalk, Conn, 1987, Appleton & Lange.

Maxwell M and others: Clinical disorders of fluid and electrolyte metabolism, New York, 1987, McGraw-Hill Book Co.

McIntyre K and Lewis A: Textbook of advanced cardiac life support, Dallas, 1983, American Heart Association.

Narins R and Emmett M: Simple and mixed acid-base disorders: a practical approach, Medicine 59(3):161-185, 1980.

Rose B: Clinical physiology of acid-base and electrolyte disorders, ed 3, New York, 1989, McGraw-Hill Book Co.

Shapiro B and others: Clinical application of blood gases, Chicago, 1989, Year Book Medical Publishers, Inc.

Valtin H and Gennari F: Acid-base disorders, Boston, 1987, Little, Brown & Co, Inc.

York K: The lung and fluid-electrolyte and acid-base imbalances, Nurs Clin North Am 22(4):805-814, 1987.

Chapter

14

Arterial Blood Gas Analysis

Arterial blood gas (ABG) values are used to determine acid-base status and oxygenation. ABGs provide information about the amount of oxygen, CO_2, **oxygen saturation,** arterial pH, HCO_3^-, and in most institutions, the base excess. Blood gas analysis is based on arterial sampling because venous blood is not suitable for assessment of oxygen tension. (Normal arterial and venous blood gas values are presented in Table 14-1.) Capillary blood, however, can be used for pH and P_{CO_2} analysis, and the earlobe is frequently used to collect the capillary specimen. Use of capillary blood is not recommended if the systolic blood pressure is <95 mm Hg or if the area has a poor blood supply.

OBTAINING ARTERIAL BLOOD

Without exposing it to air (i.e., anaerobically), arterial blood is obtained in a heparinized syringe from an appropriate artery (i.e., the brachial, radial, or femoral). Immediately after 2 to 4 ml of blood are withdrawn, the syringe is capped and placed on ice for transportation to the laboratory. If the analysis will be delayed for more than 30 minutes, the specimen is refrigerated because at room temperature the pH decreases at a rate of approximately 0.015 pH units every half hour.

Specific arterial puncture procedures will not be discussed in this book. Refer to other textbooks or your institution's policies and procedures manual for these procedures.

Table 14-1 Normal arterial and venous blood gas values

Arterial Values			Venous Values	
	Perfect	**Range**		
pH	7.40	7.35-7.45	pH	7.32-7.38
$Paco_2$	40 mm Hg	35-45 mm Hg	Pco_2	42-50 mm Hg
Pao_2	95 mm Hg	80-100 mm Hg	Po_2	40 mm Hg
Saturation	95%-99%		Saturation	75%
Base excess	+ or −2			
Total CO_2content	23-27 mm/L			
HCO_3^-	24 mEq/L	22-26 mEq/L	HCO_3^-	23-27 mEq/L

ARTERIAL BLOOD GAS (ABG) COMPONENTS

pH

pH measures the hydrogen ion (H^+) concentration and is the most important component of the ABG analysis when assessing acid-base status. pH directly reflects the acid-to-base ratio and signals the presence and severity of any acid-base derangement, although the pH alone does not identify the cause of the disorder. The perfect pH is 7.40, with a normal range of 7.35 to 7.45. It is important to remember that subtle changes in pH can reflect an acid-base disorder even when the pH is in the normal range. A *normal* pH in the presence of an acid-base disorder may be caused by a mixed disorder, with each imbalance actually cancelling the other (i.e., respiratory alkalosis and metabolic acidosis) or it may be caused by compensatory mechanisms that return the pH to near normal.

$Paco_2$

$Paco_2$ refers to the partial pressure of CO_2 in arterial blood. It is the most accurate reflection of alveolar function. $Paco_2$ is the respiratory component of acid-base regulation, and it is adjusted by changes in the rate and depth of alveolar ventilation. While factors other than alveolar function can affect Pao_2, they do not directly alter the $Paco_2$.

Hypercapnia ($Paco_2$ >45 mm Hg) signals alveolar hypoventilation and respiratory acidosis or respiratory compensation for metabolic alkalosis. **Hypocapnia** ($Paco_2$ <35 mm Hg) stems from alveolar hyperventilation and results in respiratory alkalosis. A decreased $Paco_2$ also can be the result of respiratory compensation for metabolic acidosis. When there is an alteration in pH secondary to metabolic disturbances, respiratory compensation occurs rapidly, within minutes.

HCO_3^-

Bicarbonate is the major base component in the acid-base system and it is regulated by the kidneys, where it is reabsorbed and regenerated to maintain a normal acid-base environment. Decreased HCO_3^- levels (<22 mEq/L) are usually indicative of metabolic acidosis. Elevated HCO_3^- levels reflect metabolic alkalosis, either as a primary metabolic disorder or as a compensatory alteration in response to respiratory acidosis. HCO_3^- is frequently calculated from the pH and $Paco_2$ and is obtained from the ABG value using the Henderson-Hasselbalch equation (see Chapter 13). Serum HCO_3^-, which is the sum of HCO_3^-, dissolved CO_2, and carbonic acid, is often reported as serum CO_2 when it is listed with other electrolytes on routine electrolyte chemistry reports. The terminology is sometimes confusing, but serum CO_2, serum HCO_3^-, and total CO_2 content are all measures of the total concentration of bicarbonate in the blood. It is imperative to analyze the pH and HCO_3^- closely, along with the $Paco_2$. Only by analyzing all three components will the nurse be able to determine if the $Paco_2$ value is normal, signals a primary respiratory disturbance, or is the result of a compensatory response to a metabolic acid-base disorder.

Pao_2

Pao_2 is the partial pressure of arterial oxygen. The normal value of Pao_2 varies slightly with age. In general, Pao_2 decreases from the minimum 80 mm Hg by 1 mm Hg for every year over age 60 (not applicable in individuals older than age 90). In the healthy newborn the acceptable Pao_2 ranges from 40 to 70 mm Hg on room air. A Pao_2 within normal limits has no primary role in acid-base regulation. A low Pao_2 *(hypoxemia)* results either from breathing air with a low-oxygen concentration (e.g., at high altitude) or from alterations in the function of the respiratory system (i.e., hypoventilation disorders or diffusion abnormalities). Hypoxemia is defined as a Pao_2 below the range of normal for the individual's age group. *Hypoxia,* which is inadequate oxygenation at the tissue level, can occur as a result of hypoxemia. Hypoxia is most common when the Pao_2 decreases to below 50 to 60 mm Hg. Lack of oxygen at the tissue level results in anaerobic metabolism, which causes lactic acid production and metabolic acidosis. While hypoxemia is determined by measuring the Pao_2, hypoxia is not measured directly, but it is inferred from signs and symptoms (e.g., increased heart rate, increased respiratory rate, decreased blood pressure, diaphoresis, breathlessness, weakness, headache, and lethargy) and supporting laboratory data (i.e., lactate levels).

Saturation

Saturation (Sao_2) measures the degree to which hemoglobin is saturated by oxygen. Saturation can be affected by changes in temperature, pH, and $Paco_2$. Oxygen saturation is often determined through calculations using standard hemoglobin values. Inaccuracies occur if the patient's actual hemoglobin differs markedly from the values used by the laboratory for calculations, as with anemia.

Oxy-hemoglobin dissociation curve

The oxy-hemoglobin dissociation curve demonstrates the relationship between Po_2 and oxygen saturation. Simply stated, the oxy-hemoglobin dissociation curve shows that when the Po_2 is high, such as in the capillaries of the lung, the hemoglobin is fully saturated with oxygen (Figure 14-1). When the Po_2 is lower, as in the capillaries of the tissue, the hemoglobin is less saturated, having freed oxygen for use at the tissue level. This relationship between Po_2 and O_2 saturation

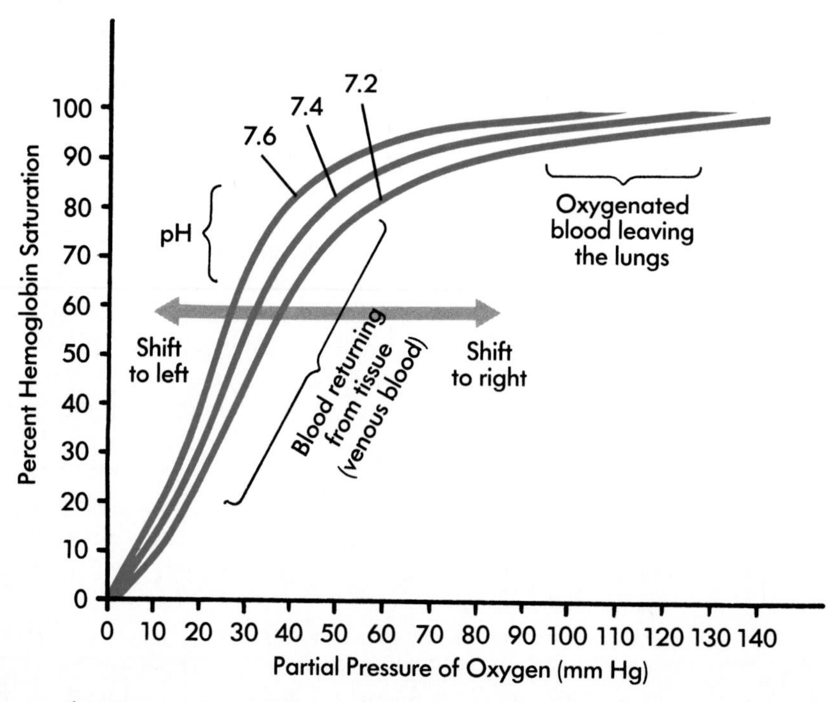

Figure 14-1 The oxy-hemoglobin dissociation curve demonstrates the affinity of hemoglobin for oxygen at a normal pH of 7.40. Saturation of hemoglobin is greatest in oxygenated blood leaving the lungs and decreases as oxygen is released at the tissue level. Shifts in the curve can occur with numerous conditions. A shift to the right occurs with acidemia (increased $Paco_2$) and hyperthermia and increases the movement of O_2 from the blood to the tissue. A shift to the left occurs with alkalemia (decreased $Paco_2$) and hypothermia and results in decreased movement of O_2 from the blood to the tissues.

(Modified from Guyton AC: Textbook of medical physiology, ed 7, Philadelphia, 1986, WB Saunders Co.)

is not linear. There is not always the same degree of change in O_2 saturation based on a similar degree of change in the Po_2.

When the Pao_2 decreases to below 60 mm Hg, there is a sharp decline in the saturation, and the decline continues with each small decrease in Pao_2. On the other hand, when the Pao_2 is > 60 mm Hg, there is a much smaller change in the saturation with each subsequent increase in Pao_2 (i.e., for a Pao_2 of 60 mm Hg, the saturation is 90%, while an increase in Pao_2 to 70 mm Hg only increases the saturation to 94%). Therefore it is important to keep the $Pao_2 \geq 60$ mm Hg because at this point the hemoglobin is saturated. This dissociation curve is applicable only under normal conditions. Alterations in acid-base balance or thermoregulation can affect the dissociation curve as well. For example, during acidosis, hypercapnia, and febrile states, hemoglobin has a decreased affinity for oxygen (i.e., saturation is less for a given Pao_2). This is called a shift of the oxygen dissociation curve to the right. A *shift to the right* aids the movement of oxygen from the blood to the tissues. When acidosis is present, for example, in instances of decreased tissue perfusion, the increased delivery of oxygen helps prevent further lactic acid accumulation. Conversely, with alkalemia, hypocapnia, and hypothermia, the oxygen dissociation curve shifts to the left, resulting in an increased hemoglobin affinity for oxygen. In other words, the saturation is greater for a given Pao_2. Thus oxygen movement from the blood to the tissue decreases. Other factors, such as hypophosphatemia (see Chapter 11), can affect the affinity of hemoglobin for oxygen, but the aforementioned are the most common causes for shifts in the oxy-hemoglobin dissociation curve.

Fraction of inspired oxygen (Fio_2)

Oxygen saturation is also related to the fractional percentage of oxygen (Fio_2) in the inspired air. Normal Fio_2 of room air is 21%, which correlates with a 96% oxygen saturation in a healthy adult. The O_2 saturation should increase as the Fio_2 increases.

Base Excess

Base excess is an excess or deficit of buffer base from what was calculated and expected for a specific patient based on pH, HCO_3^-, and hemoglobin concentration. Buffer base refers to all of the substances in the buffering system of whole blood that are able to bind excess hydrogen ions (H^+). The buffers primarily used to make these determinations are HCO_3^- and hemoglobin. The normal buffer base values for any patient are calculated based on the patient's actual hemoglobin concentration and the normal figures for pH and HCO_3^-. The normal base excess value is +2 to −2. The terms *base excess* and *base deficit* are derived from the difference in the measured total quantity of blood base buffers from what was calculated as normal for that patient. For example, if there is an increase in total buffer base (metabolic alkalosis), it is considered a positive base excess (> +2). On the other hand, if there is a decrease in total buffer base from the expected normal (such as in decreased HCO_3^-), it is considered a negative base excess or base deficit (< -2). Base excess refers to both the HCO_3^- and the hemoglobin, so it provides more information than just the HCO_3^-. The term *base excess* can be confusing, and while the information it relays is important, it is not critical in determining a patient's acid-base status.

STEP-BY-STEP GUIDE TO ABG ANALYSIS

A systematic step-by-step analysis is crucial to the accurate and quick interpretation of ABG values. ABG analysis includes identifying pH abnormalities and causative factors and assessing the degree of compensation present, as well as oxygenation. All these factors must be viewed in light of the clinical situation, taking the patient's history into consideration. Armed with knowledge about the patient's specific acid-base imbalance, nurses can make appropriate nursing diagnoses and develop a plan of care to meet the patient's needs. For more information on ABG analysis see Table 14-2.

Step One

Determine the pH value. If it deviates from 7.40, note how much it deviates and in which direction. For example, pH >7.40 equals alkalosis; pH <7.40 equals acidosis. Is the pH in the normal range of 7.35 to 7.45 or is it in the critical range of >7.55 or <7.20?

Step Two

Check the $Paco_2$. If it deviates from 40 mm Hg, how much does it deviate and in which direction? Does the change in $Paco_2$ correspond to the direction of the change in pH? For example, a decrease in $Paco_2$ would indicate an alkalosis. Does the pH reflect this? (The pH would increase with an alkalosis.) An increase in $Paco_2$ would indicate an acidosis. Does the pH reflect this? (The pH would be decreased with an acidosis.)

Step Three

Determine the HCO_3^- value (may be referred to as *total CO_2 content, serum CO_2, or serum HCO_3^-*). If it deviates from 24 mEq/L, note the degree and direction of deviation. Does the change in HCO_3^- correspond to the change in pH? For example, a decrease in HCO_3^- concentration would indicate an acidosis. Does the pH also reflect this? (The pH would be decreased with acidosis.) An increase in HCO_3^- would signal an alkalosis. Does the pH reflect this? (The pH would increase with alkalosis.)

Table 14-2 Brief overview of simple acid-base disturbances and compensatory responses

Disturbance	pH	$Paco_2$	HCO_3^-	Compensatory Response	Results of Compensation
Respiratory acidosis					
Acute	↓	↑	no change or very slight increase	Immediate release of tissue buffers (HCO_3^-)	1.0 mEq/L increase in HCO_3^- from patient's baseline for every 10 mm Hg increase in $Paco_2$
Chronic	slightly decreased	↑	↑	Increased renal reabsorption of HCO_3^-. Renal compensation begins in 8 hrs and reaches maximal effect in 3-5 days	3.5 mEq/L increase in HCO_3^- for every 10 mm Hg increase in $Paco_2$
Respiratory alkalosis					
Acute	↑	↓	no change or very slight decrease	Immediate release of tissue buffers (H^+)	2.0 mEq/L fall in HCO_3^- from patient's baseline for every 10 mm Hg decrease in $Paco_2$
Chronic	slightly increased or normal*	↓	↓	Decreased renal reabsorption of HCO_3^-. Initial effects of renal compensation takes about 8 hrs with a maximal effect in 3-5 days	5.0 mEq/L fall in HCO_3^- for each 10 mm Hg decrease in $Paco_2$
Metabolic acidosis	↓	↓	↓	Hyperventilation occurs immediately	1.2 mm Hg decrease in $Paco_2$ for each 1 mEq/L decrease in HCO_3^-
Metabolic alkalosis	↑	↑	↑	Hypoventilation occurs immediately	0.7 mm Hg increase in $Paco_2$ for every 1 mEq/L increase in HCO_3^-

*The only acid-base disturbance in which compensation can return the pH to normal.

Step Four

If both the $Paco_2$ and HCO_3 deviate from normal, which value corresponds more closely to the pH value? For example, if the pH reflects acidosis, which value (e.g., increased $Paco_2$ or decreased HCO_3^-) also reflects acidosis? The value that most closely corresponds to the pH and deviates the most from normal points to the primary disturbance responsible for the alteration in pH. There are 2 rules that help assess the effect that changes in $Paco_2$ and HCO_3^- have on pH. The first rule states that a change in $Paco_2$ by 10 mm Hg, whether up or down, will increase or decrease pH by 0.08. The second rule states that for every change in HCO_3^- by 10 mEq/L, there will be a corresponding change in pH of 0.15 units. Applying these rules will help in determining if the abnormal pH is caused by a respiratory or metabolic disorder or a combination of the two (Table 14-3).

A mixed metabolic-respiratory disturbance or compensatory elements may be present when the HCO_3^- and $Paco_2$ are both abnormal. With certain mixed disturbances, both the $Paco_2$ and the HCO_3^- may be abnormal with an abnormal pH. These mixed disturbances will be discussed in Chapter 19.

Step Five

Check the Pao_2 and the oxygen saturation to determine if they are within normal limits, increased, or decreased. A decrease in Pao_2 to 60 mm Hg or less will result in a severe decrease in oxygen saturation and a decrease in the delivery of oxygen at the tissue level, which is referred to as hypoxia. A decrease in Pao_2 and oxygen saturation may signal the need to increase the oxygen concentration the patient is receiving. Conversely, a high Pao_2 in a patient receiving oxygen may indicate a need to decrease the concentration of the delivered oxygen (Fio_2).

For a comparison of typical ABG values with acid-base disturbances, see Table 14-4.

Table 14-3 Rules for predicting pH based on $Paco_2$ and HCO_3^- changes

Alteration	Change in pH	End Result
$Paco_2$ ↑ 10 mm Hg 40→50	pH ↓ 0.08 7.40→7.32	Respiratory acidosis
$Paco_2$ ↓ 10 mm Hg 40→30	pH ↑ 0.08 7.40→7.48	Respiratory alkalosis
HCO_3^- ↓ 10 mEq/L 24 mEq/L→14 mEq/L	pH ↓ 0.15 7.40→7.25	Metabolic acidosis
HCO_3^- ↑ 10 mEq/L 24→34	pH ↑ 0.15 7.40→7.55	Metabolic alkalosis

ANALYZING OTHER LABORATORY TESTS
Total CO_2 Content

Total CO_2 measures the sum of HCO_3^-, H_2CO_3, and dissolved CO_2 in the blood. The amount of dissolved CO_2 and H_2CO_3 in the blood is very small, so the total CO_2 is basically a measure of the HCO_3^-. The total CO_2 content will be slightly higher (1 to 3 mmol/L) than the bicarbonate concentration calculated using the Henderson-Hasselbalch equation (see Chapter 13) and the pH and $Paco_2$ values from the blood gas sample. The reason for the difference in these calculated values is the small amount of H_2CO_3 and dissolved CO_2 that are measured with the HCO_3^- to determine total CO_2 content. The total CO_2 content is usually determined along with the electrolyte profile.

Anion Gap

The anion gap is the difference between the commonly measured cations and anions of the plasma. The electrolytes used to determine the anion gap are Na^+, Cl^-, and HCO_3^-. Thus the anion gap equals Na^+ minus (Cl^- plus HCO_3^-). Normal anion gap is 12 (+ or −2). If the gap deviates from normal, this indicates an abnormal quantity of either unmeasured anions or cations. The unmeasured cations that may decrease the anion gap are K^+, Ca^{2+}, and Mg^{2+}. These cations rarely cause a significant decrease in the anion gap because of their small concentration. The unmeasured anions that can cause an increase in the anion gap are protein-, PO_4^-, SO_4^-, and organic acids, including lactic acid and ketoacid. An increased anion gap is seen in numerous clinical situations involving acid-base disturbances. The anion gap is very helpful when determining the cause of metabolic acidosis. Metabolic acidosis is described according to the categories of normal or increased anion gap. Anion gap will be discussed in greater detail in Chapter 17.

Table 14-4 ABG comparisons of acid-base disorders

		Acidosis			Alkalosis		
		$Paco_2$	pH	HCO_3^-	$Paco_2$	pH	HCO_3^-
Simple	Respiratory	50	7.15*	25	25	7.6*	24
	Metabolic	38	7.20	15	44	7.54	36
Compensated	Respiratory	66	7.37	34	25	7.54	21
	Metabolic	23	7.28	9	50	7.42	31
Mixed Disorder		50	7.26	20	40	7.56	38

*Note the greater changes in pH with acute respiratory disorders because renal compensation is delayed.

Potassium

Patients with acid-base disturbances often have potassium disturbances, either hyperkalemia or hypokalemia, depending on the specific disorder. In acidosis the excess extracellular hydrogen ions (H^+) move into the cells to be buffered by intracellular buffers. For electroneutrality to be maintained, K^+ must then move out of the cells and into the extracellular fluid (ECF). This shift of ions results in a transient buffering of the intracellular pH and a rise in serum K^+. Although the elevated serum K^+ does not reflect an increase in overall K^+ stores, the hyperkalemia can cause serious disturbances. While hyperkalemia does not always occur with acidosis, it can be a helpful diagnostic sign (see Chapter 9).

A shift of H^+ and K^+ also occurs in alkalosis, but in this instance, hydrogen ions (H^+) are released from intracellular buffers to the extracellular space to restore the pH. To maintain electroneutrality, K^+ must move into the cells from the ECF. The degree of hypokalemia with alkalemia is usually very mild (see Chapter 9).

Concentrations of Ca^{2+} and Mg^{2+} also can be affected when the acid-base balance is deranged. See Chapters 10 and 12 for further discussion.

OTHER DIAGNOSTIC INFORMATION

Arterial-venous Difference (A-VO_2)

The difference between arterial oxygen content and the mixed venous oxygen content reflects how much O_2 has been utilized by the tissues (called tissue extraction of oxygen). Mixed venous sampling is usually done in the pulmonary artery via a pulmonary artery catheter, such as the Swan-Ganz. When the delivery of oxygen to the tissues is inadequate, as in heart failure, the normal difference between arterial and mixed venous oxygen content widens because tissue oxygen demand has exceeded oxygen delivery. Simultaneous analysis of arterial and pulmonary artery blood sampling thus provides an accurate and reliable index of ventricular function.

Alveolar-arterial Oxygen Difference (A-aO_2 Difference)

The normal difference between alveolar (A) and arterial (a) oxygen is 10 to 15 mm Hg on room air. This difference increases when there is a disorder in gas exchange in the lung, such as pneumonitis or adult respiratory distress syndrome (ARDS). The A-a oxygen difference is determined by using the Pao_2 from the ABG analyses. The Pao_2 is based on a formula that takes into account the inspired O_2 and the presence of CO_2 in the alveoli.

Case Study 14-1

In the following case studies, we will be analyzing only the components of the ABG that reflect the acid-base status. All interpretations and calculations are based on the *perfect* value for each component rather than the normal range.

The patient is admitted with the following laboratory data:

ABG Values	Normal (*perfect*) Values
pH 7.16	7.40
$Paco_2$ 70 mm Hg	40 mm Hg
HCO_3^- 25 mEq/L	24 mEq/L

1. Question: Is the pH normal?

Answer: No, the pH is low, signaling the presence of acidosis.

2. Question: Does the $Paco_2$ deviate from normal?

Answer: Yes, it is 30 mm Hg higher than normal.

3. Question: Does the change in $Paco_2$ correlate with the direction of the change in pH?

Answer: Yes, the pH reflects acidosis and the increase in $Paco_2$ also signals acidosis.

4. Question: What is the patient's disorder?

Answer: Acute respiratory acidosis. The pH is decreased, the $Paco_2$ is increased, and the HCO_3^- is normal. This respiratory acidosis could be the result of any process that impairs ventilation, such as sedative overdose, severe pulmonary edema, obesity. Specific causes of respiratory acidosis will be discussed in Chapter 15.

5. Question: Has there been any compensation by the kidneys?

Answer: No, with an acute rise in $Paco_2$ (respiratory acidosis), renal compensation does not begin for approximately 8 hours and does not reach maximal effect for 3 to 4 days. Over this time, you will see increasing acidification of the urine.

Case Study 14-2

ABG Values	**Normal (*perfect*) Values**
pH 7.39	7.40
$Paco_2$ 32 mm Hg	40 mm Hg
HCO_3^- 22 mEq/L	24 mEq/L

1. **Question:** Is the pH normal?

   Answer: No, it is slightly low, reflecting acidosis.

2. **Question:** Does the $Paco_2$ deviate from normal?

   Answer: Yes, the $Paco_2$ is 8 mm Hg less than normal.

3. **Question:** Does the change in $Paco_2$ correspond to the decrease in pH?

   Answer: No, the $Paco_2$ decrease reflects alkalosis, while the decreased pH signals acidosis.

4. **Question:** Does the HCO_3^- deviate from normal?

   Answer: Yes, the HCO_3^- is 2 mEq/L less than normal.

5. **Question:** Does the decrease in HCO_3^- correspond to the decrease in pH?

   Answer: Yes, both values are reflective of acidosis.

6. **Question:** What disorder does the patient have?

   Answer: Although not significant clinically, the very small change in pH and change in HCO_3^- signal a metabolic acidosis.

7. **Question:** How do you explain the decrease in $Paco_2$?

   Answer: The decreased $Paco_2$ is a respiratory compensation for the metabolic acidosis. The hyperventilation causes a decrease in $Paco_2$ to compensate for the decrease in HCO_3^-, which results from the buffering of an increased acid load. The decrease in $Paco_2$ helps maintain the 20:1 ratio of bases to acids in light of the decreased HCO_3^-. This slight metabolic acidosis could be the result of loss of base from diarrhea. Respiratory compensation occurs on a minute-by-minute basis, and the body's pH is protected from drastic changes in pH. For example, the above ABG values also could be the result of exercise. The HCO_3^- and pH decrease slightly with the production of lactic acid but the concurrent hyperventilation (as a compensatory respiratory response) keeps the pH from decreasing to a great degree.

Case Study 14-3

ABG Values	**Normal (*perfect*) Values**
pH 7.45	7.40
$Paco_2$ 47 mm Hg	40 mm Hg
HCO_3^- 34 mEq/L	24 mEq/L

1. **Question:** Is the pH normal?

   Answer: No, the pH is increased, signaling the presence of alkalosis.

2. **Question:** Is the $Paco_2$ normal?

   Answer: No, it is 7 mm Hg higher than normal.

3. **Question:** Does the change in $Paco_2$ correspond to the change in pH?

   Answer: No, an increasing $Paco_2$ (acidosis) should cause the pH to decrease (to <7.40).

4. **Question:** What pH value would you expect with a $Paco_2$ of 47?

   Answer: For every 10 mm Hg increase in $Paco_2$ there is a decrease in pH of 0.08. The $Paco_2$ has risen 7 mm Hg, which would cause approximately a 0.05 decrease in pH to 7.35 ($7.40 - 0.05 = 7.35$). The pH and the $Paco_2$ do not correspond.

5. **Question:** Is the HCO_3^- normal?

   Answer: No, the HCO_3^- is 10 mEq higher than normal (alkalosis). This change in HCO_3^- corresponds to the change in pH (alkalosis).

6. **Question:** What is the patient's acid-base disorder?

   Answer: Metabolic alkalosis.

7. **Question:** What would you expect the pH to be with this HCO_3^- value?

   Answer: For every 10 mEq increase in HCO_3^-, the pH should increase by 0.15. In this case, with a HCO_3^- of 34, the anticipated pH should be 7.55. Although the pH does reflect alkalosis at 7.45 the pH does not change to the degree anticipated by the rise in HCO_3^-.

8. **Question:** What prevented the pH from increasing as much as expected?

   Answer: The increase in $Paco_2$ from 40 mm Hg to 47 mm Hg has prevented a more serious metabolic alkalosis. The respiratory compensation for metabolic alkalosis is the retention of CO_2 (hypoventilation). Although this compensation occurs rapidly, it cannot return the pH to normal because the hypoventilation causing the retention of CO_2 also can cause a decreased Pao_2 with hypoxemia.

9. Question: Is the degree of respiratory compensation what you would expect, given the elevation of HCO_3^- to 34 mEq/L?

Answer: Yes, according to the information in Table 14-2 , for every 1 mEq/L increase in HCO_3^-, there is a 0.7 mm Hg compensatory increase in $Paco_2$. In this case, the 10 mEq/L increase in HCO_3^- would result in a 7 mm Hg increase in $Paco_2$ ($10 \times 0.7 = 7$). Adding 7 to 40, which is the *perfect* $Paco_2$ value, equals 47, which is what has occurred. Metabolic alkalosis could be the result of several causes, including NG suction or vomiting. For a complete discussion of metabolic alkalosis, see Chapter 18.

References

Adrogué H and others: Assessing acid-base status in circulatory failure: differences between arterial and central venous blood, N Engl J Med 320(20):1312-1316, 1989.

Bleich H: The clinical implications of venous carbon dioxide tension, N Engl J Med 320(20):1345-1346, 1989.

Civetta J and others: Critical care, Philadelphia, 1988, JB Lippincott Co.

Easterday U and Howard C: Respiratory dysfunctions. In Swearingen PL and others: Manual of critical care: applying nursing diagnoses to adult critical illness, St Louis, 1988, The CV Mosby Co.

Easterday U: Acid-base balance. In Horne M and Swearingen PL: Pocket guide to fluids and electrolytes, St Louis, 1989, The CV Mosby Co.

Fishbach T: A manual of laboratory diagnostic tests, Philadelphia, 1988, JB Lippincott Co.

Hricik D and Kassirer J: Understanding and using anion gap, Consultant 23(7):130-143, 1983.

Kee JL: Laboratory and diagnostic tests with nursing implications, Norwalk, Conn, 1987, Appleton & Lange.

Miller W: The ABCs of blood gases, Emerg Med 16(3):37-56, 1984.

Pfister S and Bullas J: Interpreting arterial blood gas values, Crit Care Nurs 6(4):9-14, 1986.

Shapiro B and others: Clinical application of blood gases, Chicago, 1989, Year Book Medical Publishers, Inc.

Chapter

15

Respiratory Acidosis

Respiratory acidosis, whether acute or chronic, is the result of alveolar hypoventilation, leading to an accumulation of excess CO_2 ($Paco_2$ > 40 mm Hg). Derangements in $Paco_2$ are a direct reflection of the degree of ventilation dysfunction.

Normally, CO_2 excretion equals CO_2 production. When there is an excess amount of CO_2, it is not the result of overproduction, but rather the failure of the lungs to eliminate the necessary amount to maintain the $Paco_2$ at 40 mm Hg. A rapid increase in $Paco_2$ causes acute respiratory acidosis, whereas chronically elevated $Paco_2$ causes a chronic respiratory acidosis. There are many causes of respiratory acidosis, but the underlying factor is the inability of the lungs to eliminate the appropriate amount of CO_2 to maintain the normal acid-base ratio.

Any disorder that causes a decrease in effective alveolar ventilation, thus the ability to eliminate CO_2, can result in an acute or chronic respiratory acidosis. There are two major mechanisms that decrease effective alveolar ventilation. The first is a decrease in **minute ventilation** (minute ventilation = RR × tidal volume; thus a decrease in either RR or tidal volume can cause hypoventilation). The second mechanism is an inequality in the ratio between ventilation and perfusion in the lungs (i.e., shunting of venous blood past unventilated alveoli). This phenomenon is called **ventilation-perfusion mismatch.**

The two components of ventilation that must be regulated to control levels of CO_2 are respiratory rate and depth. Numerous terms have been used when discussing ventilation. **Hyperpnea** and **tachypnea** refer to an increased respiratory rate but do not describe adequacy of ventilation. **Hypoventilation** refers to inadequate ventilation regardless of the respiratory rate. Hypoventilation with a resultant increase in $Paco_2$ can occur with increased or decreased respiratory rates. **Tidal volume,** which is the amount of air inhaled and exhaled during normal ventilation, is determined either through measurement or visual assessment of respiratory depth. The term *shallow* is often used to describe the appearance of a decreased tidal volume. *Hypercapnia* (also known as hypercarbia) is another term used to describe increased levels of $Paco_2$. During the discussion of respiratory acidosis, increased $Paco_2$ will be referred to as hypercapnia. The terms *hypoventilation* and *ventilation-perfusion mismatch* will be used to describe the processes leading to increased $Paco_2$ concentrations.

ACUTE RESPIRATORY ACIDOSIS

Acute respiratory acidosis occurs when there is a sudden failure in ventilation that results in a rapidly elevated $Paco_2$. An increase in $Paco_2$ is the normal stimulus that drives the respiratory center to increase the rate and depth of ventilation. If the respiratory center does not respond to this stimulus, or if the ventilatory response is inadequate and the excess levels of CO_2 are not eliminated, $Paco_2$ will continue to increase, pH will decrease, and respiratory acidosis will occur.

History and Risk Factors

Many disease processes can cause respiratory acidosis (see the accompanying box). Ventilatory failure resulting in respiratory acidosis usually conjures up images of individuals with acute severe pulmonary diseases. While severe pulmonary diseases, such as severe pneumonia and adult respiratory distress syndrome (ARDS), can cause acute increases in $Paco_2$, they are not the most common causes for respiratory acidosis. Usually sudden respiratory failure occurs in patients who have a depressed respiratory center—either from drugs or cerebral injury or disease—or in patients who have experienced a catastrophic event, such as cardiac arrest.

Dysfunctions of the respiratory/thoracic structures in healthy lung states

Respiratory acidosis can result from a muscular or mechanical inability of the respiratory and thoracic structures. Examples of this include flail chest, a condition in which numerous fractured ribs impede the normal movement of the thoracic cage; severe hypokalemia (potassium <2.0 mEq/L), which results in respiratory muscle weakness; and obstructive states (e.g., bronchospasm or foreign object), in which elimination of CO_2 is impeded. In these examples, the lungs are healthy and normal. When ventilation is affected by nonpulmonary causes, such as neuromuscular or mechanical problems, the patient will be tachypneic and dyspneic. However, when the hypoventilation is caused by an impaired respiratory center, for example, with cerebral trauma or sedative overdose, the respiratory rate usually will be decreased.

Potential Causes of Acute Respiratory Acidosis

Pulmonary/Thoracic Disorders
Severe pneumonia
ARDS
Flail chest
Pneumothorax
Hemothorax
Smoke inhalation

Airway Obstruction
Aspiration
Laryngospasm (anaphylaxis, severe hypocalcemia)
Severe bronchospasm
Severe, prolonged acute asthma attack

CNS Depression
Sedative overdose
Anesthesia
Cerebral trauma
Cerebral infarct

Metabolic Causes
High-carbohydrate diet

Neuromuscular Abnormalities
Guillain-Barré syndrome
Hypokalemia
High-cervical cordotomy
Drugs (e.g., curare)
Toxins

Systemic Causes
Cardiac arrest
Massive pulmonary embolus
Severe pulmonary edema

Mechanical Ventilation
Fixed minute ventilation with increased CO_2 production
Inappropriate dead space
Equipment failure

Acute lung disorders

When an acute lung disorder is present, hypoxemia will occur before hypercapnia. The delay in developing hypercapnia occurs because CO_2 is 20 times more diffusable across the alveolar capillary membrane than is oxygen. Because CO_2 is so readily diffusable, even small increases in $Paco_2$ in a previously healthy individual is a very serious sign of respiratory muscle fatigue, severe ventilatory dysfunction, and perhaps impending respiratory arrest. When hypercapnia occurs in acute pulmonary disease, such as pulmonary edema, pneumonia, ARDS, or bronchial asthma, it is a very grave prognostic indicator.

If hypercapnia is present and the individual is breathing room air, hypoxemia also will be noted. In the early stages of acute pulmonary disease the hypoxemia that is present will cause the respiratory rate and depth to increase, leading to respiratory alkalosis (see Chapter 16). It is only as the pulmonary disease progresses and the elimination of CO_2 is impaired that the individual will exhibit an increased $Paco_2$ and acute respiratory acidosis along with the hypoxemia. For example, patients with symptomatic asthma typically have a low $Paco_2$ (respiratory alkalosis) and hypoxemia.

Only when the obstruction is severe and prolonged and the respiratory muscles fatigue will the $Paco_2$ levels increase (respiratory acidosis). In acute events, such as cardiac arrest, that lead to dramatic and sudden increases in $Paco_2$, the accompanying hypoxia is a more life-threatening concern than the increasing $Paco_2$. $Paco_2$ rises steadily in complete ventilatory failure, with an increase of 3 to 5 mm Hg of CO_2 every minute, so 10 to 15 minutes of apnea would be necessary before severe hypercapnia would be noted. Conversely, hypoxia that is life threatening occurs in much less time, so oxygenating the patient is the most important goal.

Respiratory center depression

Drug-related respiratory center depression is one of the most common causes of respiratory acidosis. The causative agents most closely associated with drug-induced hypercapnia include general anesthetics (an exception is nitrous oxide), the anesthetic induction agent fentanyl citrate, and overdoses of opiates, barbiturates, and ethanol.

Narcotics

Narcotics, even in therapeutic doses, depress respiratory activity to some degree by decreasing the responsiveness of the respiratory center to increased CO_2 levels. It is impossible to predict how an individual will respond to a specific narcotic dose, and thus it is critical that the patient's response to the narcotic be monitored closely, especially the first dose (McCaffery and Beebe, 1989). In most patients, however, the degree of respiratory depression is not clinically significant, but nurses are often reluctant to increase the narcotic dose to the maximum amount prescribed, even if the patient has not experienced adequate pain relief at the lower dosage. Nurses' fears of producing respiratory center depression should not prevent a patient from receiving adequate pain relief. Severe pain is the body's natural antidote or antagonist to the respiratory depression caused by narcotics (Jaffe and Martin, 1985). Patients in severe pain usually can tolerate much higher doses of narcotics without respiratory depression. However, if the pain suddenly stops (e.g., in the patient who suddenly passes a kidney stone) this patient must be monitored closely for respiratory depression.

Patients' responses to a narcotic must be monitored according to their baseline status. For example, a respiratory rate that decreases from 16 to 10 breaths/min is more significant than a respiratory rate that decreases from 12 to 10 breaths/min. Maximal respiratory depression occurs within approximately 7 to 10 minutes after intravenous (IV) narcotics, 30 minutes after intramuscular doses, and within 90 minutes after subcutaneous doses.

Equianalgesic doses provide approximately the same relief of pain. For example, morphine 10 mg IM equals meperidine (Demerol) 75 mg IM. With a few exceptions*,

* The narcotic agonist-antagonists buprenorphine (Buprenex), butorphanol (Stadol), and nalbuphine (Nubain) are not as likely to cause severe respiratory depression.

equianalgesic doses of narcotics cause the same degree of respiratory depression. The individual's response to any change in narcotic or dose, even if equianalgesic, must be assessed continuously. If clinically significant respiratory depression occurs after a narcotic has been administered, naloxone (Narcan), a narcotic antagonist, can be administered IV to reverse this state. Narcan, however, is not effective in reversing the respiratory depression that can be caused by buprenorphine.

Neuromuscular dysfunctions

Paralyzing drugs, such as curare and succinylcholine, produce respiratory muscle paralysis, which leads to increased $Paco_2$ levels and respiratory acidosis unless the patient is mechanically ventilated. Other drugs that may contribute to the development of respiratory acidosis due to their ability to block transmission of the nerve impulse to the muscles of respiration include aminoglycosides, bacitracin, polymycin, and colistin sulfate.

Individuals with neuromuscular disorders, such as amyotrophic lateral sclerosis (ALS), Guillain-Barré syndrome, and myasthenic crisis, are also at risk for developing respiratory failure. The respiratory failure is a result of impaired muscles of respiration. The weakened or paralyzed respiratory muscles cannot provide the energy required to eliminate the necessary CO_2, and therefore respiratory acidosis develops.

Mechanical ventilation

If an individual is being ventilated mechanically at a fixed respiratory rate and volume and the production of CO_2 rises without a corresponding change in the patient's ability to eliminate it, respiratory acidosis can result. When fixed ventilatory rates are used, adjustments must be made in minute ventilation (i.e., RR or tidal volume) so that excess CO_2 can be eliminated. These ventilator adjustments are followed by ABG analysis to determine the adequacy of CO_2 elimination.

High-carbohydrate feedings

When parenteral or enteral nutrition with high concentrations of carbohydrates is given, O_2 consumption and CO_2 production increase as the carbohydrates are metabolized to CO_2 and H_2O. This increase in CO_2 production has been associated with increased $Paco_2$ in individuals on controlled mechanical ventilation. These feedings also may increase the ventilatory requirements of patients on other modes of ventilation. When patients in respiratory failure receive glucose loading or high-carbohydrate feedings, the hypercapnia may worsen, and weaning from the ventilator can be prolonged. Caloric needs should be met by providing a diet high in fat and low in carbohydrates. There are commercially prepared enteral feedings that are high in calories, high in fat, and low in carbohydrates that can be used with this patient population. Noncarbohydrate calories also can be provided by using IV fat emulsions.

Physiologic Responses

Chemical buffering

Acute rises in $Paco_2$ precipitate a rise in extracellular bicarbonate (HCO_3^-) through the action of nonbicarbonate buffers (primarily hemoglobin and proteins). The buffering begins immediately, reaches its maximum in 15 to 30 minutes, and persists for up to 6 hours. The resultant rise in serum HCO_3^- from this chemical buffering is small, approximately 1 mEq/L for every 10 mm Hg rise in $Paco_2$. The buffering is insufficient in maintaining a normal pH in the presence of an elevated $Paco_2$, and so there will be some degree of acidemia (pH <7.40).

Respiratory

The problem in respiratory acidosis is inadequate elimination of CO_2 by the lungs, and thus there is no available respiratory compensation.

Renal

Renal compensation is stimulated by the presence of increased CO_2 levels, but it takes hours to days for it to have a significant impact on the pH. It will take 3 to 5 days for the pH to return to near normal.

Assessment

Central nervous system manifestations

In general, the signs and symptoms of acute respiratory acidosis are related to the effect increased CO_2 has on the central nervous system (CNS), regardless of the cause. The CNS manifestations result from the fall in pH of the cerebrospinal fluid (CSF) as CO_2 increases. The decrease in CSF pH is greater in respiratory acidosis than in metabolic acidosis because CO_2 crosses the blood-brain barrier quickly, while HCO_3^- crosses it slowly. As expected, the CNS effects are also greater in acute respiratory acidosis than in acute metabolic acidosis. When the $Paco_2$ increases rapidly to 60 mm Hg or greater, numerous CNS manifestations are present, including restlessness, confusion, headache, nausea, vomiting, fine tremors to frank asterixis, lethargy, loss of consciousness, and seizures. The early signs of restlessness and slight confusion are subtle but very important clues to a serious problem.

Cardiopulmonary manifestations

Generally, acidemia diminishes the response of body tissues to sympathetic nervous system (SNS) stimulation, but this is offset because hypercapnia causes a great release of epinephrine and norepinephrine (called catecholamines) from the SNS. The increased catecholamine levels cause increased heart rate and increased cardiac output. The pulses bound, and because of peripheral vasodilatation, the skin becomes warm. The respiratory rate is usually increased initially (except when disorders involving the respiratory center are involved), but the respirations may be shallow and ineffective due to the rapid rate. As muscle fatigue occurs, the respiratory rate may decrease. With acute respiratory acidosis, the patient may complain of breathlessness and discomfort. Cyanosis may or may not be present, and it is not a reliable clue to the presence or severity of acute respiratory acidosis.

Vascular manifestations

Hypercapnia causes two opposing effects on the vasculature: vasodilatation and vasoconstriction. These effects usually balance each other out so that neither effect predominates except in the cerebral vessels. Increased $Paco_2$ results in increased cerebral blood flow and causes marked vasodilatation in cerebral blood vessels, leading to headache, papilledema, and increased intracranial pressure (ICP). The cerebral vessel vasodilatation caused by increased $Paco_2$ levels provides the physiologic basis for hyperventilating patients with increased ICP. When the $Paco_2$ is lowered, as occurs with hyperventilation, its effect on cerebral blood flow and dilatation is reduced and the ICP decreases.

Acute hypoxia

It is important to be alert to the signs and symptoms of hypoxia since hypoxemia is always present to some degree with hypercapnia. The most frequently occurring and important manifestations of hypoxia are related to its effect on the nervous system, including impaired judgment, motor incoordination, and irritability. As hypoxia worsens, the brainstem is affected, leading to respiratory arrest. Severe and prolonged hypoxia can impair hepatic and renal function. Signs of renal impairment, such as a decrease in urinary output (<30 ml/hr) and a rise in creatinine, may occur. See Table 15-1 for quick assessment guidelines for individuals with acute respiratory acidosis.

Diagnostic Studies

Arterial blood gas (ABG) values

Arterial blood gas (ABG) analysis is a necessity in diagnosing the presence of respiratory acidosis and determining its severity. A $Paco_2$ greater than 40 mm Hg and a pH less than 7.40 are diagnostic of this state. The normal regulatory mechanisms for CO_2 ($Paco_2$, pH, hypoxia) are ineffective, and the $Paco_2$ increases above 40 mm Hg. All individuals with hypercapnia will experience some degree of hypoxemia while breathing room air. Often the hypoxemia is severe, leading to hypoxia, and depending on the underlying pathology, may not respond to increasing concentrations of supplemental oxygen.

Table 15-1 Quick assessment guidelines for individuals with acute respiratory acidosis

Signs and Symptoms	Physical Assessment	Hemodynamic Parameters
Restlessness, confusion, dyspnea, headache, nausea, vomiting, fine tremors to frank asterixis, lethargy, loss of consciousness, and seizures	↑ HR and RR, diaphoresis, and cyanosis. If severe, may result in ↑ ICP with papilledema and dilated conjunctival and facial blood vessels	↑ ICP, ventricular dysrhythmias

History and Risk Factors

Acute respiratory disease: Acute respiratory failure from several causes, including pneumonia, ARDS

Overdose of drugs: Oversedation with drugs that cause respiratory center depression, such as opiates, barbiturates, ethanol

Chest wall trauma: Flail chest, pneumothorax

CNS trauma/lesions: Can lead to respiratory center depression

Asphyxiation: Mechanical obstruction, anaphylaxis

Impaired respiratory muscles: For example, with hypophosphatemia, hyperkalemia, polio, malnutrition, myasthenia gravis, muscular dystrophy, and Guillain-Barré syndrome

Iatrogenic: Inappropriate mechanical ventilation (increased dead space, insufficient rate or volume); high Fio_2 in the presence of CO_2 retention

Bicarbonate (HCO_3^-)

Initially, HCO_3^- values will be normal or slightly elevated. The slight increase is a result of the release of tissue buffers, and the increase will be approximately 1 mEq/L for every 10 mm Hg increase in $Paco_2$. Normally, the HCO_3^- does not exceed 32 mEq/L in uncomplicated acute respiratory acidosis. If it is higher than this amount, a concurrent metabolic alkalosis may be present. If the HCO_3^- is less than predicted, a concurrent metabolic acidosis may be present. Remember, during acute respiratory acidosis there is no significant renal compensation. Renal compensation takes days to affect the HCO_3^- concentration significantly.

Serum electrolytes

Serum electrolytes usually will not be altered greatly with respiratory acidosis. The serum Na^+ and K^+ will be mildly increased, and the anion gap will be normal. If the electrolytes or the anion gap is abnormally altered, another acid-base disturbance is present. The most common disorder that occurs with respiratory acidosis is metabolic acidosis. It is the most common because the hypoxemia that is present can lead to lactic acidosis.

Chest x-ray

A chest x-ray can determine the presence and severity of pulmonary diseases (such as ARDS and pneumonia) that can lead to respiratory acidosis.

Medical Management

The goal of medical management is to restore effective alveolar ventilation, which will return the $Paco_2$ to normal. In addition, prompt identification and treatment of the underlying disorder is critical. These interventions may include antibiotics, bronchodilatation, and specific breathing treatments.

Oxygen therapy

Oxygen therapy is often indicated because severe hypoxia often accompanies hypercapnia in acute respiratory failure. A Pao_2 less than 60 mm Hg can lead to tissue damage, while a Pao_2 of 30 mm Hg or less can result in death. When oxygen is used to treat uncomplicated acute respiratory acidosis, it carries no risk of making the hypercapnia worse. However, because of the potentially toxic effects, the concentration of oxygen used to treat hypoxemia should be the lowest concentration possible to maintain the Pao_2 between 50 to 60 mm Hg. A fraction of inspired oxygen (Fio_2) of 0.40 or less is desired. Even this concentration of oxygen can produce changes in lung cells, although no long-term negative effects have been documented at oxygen concentrations of 40% or less. Oxygen concentrations greater than 80% are toxic to the lung in 5 to 6 days. Oxygen therapy for acute respiratory acidosis, when it is superimposed on a chronic respiratory acidosis with CO_2 retention, will be discussed later (see the section on chronic respiratory acidosis later in this chapter).

Intubation and mechanical ventilation

If the $Paco_2$ is greater than 50 to 60 mm Hg, the Pao_2 is less than 50 mm Hg, and signs of ventilatory failure are present (e.g., confusion and lethargy), the patient usually requires intubation and mechanical ventilation. The goal of ventilation is to maintain the Pao_2 at 60 to 70 mm Hg and return the $Paco_2$ to normal. In acute respiratory failure, returning the $Paco_2$ to within normal range does not pose the risk of a life-threatening metabolic alkalosis as it would in chronic respiratory acidosis (see the section on chronic respiratory acidosis later in this chapter).

Sodium bicarbonate ($NaHCO_3$)

Rapid correction of pH in acute respiratory acidosis with $NaHCO_3$ is usually not indicated because it can result in fluid volume excess by increasing osmolar concentration of the blood, thereby causing fluid to shift into the intravascular space, which could further compromise respiratory function. Treatment with $NaHCO_3$ also can result in a metabolic alkalosis as the $Paco_2$ returns to normal with ventilatory support. Occasionally, $NaHCO_3$ is indicated when the pH is less than 7.0 in acute respiratory acidosis if the cause is bronchial asthma or bronchospasm. In these circumstances if epinephrine is to be used, treatment with $NaHCO_3$ first to increase the pH to about 7.20 will make the tissues more responsive to the effects of epinephrine (bronchodilatation). This is necessary because acidemia diminishes the tissues' response to SNS stimulation (i.e., to catecholamines, such as epinephrine). In general, however, the use of bicarbonate or nonbicarbonate buffers is not indicated in the treatment of respiratory acidosis.

Nursing Diagnoses and Interventions

Impaired gas exchange related to alveolar hypoventilation secondary to underlying disease process

Desired outcome: Patient has adequate gas exchange as evidenced by $Pao_2 \geq 60$ mm Hg, $Paco_2 \leq 45$ mm Hg, pH 7.35-7.45, RR 12 to 20 breaths/min with a normal respiratory pattern and depth (eupnea), and absence of adventitious (abnormal) breath sounds.

1. Monitor serial ABG results to detect continued presence of hypercapnia or hypoxemia. Report significant findings (i.e., variance of 10 mm Hg in $Paco_2$ or Pao_2).
2. Assess and document character of respiratory effort: rate, depth, rhythm, and use of accessory muscles of respiration.
3. Assess patient for signs and symptoms of respiratory distress: restlessness, anxiety, confusion, and tachypnea (RR >20 breaths/min).
4. Position patient for comfort and to ensure optimal gas exchange. Usually, semi-Fowler's position allows for adequate expansion of the chest wall, but the specific pathologic process must be considered when positioning patients.
   - For unilateral lung disease in the patient who is ventilated mechanically, a side-lying (lateral decubitus) position may increase perfusion in the dependent (healthy) lung and increase ventilation to the upper (diseased) lung.
   - For patients with bilateral lung disease resulting from ARDS, requiring mechanical ventilation, the prone position may improve gas exchange by decreasing the edema and increasing the ventilation to the dependent lung area (Langer, 1988). Hypoventilation can occur in dependent areas with mechanical ventilation, resulting in atelectasis. Not all patients with ARDS will benefit from the prone position and a brief trial (30 minutes) in this position is recommended to identify patients who can benefit from it. Improved ventilatory status will be noted by improvement in breath sounds and ABG values (increased Pao_2 on same oxygen concentration and ventilator settings).

Potential sensory/perceptual alterations related to changes in sensorium secondary to disturbance in acid-base regulation

Desired outcome: Patient verbalizes orientation to person, place, and time and remains free of injury.

1. Monitor ABG and serum CO_2 results. Notify MD of abnormal values and significant changes (i.e., variances of $\geq$ 10 mm Hg in Pao_2 and $Paco_2$ and changes of 4 to 5 mEq/L in serum HCO_3^-).
2. At frequent intervals assess and document patient's LOC and orientation to person, place, and time. Notify MD of significant findings immediately. Subtle changes, such as restlessness and mild confusion, can herald an increase in $Paco_2$.
3. Use reality therapy, such as familiar photos, a calendar, and a clock with a face large enough for patient to see. Keep these items at the bedside and within the patient's visual field. If the patient normally wears eyeglasses or hearing aids, make sure these items are accessible, and help patient use them, if necessary.
4. Keep the bed in its lowest position, with all siderails up and the wheels locked.
5. If the patient is allowed out of bed, remind patient to ask for assistance before getting up.
6. Offer confused patients the opportunity to toilet at frequent intervals. Many falls result from a disoriented or unsteady patient's attempt to use the toilet.
7. Use a night light to minimize patient's confusion in an unfamiliar and unlit environment.
8. If the patient's confusion persists despite reorientation, increase the frequency of observations, with concomitant reorientation and documentation. Alert MD to continued or increasing confusion.
9. Reassure significant others that patient's confusion will abate with treatment.
10. If patient remains at risk for injury, obtain a prescription for a protective restraining device per agency policy.
    - Document need for restraining device based on patient's behavior upon initiation and q8h thereafter.
    - Choose the least restrictive device that will protect patient. Add additional device(s) as needed, based on documented patient behaviors.
    - Document type of device used. A protective vest is applied to the upper torso. Most vests criss cross or form a *V* in the front. To avoid potential injury to the patient, read the manufacturer's recommendations for proper application. Soft limb protectors are applied to the upper extremities (may be applied to upper and lower extremities in some cases).
    - Monitor patient q15 to 30 min while restraints are in use, documenting mental status and response to protective device(s).
    - Document frequency (at least q2h) of device removal, repositioning of patient, and reapplication of the device.

Potential alteration in oral mucous membrane related to abnormal breathing pattern

Desired outcome: Patient's oral mucosa, lips, and tongue remain moist and intact.

1. Assess patient's oral mucous membrane, lips, and tongue q2h, noting presence of dryness, exudate, swelling, blisters, and ulcers.
2. If the patient is alert and able to take fluids orally, offer frequent sips of water or ice chips to alleviate dryness.
3. Perform mouth care q2 to 4h, using a soft-bristled toothbrush to cleanse the teeth and a moistened cloth or Toothette (small sponge on a stick) to moisten crusty areas or exudate on the tongue and oral mucosa. If patient is intubated, suction the mouth to remove fluid and debris, and then use a toothbrush or Toothette as appropriate to cleanse the mouth and teeth.
4. If indicated, use an artificial saliva preparation to assist in keeping the mucous membrane moist. Avoid use of lemon glycerine swabs, which can contribute to dryness.

Sleep pattern disturbance related to frequent treatments and procedures

Desired outcome: Patient sleeps undisturbed for at least 90 minutes at a time and relates a feeling of well being.

1. Gather information about patient's normal sleep habits: bedtime rituals; usual position; hours of sleep required; number of pillows used; sensitivity to light, noise, and touch. Based on data gathered, attempt to incorporate patient's needs into his or her care plan.
2. Cluster activities and procedures to provide 90-minute periods of uninterrupted sleep. Even brief (30-second) interruptions can result in feelings of fatigue.
3. Attempt to administer unpleasant or uncomfortable procedures or treatments at least 1 hour before bedtime to allow time for relaxation before patient attempts to sleep.
4. Offer backrubs, repositioning, and relaxation techniques (quiet music, instructions for imagery) at bedtime.
5. If patient requires a daytime nap because of fatigue, provide opportunities to sleep between 1PM and 3PM. Sleep attained at this time is most restful and least likely to disrupt nighttime sleep. Discourage napping after 7PM, as it may prevent patient from falling asleep and maintaining a normal sleep cycle.

Potential for ineffective family coping related to stress reaction secondary to catastrophic illness of family member

Desired outcome: Family members exhibit effective coping mechanisms, seek support from others, and discuss concerns among the family unit.

1. Establish an open line of communication with the family, providing an atmosphere in which family members can ask questions, ventilate feelings, and discuss concerns.

2. Assess family members' knowledge about patient's therapies and treatments. Provide information, as needed, and reinforce information family has received from other health care members.
3. Provide opportunities and areas for family members to talk privately as well as share concerns with health care members.
4. Determine effective coping strategies that have been used by the family in other stressful situations and encourage the family to use them.
5. Enable family members to spend time alone with patient for short, frequent intervals.
6. Encourage family members to pursue diversional activities outside the hospital to help alleviate stress. Families often feel the need for *permission* from health care members to leave the waiting room or hospital.
7. Offer realistic hope.

CHRONIC RESPIRATORY ACIDOSIS (COMPENSATED)

Chronic respiratory acidosis occurs in individuals with pulmonary disorders, in which there is a decrease in the effective alveolar ventilation or a mismatched ratio of ventilation to perfusion (ventilation-perfusion mismatch), for example, chronic bronchitis or emphysema. Over time the amount of CO_2 eliminated is less than the amount generated, and thus the $Paco_2$ levels increase. If the elevated $Paco_2$ levels exist for 3 to 5 days or more, maximal renal compensation will occur. The $Paco_2$ will remain elevated, but because of the increased HCO_3^- concentration (caused by renal compensation), the pH, while still reflecting a small degree of acidosis, will be nearly normal. Individuals whose acute respiratory acidosis lasts 3 to 5 days or more will evoke the maximal renal compensatory response and their acid base status should now be viewed as chronic respiratory acidosis. The HCO_3^- concentration will increase 3 to 4 mE/L for every 10 mm Hg rise in $Paco_2$ in individuals with chronic CO_2 retention. Patients with chronic lung disease also can experience acute rises in $Paco_2$ secondary to superimposed disease states, such as pneumonia. In this case, the $Paco_2$ rises acutely from the patient's baseline. The degree of renal compensation present usually is inadequate to meet the increased demands for buffering, and a decrease in pH will occur until a new steady state can be reestablished, with further increases in serum HCO_3^- concentration to offset the new increase in $Paco_2$ levels.

As mentioned earlier, a decrease in CSF pH resulting from an increase in $Paco_2$ is the normal stimulus for the respiratory center to increase the rate and depth of respirations. When hypercapnia is sustained (e.g., with chronic bronchitis), the respiratory center becomes less sensitive to further increases in $Paco_2$, and hypoxemia becomes the stimulus to breathe. To some degree hypoxemia is always present with chronic hypercapnia. In fact, the hypoxemia precedes the hypercapnia. The exchange of oxygen is impaired first because of the pulmonary disease, followed by impairment in the exchange of CO_2 as the disease progresses. Sustained hypercapnia results in diminished sensitivity of the respiratory center to further increases in $Paco_2$. This decreased sensitivity to increased levels of CO_2 results from the increased concentration of HCO_3^- in the CSF. The HCO_3^- concentration in the CSF and other body fluids increases as the kidneys compensate for the respiratory acidosis. With the presence of increased HCO_3^- concentration in the CSF, the pH of the CSF fluids does not decrease even though the CO_2 concentration increases. Without the change in CSF pH the respiratory center is not stimulated.

When increased $Paco_2$ concentration no longer stimulates the respiratory center, hypoxemia becomes the stimulus to breathe. If the hypoxemia is corrected by the administration of oxygen, the stimulus to breathe may be removed, resulting in a respiratory arrest. It is important to remember, however, that oxygen is not always harmful to patients with chronic obstructive pulmonary disease (COPD). Not all patients with COPD retain CO_2, and thus they do not all rely on hypoxemia to stimulate respirations. It is crucial that nurses know the individual's history to determine the potential risk when administering oxygen to a patient with COPD.

History and Risk Factors

There are many causes of chronic respiratory acidosis, chronic obstructive pulmonary disease (COPD) being the most common (see the accompanying box).

Chronic obstructive pulmonary disease (COPD)

COPD refers to two distinct, though related processes: chronic bronchitis and emphysema.* An individual with COPD rarely has *pure* emphysema or *pure* chronic bronchitis—components of both are usually present. While these diseases may coexist, usually one is predominant. It is important to be aware of the pathophysiology and clinical progress for each of these disorders since this information helps nurses understand treatment differences, expected complications, and outcomes for these patients.

Chronic bronchitis

Chronic bronchitis causes airway obstruction because of increased mucus production, inflammation, and bronchospasm. Patients who have chronic bronchitis often have a long history of smoking and a productive cough with episodes of purulent sputum. Hypoxemia is present (Pao_2 45 to 60 mm Hg) and is the result of the shunting of venous blood past unventilated alveoli (ventilation-perfusion mismatch). This poorly oxygenated blood mixes with blood oxygenated in healthier regions of the lung, resulting in a decreased Pao_2. The hypoxemia that results plays a critical role in disease progression. The hypoxemia becomes the stimulus to breath because of the chronic CO_2 retention common with chronic bronchitis. If the hypoxemia is corrected by

* Although asthma is classified as a chronic obstructive pulmonary disease, it does not cause chronic respiratory acidosis, so it will not be discussed here. For further information about asthma, see Chapter 16.

Potential Causes of Chronic Respiratory Acidosis

Obstructive Diseases	Restriction of Ventilation	Neuromuscular Abnormalities	Depression of the Respiratory Center
Emphysema	Kyphoscoliosis	Poliomylitis	Brain tumor
Chronic bronchitis	Hydrothorax	Muscular dystrophy	Bulbar poliomyelitis
Cystic fibrosis	Fibrothorax	Multiple sclerosis	Chronic sedative overdose
Obstructive sleep apnea	Severe chronic pneumonitis	Amyotrophic lateral sclerosis (ALS)	Obesity-hypoventilation (Pickwickian) syndrome
	Obesity-hypoventilation (Pickwickian) syndrome	Diaphragmatic paralysis	

supplemental oxygen delivery, the stimulus to breath is eliminated and a respiratory arrest can occur. When hypoxemia requires treatment with supplemental oxygen, cautious administration is necessary. Chronic hypoxemia (Pao_2 <60 mm Hg) causes vasoconstriction, which produces pulmonary hypertension, which in turn results in right heart failure. Numerous clinical signs of heart failure then appear: edema, ascites, hepatomegaly, jugular venous distention. The decreased Pao_2 also results in desaturation of hemoglobin, which causes cyanosis, and stimulates production of red blood cells (RBCs), leading to polycythemia (hematocrit 45% to 60%). The combination of cyanosis and edema from the right-sided heart failure gives rise to the term *blue bloaters* to describe patients with predominant chronic bronchitis. These patients may have repeated episodes of ventilatory failure, but because there are reversible components of the disease (i.e., inflammation), they tend to respond favorably to treatment.

Emphysema

Predominant emphysema is characterized by the destruction of alveolar walls. Destruction of the alveoli leads to air trapping, which results in hyperinflation. As the alveoli are destroyed, the number of capillaries in the remaining alveoli are decreased. All these pathophysiologic changes lead to a decrease in the number of alveoli available for gas exchange and to abnormal ventilation and perfusion (e.g., some areas of the lung are ventilated but not perfused, while other areas are perfused but not ventilated). Also, as lung tissue is destroyed, there is less support of airways, causing them to collapse on inspiration. This increases the work of breathing (WOB).

Individuals with predominant emphysema usually have a history of gradual and progressive dyspnea, minimal cough, and weight loss. The respiratory rate is increased and accessory muscles of respiration (i.e., scaleni and sternocleidomastoids) are used to enhance the breathing effort even at rest. The increased respiratory rate increases the minute volume, and individuals with emphysema are able to maintain their Pao_2 and $Paco_2$ within an acceptable range until late in the disease. Because the hemoglobin is nearly saturated, cyanosis does not occur. The combination of increased respiratory rate, increased WOB, and lack of cyanosis is the reason these individuals are called *pink puffers.* COPD patients with predominant emphysema do not retain CO_2, are sensitive to increases in $Paco_2$, and usually can respond appropriately (unless a superimposed respiratory infection is present) to a respiratory stimulus. Rather than hypoxemia, increased $Paco_2$ continues to be their stimulus to breathe, so oxygen delivery does not pose the risk of inducing a respiratory arrest in these individuals since it does not abolish the drive to breathe.

Individuals with emphysema usually do not have as many bronchopulmonary infections as do those with bronchitis. When they do have a pulmonary infection, they respond poorly to treatment. In fact, since emphysema has no reversible components, there is no effective treatment to reduce the symptoms or slow the progress of the disease. Chronic, severe hypoxia and cor pulmonale usually do not occur until late in the disease process. The normal course with emphysema involves severe and progressive dyspnea.

Predisposing factors for emphysema are smoking and an inherited disorder (homozygotic α-1 antitrypsin deficiency). It is not known why some smokers develop bronchitis while others develop emphysema, although both processes are usually present to some degree in each patient population. Table 15-2 summarizes the distinguishing features of chronic bronchitis and emphysema.

Other causes

The restrictive pulmonary diseases result in decreases in alveolar ventilation and lung volume, either from anatomic restriction (e.g., kyphoscoliosis or obesity) or decreased lung compliance (e.g., interstitial fibrosis). Other causes of alveolar hypoventilation that can lead to increased $Paco_2$ include chronic respiratory center depression (e.g., tumors) and abnormal neuromuscular functioning (e.g., polio or ALS). No primary bronchopulmonary disease exists in some of the restrictive disorders, such as neuromuscular abnormalities (polio), or anatomic abnormalities (obesity).

Assessment

Signs and symptoms of chronic respiratory acidosis vary widely, based on the causes and the patient's individual response to the disease process. In general, many patients with chronic respiratory acidosis from causes listed in Table 15-3 are asymptomatic when renal compensation occurs (remember that the pH is near normal because of increased HCO_3^- concentration) and when they are free of superimposed acute disturbances, such as pneumonia. Individuals with chronically elevated $Paco_2$ may experience symptoms from their underlying disease process, but the signs and symptoms are not related specifically to the chronic respiratory acidosis. Even a chronically elevated $Paco_2$ that exceeds 70 mm Hg usually will not cause symptoms.

Table 15-2 Distinguishing features of chronic bronchitis and emphysema

Features	Chronic Bronchitis	Emphysema
$Paco_2$	50-60 mm Hg	35-40 mm Hg
Pao_2	45-60 mm Hg	65-75 mm Hg
Dyspnea	Severe	Mild
Cough	Before dyspnea occurs	After dyspnea occurs
Sputum	Copious, frequently purulent	Scant, mucoid
Etiology	Smoking, environmental and occupational exposure	Smoking, inherited
Hematocrit	50%-55%	35%-45%
Right-sided heart failure	Common	Rare, except with terminal illness
Recurrent bronchial infection	Frequent	Less frequent
Respiratory insufficiency episodes	Repeated	Frequently terminal

CNS manifestations of hypercapnia will be present if an acute pulmonary insult occurs and causes the $Paco_2$ to rise rapidly above the patient's baseline or if the stimulus to breathe is hypoxemia and it is eliminated by the delivery of high concentrations of oxygen. If the baseline $Paco_2$ increases rapidly, the patient may experience a dull headache, weakness, **asterixis**, agitation, and insomnia, progressing to somnolence and coma. The somnolence and coma resulting from the rapid rise in CO_2 is often referred to as *carbon dioxide narcosis.* If chronic hypoxemia is present and it leads to cor pulmonale, signs of right-sided heart failure may occur (i.e., ascites, edema, and hepatomegaly). See Table 15-3 for quick assessment guidelines for individuals with chronic respiratory acidosis.

Diagnostic Studies

Arterial blood gas (ABG) values

ABG values provide the data necessary for determining the presence and severity of chronic respiratory acidosis. Although the $Paco_2$ will be elevated (>40 mm Hg), the pH will be close to normal (though less than 7.40) because of renal compensation. In patients experiencing an acute pulmonary infection or other compromising disorder, the $Paco_2$ may be elevated above the patient's normal baseline and beyond what the kidneys will have compensated for, resulting in a decreased pH. The Pao_2 will be less than normal. The degree of hypoxemia will depend on the patient's chronic disorder and the presence of an acute illness.

Table 15-3 Quick assessment guidelines for individuals with chronic respiratory acidosis

Signs and Symptoms	Physical Assessment
If renal compensation occurs, no specific symptoms will be noted. With a rapid rise in $Paco_2$, the following may be present: dull headache, weakness, dyspnea, asterixis, agitation, and insomnia, progressing to somnolence and coma	Tachypnea, cyanosis. Severe hypercapnia ($Paco_2$ >70 mm Hg) may cause cerebral vasodilatation resulting in ↑ ICP, papilledema, and dilated conjunctival and facial blood vessels. Edema may be present secondary to right ventricular failure

History and Risk Factors

COPD: Emphysema and chronic bronchitis
Development of superimposed acute respiratory infection in a patient with COPD
Exposure to pulmonary toxins: Occupational risk; pollution
Restrictive diseases: Kyphosis, extreme obesity (Pickwickian syndrome)

Bicarbonate (HCO_3^-)

HCO_3^- is especially helpful in determining the level of renal compensation that has occurred (i.e., increased HCO_3^- with a near normal pH if fully compensated). This information is particularly useful in identifying *mixed* acid-base disturbances (see Chapter 19) because the HCO_3^- is expected to be elevated by 3 to 4 mEq/L for each 10 mm Hg increase in $Paco_2$. If the HCO_3^- concentration is less than what is anticipated for the degree of hypercapnia, this could signal the presence of another pathologic process (i.e., metabolic acidosis). Conversely, if the HCO_3^- is greater than 45 mEq/L, another underlying problem is present (i.e., metabolic alkalosis) because HCO_3^- rarely exceeds 45 mEq/L as renal compensation for respiratory acidosis.

Serum electrolytes

The Na^+, K^+, and Cl^- all should be within normal limits unless other acid-base or electrolyte imbalances are present.

Chest x-ray

A chest x-ray will determine the extent of the underlying pulmonary disease and identify further pathologic changes (e.g., pneumonia) that may be responsible for the presence of an acute exacerbation.

Electrocardiogram (EKG)

An EKG identifies cardiac involvement from COPD. The combination of an isoelectric P wave and an isoelectric QRS in lead I in a patient who is middle-aged or older strongly suggests the presence of chronic bronchitis or emphysema. These changes are the result of the shift in position of the heart due to hyperinflation in the lungs. In addition, supraventricular tachycardia and ventricular tachycardia are common in an individual with chronic respiratory acidosis who is experiencing an acute exacerbation.

Sputum culture

A sputum culture is done to determine the presence of pathogens (e.g., with pneumonia) that may be causing acute exacerbations of the pulmonary disease in a patient with COPD.

Medical Management

Oxygen therapy

In patients with COPD, hypoxemia usually responds well to supplemental oxygen treatment, but it must be delivered cautiously to those individuals in whom hypoxemia rather than hypercapnia stimulates ventilation. Most patients with COPD receive adequate oxygenation with 1 to 2 L of oxygen by nasal cannula or mask delivery, with an FiO_2 at 0.24 to 0.28, but the adequacy of oxygen therapy must be based on ABG findings. An FiO_2 greater than 0.40 is usually not required.

Intubation and mechanical ventilation

Patients may require intubation and mechanical ventilation if the drive to breathe is eliminated by the oxygen therapy or if the individual is experiencing acute exacerbations, such as pneumonia, causing a further decline in oxygenation and an acute rise in $PaCO_2$. The goal of mechanical ventilation is to maintain the PaO_2 at 50 to 60 mm Hg and the $PaCO_2$ at less than 50 mm Hg. In general, the $PaCO_2$ should not be decreased by more than 10 mm Hg each hour (Effros, 1988). If the $PaCO_2$ is corrected rapidly, a dangerous metabolic alkalosis may develop. This metabolic alkalosis is the result of an increased HCO_3^- concentration that occurs as renal compensation for the chronic respiratory acidosis. When chronically elevated $PaCO_2$ decreases rapidly, HCO_3^- remains elevated until the excess can be eliminated by the kidneys. This alkalosis is referred to as *posthypercapneic alkalosis,* and it can cause cardiac dysrhythmias, confusion, seizures, coma, and death.

Pharmacotherapy

Bronchodilators and antibiotics are given as indicated. Narcotics and sedatives can depress the respiratory center (see the first section in this chapter on acute respiratory acidosis) and are avoided unless the patient is intubated and mechanically ventilated. Since HCO_3^- crosses the blood brain barrier slowly, and the Na^+ load can cause pulmonary edema, $NaHCO_3$ is an inappropriate treatment for chronic respiratory acidosis.

Chest physiotherapy

Chest physiotherapy may aid in the expectoration of sputum. It includes postural drainage if copious secretions are present. The patient is assessed closely during chest physiotherapy because the treatment may be poorly tolerated, especially the postural drainage component.

Nursing Diagnoses and Interventions

Impaired gas exchange related to trapping of CO_2 secondary to pulmonary tissue destruction (appropriate for the patient with COPD)

Desired outcome: ABG values reflect a $Paco_2$ and pH within acceptable range, based on patient's underlying pulmonary disease.

1. Monitor serial ABG results to assess patient's response to therapy. Report significant findings to MD: an increasing $Paco_2$ and a decreasing pH.
2. Assess and document the patient's respiratory status: respiratory rate and rhythm, exertional effort, and breath sounds. Compare pretreatment findings with posttreatment (e.g., oxygen therapy, physiotherapy, or medications) findings for evidence of improvement.
3. Assess and document patient's LOC. If $Paco_2$ increases, be alert to subtle, progressive changes in mental status. A common progression is agitation → insomnia → somnolence → coma. To avoid a comatose state secondary to rising CO_2 levels, always evaluate the *arousability* of a patient with elevated $Paco_2$ who appears to be sleeping. Notify MD if patient is difficult to arouse.
4. Ensure appropriate delivery of prescribed oxygen therapy. Assess patient's respiratory status after every change in Fio_2. Patients with chronic CO_2 retention may be very sensitive to increases in Fio_2, resulting in depressed ventilatory drive. If the patient requires mechanical ventilation, be aware of the importance of maintaining the compensated acid-base environment. If the $Paco_2$ were rapidly decreased by a high respiratory rate via mechanical ventilation, a severe metabolic alkalosis could develop. The sudden onset of metabolic alkalosis may lead to hypocalcemia, which can result in tetany (see *Hypocalcemia,* Chapter 10).
5. Assess for the presence of bowel sounds and monitor for gastric distention, which can impede movement of the diaphragm and restrict ventilatory effort even further.
6. Monitor the patient's stool and NG drainage for the presence of occult blood. Peptic ulcers are common complications in COPD patients with respiratory failure.
7. If patient is not intubated, encourage use of pursed-lip breathing (inhalation through nose, with slow exhalation through pursed lips), which helps airways remain open and allows for better air excursion. Optimally, this technique will diminish air entrapment in the lungs and make respiratory effort more efficient.

Ineffective airway clearance related to viscous secretions and fatigue

Desired outcome: Auscultation of patient's airway reveals the absence of adventitious breath sounds.

1. Be alert to increased fatigue or lethargy as a potential sign of increasing $Paco_2$.
2. If the patient is unable to raise secretions independently, suction as often as need is determined by assessment findings.
3. If prescribed, administer chest physiotherapy. (Chest physiotherapy may be contraindicated in some patients with chronic CO_2 retention.) Evaluate effectiveness of therapy by assessing breath sounds, ABG results, and patency of airway both before and after treatment.
4. Ensure adequate fluid intake to compensate for increased insensible losses caused by increased respiratory rate, febrile state, and diaphoresis.
5. Encourage nonintubated patient to continue with pursed-lip breathing (inhalation through nose, with slow exhalation through pursed lips), which will increase efficiency and effectiveness of respiratory effort.

For other nursing diagnoses and interventions, see the first section in this chapter on acute respiratory acidosis for **Potential alteration in oral mucous membrane** related to abnormal breathing pattern and **Sleep pattern disturbance** related to frequent treatments and procedures.

Case Study 15-1

J.H. is a 23-year-old woman who is brought to the emergency room because of decreased responsiveness. Her roommate found her with an empty bottle of secobarbital (Seconal) and a half empty bottle of wine near her. Her roommate states that J.H. has been depressed about her unemployment. Her VS are as follows:

HR 120 bpm
RR 28 breaths/min and shallow
BP 70/40 mm Hg.

J.H.'s ABG and Electrolyte Values	**Normal (*perfect*) Values**
pH 7.20	7.40
$Paco_2$ 65 mm Hg	40 mm Hg
Pao_2 45 mm Hg	range: 80-95 mm Hg
HCO_3^- 26 mEq/L	24 mEq/L
K^+ 3.8 mEq/L	range: 3.5-5.0 mEq/L
Na^+ 140 mEq/L	range: 137-147 mEq/L
Cl^- 104 mEq/L	range: 95-108 mEq/L

1. **Question:** What primary acid-base disturbance is present?

   Answer: The decreased pH reflects a severe acidosis; the $Paco_2$ is increased, which causes acidosis and corresponds with the pH. The bicarbonate is normal. Thus ABGs reflect a primary respiratory acidosis.

2. **Question:** Is the pH value what you would expect to see, given the other ABG values?

   Answer: Yes. The pH should decrease by 0.08 units for every 10 mm Hg increase in $Paco_2$. The $Paco_2$ value is 25 mm Hg greater than normal (40 mm Hg). $25 \times 0.08 = 0.20$. $7.40 - 0.20 = 7.20$.

3. **Question:** Has any degree of compensation occurred?

   Answer: Yes. Immediate tissue buffering (with nonbicarbonate buffers) has occurred, which increases plasma HCO_3^- by 2 to 3 mEq/L. No renal compensation has occurred because not enough time has elapsed.

4. **Question:** What is the apparent cause of the acute respiratory acidosis?

   Answer: Barbiturate overdose is the primary cause, with ethanol alcohol (ETOH) enhancing the CNS depression. Barbiturates lead to respiratory center depression, resulting in hypoventilation (RR 28 breaths/min and $Paco_2$ 65 mm Hg). The respiratory rate appears adequate but the breaths are very shallow, leading to decreased minute volume and hypoventilation, which results in accumulation of CO_2.

5. **Question:** Is the anion gap normal?

   Answer: Yes. Na^+ 140 − (Cl^- 104 + HCO_3^- 26) = 10. Normal anion gap is 12 (+ or − 2).

6. **Question:** Given the ABG values, which is the value that causes the most concern?

   Answer: The Pao_2 of 45 mm Hg causes the most concern, since it is in the life-threatening range. Tissue hypoxia can occur when Pao_2 values fall below 60 mm Hg. When adequate oxygen is unavailable at the cellular level, anaerobic metabolism occurs and lactic acidosis is produced, which can lead to metabolic acidosis. The pH (7.20) is not incompatible with life, but interventions for returning it to within normal range are necessary. The increased $Paco_2$ of 65 mm Hg is the source of the decreased pH. When the $Paco_2$ is lowered and the acid-to-base ratio of 1:20 is restored, the pH will return to normal and further treatment to raise the pH will be unnecessary.

Case Study 15-2

K.E. is a 74-year-old man who is being seen in his physician's office because of shortness of breath (SOB) that has become worse over the last several months. He has a cough that is productive of thick, tenacious sputum and he states that his feet swell occasionally, preventing him from getting his shoes on. He does not have cyanosis or digital clubbing. He states that he is a heavy smoker.

K.E.'s ABG Results on Room Air	**Normal (*perfect*) Values**
pH: 7.39	7.40
$Paco_2$: 62 mm Hg	40 mm Hg
Pao_2: 54 mm Hg	range: 80-95 mm Hg
Saturation: 87%	range: 95%-99%
HCO_3^-: 34 mEq/L	24 mEq/L
Base excess: + 4	+ or − 2

1. Question: Based on the blood gas values, what acid-base disturbance is present?

Answer: The pH is decreased, which reflects a mild acidosis. The $Paco_2$ is greatly elevated and corresponds to the acidosis (decreased pH). The increased serum HCO_3^- reflects an alkalosis and does not correspond to the pH. The primary disturbance is a respiratory acidosis (increased $Paco_2$).

2. Question: Based on the patient's signs and symptoms and ABG values, what is the patient's predominant chronic lung disease?

Answer: The patient has predominant chronic bronchitis based on the history of severe dyspnea; copious, thick sputum production; and ABGs that reflect hypercapnia and hypoxemia. Smoking is common with both emphysema and chronic bronchitis, but patients with prodominant emphysema usually are able to maintain their CO_2 at a normal level and keep their Pao_2 in an acceptable range by increasing their minute volume via an increase in respiratory rate and respiratory effort until the end stages of their disease.

3. Question: Has any compensation occurred?

Answer: The elevated HCO_3^- reflects the renal compensation that has occurred. If renal compensation had not occurred, the $Paco_2$ of 62 would cause the pH to be $\leq$7.23, but the increased HCO_3^- concentration results in a near normal pH. The presence of significant renal compensation suggests that this is a chronic respiratory acidosis.

4. Question: What patient education intervention is especially important for a patient with chronic bronchitis?

Answer: Patients with chronic bronchitis are predisposed to bronchopulmonary infection, which can severely compromise their respiratory status. Patients need to be informed of the signs and symptoms of a respiratory infection (sputum changing to a yellow or green color, increased temperature, increasing dyspnea, chills) and told to notify their caregiver immediately. They also should be informed that cessation of smoking may improve some of the symptoms of their chronic disease. A possible nursing diagnosis for this patient is **Knowledge deficit:** Signs and symptoms of pulmonary infections and the importance of prompt notification of a health care professional.

One week later, K.E. is seen in the emergency room, confused, extremely SOB, and cyanotic. His sputum is thick and yellow and his temperature is 38.5° C (101.4° F), his HR is 112 bpm, and his RR is 20 breaths/min. The chest x-ray shows evidence of chronic changes as well as diffuse, patchy infiltrates indicating a pneumonia.

K.E.'s ABG Values on Room Air	Normal (*perfect*) Values
pH: 7.25	7.40
$Paco_2$: 92 mm Hg	40 mm Hg
Pao_2: 33 mm Hg	range: 80-95 mm Hg
O_2 saturation: 58%	range: 95%-99%
Base excess: + 5	+ or − 2
HCO_3^-: 38 mEq/L	24 mEq/L

5. Question: Based on the ABG values, which acid-base disturbance is present?

Answer: The decreased pH reflects an acidosis; the $Paco_2$, which is markedly elevated, correlates with the acidic pH; the HCO_3^- remains elevated, which is the renal compensatory response. Remember that as the $Paco_2$ increases, the pH decreases. The HCO_3^- remains elevated as a compensatory response to the steady state of chronic respiratory acidosis.

6. Question: How do you explain the decrease in pH when renal compensation has occurred?

Answer: The renal compensation maintained a steady pH of 7.39 when the $Paco_2$ was 62 mm Hg. When the $Paco_2$ suddenly increased to 92 mm Hg, the compensatory response was inadequate to meet the increased demands for buffering and the pH decreased. This patient is experiencing an acute respiratory acidosis superimposed on his chronic respiratory acidosis.

7. **Question:** What component of the ABG is the most life-threatening?

   Answer: The PaO_2 of 33 mm Hg is the most life-threatening. Tissue hypoxia may lead to metabolic acidosis through the production of lactic acids if oxygenation is not improved. The combination of increased $PaCO_2$ and decreased PaO_2 have led to K.E.'s confusion, which will lessen as his ventilation improves. The pH, although requiring attention, is not life-threatening and will improve as the $PaCO_2$ declines.

8. **Question:** Is this patient at risk for oxygen-induced apnea?

   Answer: Yes. The patient has a long history of hypercapnia and his respiratory center has become insensitive to further increases in $PaCO_2$. The drive to breathe is now stimulated by hypoxemia, so if the hypoxemia is suddenly corrected, the patient may stop breathing. This is commonly called carbon dioxide narcosis. This patient requires oxygen because of the dangerously low PaO_2, but it must be delivered cautiously (usually 1 to 2 L/min), with careful monitoring of the patient's breathing effort, sensorium, and serial ABG values, which will determine the appropriate oxygen concentration.

9. **Question:** The $PaCO_2$ is markedly elevated. Should it be corrected to normal?

   Answer: The $PaCO_2$ of 92 mm Hg requires attention. Because of the severe hypoxemia and the breathing effort required by the patient, ventilatory support is indicated before the patient has a respiratory arrest. The patient requires endotracheal intubation and mechanical ventilation until the underlying disease process can be treated. The $PaCO_2$ should be lowered slowly and only as far as the patient's baseline. Even though no further renal compensation has occurred secondary to the acute respiratory acidosis, the patient had an elevated HCO_3^- (34 mEq/L) when the $PaCO_2$ was 62. If the $PaCO_2$ is reduced to normal (40 mm Hg) a metabolic alkalosis will occur because the HCO_3^- will remain elevated until the kidneys can excrete the excess. This overcorrection of a compensated chronic respiratory acidosis is called *post-hypercapneic alkalosis.* Mechanical ventilation can bring the PaO_2 level up to baseline (54 to 60 mm Hg) and the $PaCO_2$ can be lowered to 60 to 65 mm Hg. The decrease in $PaCO_2$ should not exceed approximately 10 mm Hg/hr.

References

Ackerman G and Arruda J: Acid-base and electrolyte imbalance in respiratory failure. In Bone R: The Medical Clinics of North America Symposium on Respiratory Failure 67(3):645-656, 1983.

Easterday U: Acid-base balance. In: Horne M, Swearingen PL: Pocket guide to fluids and electrolytes, St Louis, 1989, CV Mosby Co.

Easterday U and Howard C: Respiratory dysfunctions. In Swearingen PL and others: Manual of critical care: applying nursing diagnoses to adult critical illness, St Louis, 1988, The CV Mosby Co.

Effros R: Acid-base balance. In Murray J and Nadel J: Textbook of respiratory medicine, Philadelphia, 1988, WB Saunders Co.

Fishman A: Pulmonary diseases and disorders, ed 2, New York, 1988, McGraw-Hill Book Co.

Hahn A and others: Mosby's pharmacology in nursing, St Louis, 1986, The CV Mosby Co.

Ibanez J and others: The effect of lateral position on gas exchange in patients with unilateral lung disease during mechanical ventilation, Intensive Care Med 7:231-234, 1981.

Jaffe J and Martin W: Opioid analgesics and antagonists. In Gilman A and others (editors): Goodman and Gilman's the pharmacological basis of therapeutics, ed 7, New York, 1985, Macmillan Publishing Co.

Langer M and others: The prone position in ARDS patients: a clinical study, Chest 94(1):103-107, 1988.

McCaffery M and Beebe A: Pain: clinical manual for nursing practice, St Louis, 1989, The CV Mosby Co.

Murray J and Nedel J: Textbook of respiratory medicine, Philadelphia, 1988, WB Saunders Co.

Pingleton S: Complications in acute respiratory failure. In Bone R: The Medical Clinics of North America Symposium on Respiratory Failure, 67(3):725-746, 1983.

Rose B: Clinical physiology of acid-base and electrolyte disorders, ed 3, New York, 1989, McGraw-Hill Book Co.

Schmidt G and Hall J: Acute on chronic respiratory failure, 261(23):3444-3453, 1989.

Shoemaker W and others: Textbook of critical care, Philadelphia, 1989, WB Saunders Co.

Valtin H and Gennari F: Acid-base disorders: basic concepts and clinical management, Boston, 1987, Little, Brown & Co.

York K: The lung and fluid-electrolyte and acid-base imbalances, Nurs Clin North Am 22(4):805-814, 1987.

Chapter

16

Respiratory Alkalosis

Respiratory alkalosis occurs when the respiratory center has been stimulated to increase ventilation, resulting in the elimination of more CO_2 than has been generated by metabolic processes. Thus the ratio of acids to bases is altered, with a $Paco_2$ of less than 40 mm Hg and a pH of greater than 7.40. Increased ventilation can be stimulated by peripheral chemoreceptors sensitive to Pao_2 and oxygen delivery; physical receptors in the lungs, thorax, and upper airways; direct stimulation of the respiratory center from higher cortical center stimuli (e.g., pain, anxiety, voluntary control); or drugs. The respiratory center increases minute ventilation through its effects on the rate and depth of respirations. The term **hyperventilation** refers to the elimination of more CO_2 than that which is generated, and when used correctly, implies a respiratory alkalosis. However, the term is often associated with the shallow and rapid breathing that occurs with anxiety, which may or may not result in respiratory alkalosis. In true hyperventilation with increased minute ventilation leading to respiratory alkalosis, the respiratory rate may be fast or slow, with a varying tidal volume. Thus the respiratory rate and depth cannot determine the presence or severity of the respiratory alkalosis. The diagnosis must be based on $Paco_2$ and pH values. In this discussion the terms *hyperventilation* and *respiratory alkalosis* will be used interchangeably.

The alteration in CO_2 elimination leading to acute respiratory alkalosis occurs suddenly, with resolution within 6 hours. If the respiratory alkalosis is sustained for more than 6 hours, clinically evident renal compensation is seen and the acid-base status changes to chronic respiratory alkalosis. In acute respiratory alkalosis, chemical buffering occurs and prevents a life-threatening increase in pH. Chronic respiratory alkalosis evokes a renal response that facilitates a return to a normal or near normal pH. As is true with acute and chronic respiratory acidosis, acute respiratory alkalosis results in a much greater change in pH and more symptomatology than chronic respiratory alkalosis.

There are numerous causes of acute and chronic primary respiratory alkalosis. In general, the primary causes of respiratory alkalosis are hypoxemia, pulmonary diseases that stimulate physical receptors, and direct stimulation of the respiratory center (e.g., by cortical influences and salicylates). The respiratory compensatory response to metabolic acidosis that leads to a decreased $Paco_2$ will not be discussed here since it is not a primary respiratory alkalosis, but rather an adaptive response to metabolic acidosis.

ACUTE RESPIRATORY ALKALOSIS

Frequently, acute respiratory alkalosis is referred to simply as *hyperventilation,* which conjures up images of an anxious person who is overbreathing. While it is true that hyperventilation related to anxiety can lead to respiratory alkalosis, there are numerous serious and potentially life-threatening causes of this disorder. In fact, respiratory alkalosis is one of the most commonly occurring acid-base disturbances in hospitalized patients, and it is often an early, perhaps subtle sign of the onset of an impending, serious illness or complication. Anderson and Heinrich (1987) concluded that the development of respiratory alkalosis is a poor prognostic sign in medical patients but is a less reliable sign for surgical patients, whose respiratory alkalosis is usually brief and frequently related to excessive mechanical ventilation during surgery. In general, mortality rates increase as the alkalemia (both respiratory and metabolic) worsens and the length of time the patient experiences alkalemia increases.

History and Risk Factors

There are many causes of acute respiratory alkalosis. For the purposes of this discussion they will be presented under the following three broad categories: (1) hypoxemia, (2) pulmonary disorders, and (3) direct central stimulation of the respiratory center (see the accompanying box).

Hypoxemia

The carotid and aortic body chemoreceptors are very sensitive to decreases in Pao_2. They stimulate the respiratory center when the Pao_2 falls below 60 to 70 mm Hg (hypoxemia) or when oxygen delivery to the tissues is impaired (hypoxia) by severe anemia or hypotension. The stimulation of the respiratory center results in hypocapnia and an increase in both the arterial and cerebrospinal fluid (CSF) pH. Respiratory center stimulation cannot be sustained because CO_2 concentration decreases in the CSF, and this is sensed by the central chemoreceptors. Therefore hyperventilation secondary to hypoxemia is limited because of the resulting changes in CSF CO_2 and pH.

Potential Causes of Acute Respiratory Alkalosis

Hypoxemia	Pulmonary Disorders	Central (Direct) Stimulation of the Respiratory Center
Pneumonia	Pulmonary emboli	Anxiety
High altitude (>6500 feet)	Inhalation of irritants	Fever
Hypotension	Interstitial fibrosis	Pain
Severe anemia	Pneumonia	Drugs (salicylates)
Congestive heart failure	Pulmonary edema	Voluntary or mechanical hyperventilation
	Asthma	Intracerebral trauma
		Gram-negative septicemia
		Acute cerebral vascular accident

In pulmonary disorders that cause hypoxemia (e.g., pneumonia, pulmonary edema, and pulmonary embolus), the patient initially will have a respiratory alkalosis, but it can progress to respiratory acidosis if the underlying condition worsens and the exchange of CO_2 also becomes impaired. Remember that CO_2 is 20 times more diffusable than O_2 across the alveolar membrane, so hypoxemia is the first abnormality to be noted in many pulmonary diseases. Patients presenting with an acute asthma attack, for example, usually will be hypoxemic on room air and have a decreased $Paco_2$ and a simple acute respiratory alkalosis. If the $Paco_2$ is normal or elevated, a very serious degree of obstruction is present. In fact, if the patient is symptomatic (e.g., wheezing, coughing) and the $Paco_2$ is normal or elevated, respiratory failure may occur. (For further information refer to Chapter 15.)

Acute mountain sickness

Another potential cause of hypoxemia that can lead to respiratory alkalosis is high altitude. At altitudes over 6,500 feet the partial pressure of O_2 is decreased significantly, resulting in hypoxemia, which stimulates ventilation to improve tissue oxygenation, and thus causing respiratory alkalosis. The hypoxemia and other disturbances that occur because of high altitude are referred to as *acute mountain sickness.* The constellation of signs and symptoms related to the hypoxemia can include dyspnea, malaise, headache, nausea, vomiting, insomnia, and anorexia. Tachycardia and Cheyne-Stokes respirations also can occur. The symptoms improve over several days as the body adjusts to hypocapnia and hypoxemia via a decrease in H^+ excretion, leading to a loss of bicarbonate (HCO_3^-) in the urine. This compensatory response leads to a near normal pH. Normally, the respiratory alkalosis that occurs would be considered chronic because of the time required for the kidneys to compensate, but because of the acute nature of the illness, it is called an acute respiratory alkalosis.

Pulmonary disorders

Some pulmonary diseases cause respiratory alkalosis, even in the absence of severe hypoxemia. Hyperventilation in these circumstances is caused by the stimulation of receptors in the lung, thorax, or upper airways. Three separate types of receptors have been identified: juxtacapillary (J) receptors, which are activated by pulmonary capillary congestion (i.e., pulmonary edema); irritant receptors, which are activated by noxious stimuli that irritate small airways (i.e., asthma); and stretch receptors, whose role in stimulating hyperventilation is unclear. While hypoxemia may play a role in the hyperventilation stimulated by these pulmonary disorders, the hyperventilation persists, even when the hypoxemia is corrected by oxygen administration. These receptors can stimulate the respiratory center to increase ventilation independent of oxygen, pH, or voluntary control. The role of these receptors in individuals with normal lungs is not clear.

Respiratory center stimulation

Respiratory alkalosis also results from central or direct stimulation of the respiratory center. This occurs in numerous conditions with different modes of action, but with the same outcome—hyperventilation. Pain and thermal receptors can stimulate the respiratory center directly, as does intracerebral trauma that results in hemorrhage and cerebral edema, and gram-negative septicemia. Endotoxins from gram-negative bacteria are believed to be direct stimulants of the respiratory center.

Voluntary and involuntary hyperventilation

The normal ventilatory regulation by the chemoreceptors can be overridden by voluntary hyperventilation, demonstrating the effects cortical centers have on ventilation. Onset and cessation of acute respiratory alkalosis can be very abrupt. For example, with voluntary hyperventilation the pH can increase within 15 to 20 seconds after hyperventilation begins and then return to normal immediately after hyperventilation stops. Another example of cortical control on ventilation is involuntary hyperventilation secondary to anxiety and in some types of central nervous system (CNS) diseases, such as cerebrovascular accident.

Mechanical ventilation

Mechanical ventilators set at inappropriately high rates or tidal volumes can decrease CO_2, resulting in respiratory alkalosis. Correction of the error usually results in the CO_2 returning to normal if no further pathologies exist. Respiratory alkalosis purposely caused by mechanical ventilation can be initiated in patients with cerebral trauma or in those requiring neurologic surgery. The vasoconstrictive effects of a decreased $Paco_2$ assist in decreasing intracranial pressure (ICP) and cerebral bleeding.

Salicylates

Salicylates directly stimulate the central chemoreceptors, causing hyperventilation. Metabolic acidosis also can occur with salicylate intoxication, and individuals often have a mixed acid-base disturbance as a result. Respiratory alkalosis is usually the predominant acid-base disorder in adults and in children older than 8 to 10 years of age. In younger children, metabolic acidosis predominates because ketosis and dehydration are more prevalent in this age group, especially in children under 4 years old. In mixed disorders, the hyperventilation is independent of the metabolic acidosis and will continue even when the metabolic acidosis has been corrected.

Physiologic Responses

Chemical buffering

The only immediate physiologic response in respiratory alkalosis is chemical buffering. A decrease in $Paco_2$ causes an immediate rise in pH. To prevent potentially lethal alkalemia, hydrogen ions (H^+) are released within minutes from nonbicarbonate buffers and tissues, causing a decrease in the serum HCO_3^- concentration. In the acute phase of respiratory alkalosis there is a 2 mEq/L decrease in serum HCO_3^- concentration for each 10 mm Hg decrease in $Paco_2$. It takes approximately 15 minutes for a new steady state to be reached after respiratory alkalosis has developed. The new steady state can be maintained for 5 to 6 hours if hypocapnia persists.

Respiratory

Since excess elimination of CO_2 by the lungs is the primary problem in respiratory alkalosis, no respiratory compensation can occur.

Renal

The renal response to a decreased $Paco_2$ begins almost immediately, but it is not evident clinically for 6 to 8 hours. Renal compensation, given time, will result in decreased net acid excretion, leading to reduced serum HCO_3^- concentration. The serum HCO_3^- rarely falls below 18 mEq/L as compensation for acute respiratory alkalosis. If the HCO_3^- level is less than 18 mEq/L, a concurrent acid-base disorder may be present (i.e., metabolic acidosis).

Assessment

The signs and symptoms of respiratory alkalosis, especially the CNS manifestations, typically occur when the $Paco_2$ is less than 25 to 30 mm Hg. This decrease in $Paco_2$ results in a substantial rise in the cerebral pH, cerebral vessel vasoconstriction, and impaired delivery of O_2 to the cerebral tissue. See Table 16-1 for quick assessment guidelines for individuals with acute respiratory alkalosis.

Table 16-1 Quick assessment guidelines for individuals with acute respiratory alkalosis

Signs and Symptoms	Physical Assessment	EKG Findings
Lightheadedness Anxiety Paresthesias Circumoral numbness	Increased rate and depth of respirations	Cardiac dysrhythmias ST-T wave changes
In extreme alkalosis		
Confusion Tetany Syncope Seizures		

History and Risk Factors

- *Anxiety:* Patient is often unaware of hyperventilation
- *Acute hypoxemia:* Pulmonary disorders (e.g., pneumonia, pulmonary edema, and pulmonary thromboembolism) cause hypoxemia, which stimulates the ventilatory effort in the initial stages of the disease process
- *Hypermetabolic states:* Fever, sepsis, especially gram-negative induced septicemia
- *Salicylate intoxication*
- *Excessive mechanical ventilation*
- *CNS trauma:* May result in damage to respiratory center

Central nervous system (CNS)/neuromuscular manifestations

The most typical findings with respiratory alkalosis are lightheadedness, tingling or paresthesias of the extremities and circumoral area, confusion, muscle cramps, and carpopedal spasm. Hyperreflexia, tetany, and even seizures can occur. These symptoms are related to increased irritability of the central and peripheral nervous systems and are produced by the respiratory alkalosis (alkalosis increases membrane excitability). Hypocapnia causes vasoconstriction of cerebral blood vessels, resulting in a dramatic decrease in cerebral blood flow, which causes cerebral hypoxia. The cerebral hypoxia contributes to the confusion and lightheadedness present with respiratory alkalosis. For example, in patients with sickle cell anemia, it is believed that the combination of cerebral alkalosis and cerebral hypoxia produces many of the CNS symptoms.

Cardiovascular effects

In hyperventilation caused by anxiety, tachycardia is usually present, the blood pressure (BP) remains unchanged, and chest tightness is frequently reported. These same indicators are commonly seen in other causes of respiratory alkalosis. An electrocardiogram (EKG) obtained during acute respiratory alkalosis often shows ST-T wave changes consistent with myocardial ischemia. These EKG changes occur in individuals with no known coronary artery disease and are believed to be caused by coronary artery spasm induced by acute hypocapnia. Supraventricular and ventricular dysrhythmias also may be seen, especially in patients who are critically ill.

Gastrointestinal effects

Nausea and vomiting occur frequently, especially with acute mountain sickness. These symptoms are related to cerebral hypoxia and subside as the body acclimates to the new environment.

Pulmonary effects

If respiratory alkalosis is the result of anxiety, the individual may complain of dyspnea, difficulty taking a deep breath, and feeling suffocated. Dyspnea also occurs in individuals whose respiratory alkalosis is the result of hypoxia (i.e., pulmonary fibrosis) and is often related to the energy required to maintain the work of breathing and respiratory muscle fatigue. See Table 16-1 for quick assessment guidelines for individuals with acute respiratory alkalosis.

Diagnostic Studies

Arterial blood gas (ABG) values

The $Paco_2$ will be less than 35 mm Hg and the pH will be greater than 7.40. The Pao_2 will be decreased if the respiratory alkalosis is the result of hypoxemia secondary to pneumonia, early pulmonary edema, pulmonary emboli, or other acute pulmonary disorders. The alkalosis will cause a shift to the right in the oxygen-hemoglobin dissociation curve (see Figure 14-1), causing oxygen to be more tightly bound to the hemoglobin, with the possibility of hypoxia at the tissue level.

Bicarbonate (HCO_3^-)

In acute respiratory alkalosis, serum bicarbonate (HCO_3^-) will be decreased by 2 mEq/L for each 10 mm Hg drop in $Paco_2$. The decrease in HCO_3^- is the result of hydrogen ion (H^+) release from nonbicarbonate buffers and tissues, and is not caused by the renal excretion of HCO_3^-. Remember, the HCO_3^- can be calculated from the pH and $Paco_2$, as it is done when reported with ABG values, or it can be measured with the other serum electrolytes.

Serum electrolytes

The sodium may be mildly decreased (i.e., 2 to 4 mEq/L) in respiratory alkalosis. Serum potassium will decrease slightly as it shifts from the extracellular to the intracellular space in exchange for H^+ in response to the alkalosis. (The opposite occurs in acidosis.) The chloride concentration will increase slightly to maintain electroneutrality as the serum HCO_3^- decreases. The anion gap will be normal even though the HCO_3^- concentration has decreased, because the decrease in HCO_3^- is offset by the increase in Cl^- concentration. Therefore no net change in anions is seen. Hypophosphatemia (as low as 0.5 mg/dl of phosphorus) may be present because of the intracellular shift of phosphorus that occurs with respiratory alkalosis (see Chapter 11 for more information about hypophosphatemia and respiratory alkalosis). Hypophosphatemia can result in muscle weakness secondary to impaired muscle contractility. Hypocalcemia may be present in respiratory alkalosis because of increased calcium and bicarbonate binding. Signs of hypocalcemia include muscle cramps, hyperactive reflexes, carpal spasm, tetany, and convulsions. See Chapter 10 for more information about hypocalcemia and alkalosis.

Electrocardiogram (EKG)

Tachycardia and supraventricular or ventricular dysrhythmias may be present. ST-T wave changes may be present, which indicate myocardial ischemia from coronary artery spasm caused by the alkalosis. These changes may occur in patients with no underlying coronary artery disease.

Medical Management

Treatment of the underlying disorder is the only effective therapy for respiratory alkalosis.

Oxygen therapy

If hypoxemia is the causative factor in acute respiratory alkalosis, correction of the hypoxemia with supplemental oxygen may eliminate the hyperventilation. In some pulmonary disorders, such as interstitial fibrosis and hyperventilation, respiratory alkalosis may remain even if the hypoxemia is corrected.

Reassurance

If anxiety resulting in hyperventilation has caused a decreased $Paco_2$, offering reassurance and attempting to help the patient gain control of respirations (i.e., slowing the rate) may help resolve the respiratory alkalosis. Have the patient mimic your own slow breathing pattern.

Rebreathing CO_2

If symptoms are severe, it may be necessary for the patient to breathe into a paper bag. This rebreathing of the excreted CO_2 will help raise $Paco_2$ levels. Rebreathing CO_2 through an oxygen mask with an attached CO_2 reservoir (called a rebreathing mask) will produce the same effect.

Adjusting mechanical ventilators

When hypocapnia develops in patients who are being mechanically ventilated, it may be necessary to decrease the respiratory rate and/or tidal volume. If these corrections are not sufficient to raise $Paco_2$ levels, the addition of dead space to the ventilator tubing may be necessary. Adding extra ventilator tubing (i.e., dead space) creates more conducting airway space that does not participate in gas exchange. Thus a smaller volume of air reaches the alveolar space for gas exchange and less CO_2 is eliminated. $Paco_2$ concentration increases, thereby decreasing the degree of respiratory alkalosis.

Pharmacotherapy

Sedatives and tranquilizers may be given for anxiety-induced respiratory alkalosis. Prevention of acute mountain sickness may include acetazolamide (Diamox), a carbonic anhydrase inhibitor that enables better oxygen transport with less respiratory alkalosis, or dexamethasone (Decadron). The exact actions of dexamethasone in the treatment of acute mountain sickness are not known, but the drug is believed to relieve hypoxic cerebral edema by decreasing cerebral vasoconstriction, decreasing production of CSF, and stabilizing cerebral capillary membranes. Because of potential side effects (i.e., glucose intolerance, mental disturbances, and water retention) dexamethasone is not used routinely in the treatment of acute mountain sickness.

Nursing Diagnoses and Interventions

Ineffective breathing pattern related to hyperventilation secondary to anxiety

Desired outcome: Patient's breathing pattern is effective as evidenced by a $Paco_2 \geq 35$ and a pH ≤ 7.45.

1. To help alleviate anxiety, reassure patient that a staff member will remain with him or her.
2. Encourage patient to breathe slowly. Pace patient's breathing by having him or her mimic your own breathing pattern.
3. Monitor patient's cardiac rhythm, notifying MD if dysrhythmias occur. With acute respiratory alkalosis, even a modest alkalosis can precipitate dysrhythmias in a patient with preexisting heart disease who is also taking cardiotropic drugs. In part, this is caused by the hypokalemia that occurs with alkalosis.
4. Administer sedatives or tranquilizers as prescribed.
5. Have patient rebreathe into a paper bag or into an oxygen mask with an attached CO_2 reservoir, if prescribed.

6. Ensure that patient rests undisturbed after his or her breathing pattern has stabilized. Hyperventilation can result in fatigue.

   NOTE: Hyperventilation may lead to hypocalcemic tetany despite a normal or near-normal calcium level due to increased binding of calcium (see Chapter 10).

CHRONIC RESPIRATORY ALKALOSIS

Chronic respiratory alkalosis is not seen commonly in the acutely ill hospitalized patient, but when it is present, it can be indicative of a poor prognosis, as with hepatic failure and post cerebrovascular accident. As occurs with acute respiratory alkalosis, chronic stimulation of ventilation results in greater quantities of CO_2 being eliminated than the amount generated. As renal compensation progresses to completion, a new *steady* state is reached, with a balance between acids and bases. This balance is achieved by lowering the HCO_3^- concentration proportionately to match the chronically lowered CO_2, thus establishing a new acid-base balance. The new ratio results in a normal or near-normal pH.

History and Risk Factors

The same basic mechanisms that stimulate acute respiratory alkalosis are responsible for stimulating and initiating chronic respiratory alkalosis. These include hypoxemia, direct stimulation of the respiratory center, and stimulation of the respiratory center by the J receptors. Acute alkalosis can progress to chronic respiratory alkalosis if it persists for more than 6 hours and renal compensation occurs.

Restrictive pulmonary disorders

Restrictive lung disorders, such as kyphosis, can cause chronic respiratory alkalosis. Interstitial pulmonary fibrosis, the most common restrictive lung disorder, causes chronic respiratory alkalosis by two mechanisms: hypoxemia and stimulation of the J receptors, which stimulate the respiratory center. There are numerous disorders that can result in pulmonary fibrosis (e.g., systemic lupus erythematosus, sarcoidosis, and cystic fibrosis). It is interesting that the respiratory alkalosis continues throughout the course of the pulmonary fibrosis. A normal $Paco_2$ followed by hypercapnia occurs when the illness is in its terminal stages.

Hepatic insufficiency

Hepatic insufficiency is a common cause of chronic respiratory alkalosis in hospitalized patients. The respiratory center is stimulated directly by the ammonia and other metabolic byproducts that accumulate in liver disease (see Case Study 19-3).

Hyperventilation can occur in response to the hypoxemia that is often present due to the ventilation-perfusion mismatch and alterations in pulmonary vasculature that are seen in liver disease. Hypoxemia leading to hyperventilation also may develop secondary to poor diaphragmatic excursion because of the ascites. Numerous concurrent acid-base disturbances may be present in patients with hepatic insufficiency (see Chapter 19). Patients with liver disease who have a sustained respiratory alkalosis have a very poor prognosis.

Central nervous system (CNS) disorders

Chronic respiratory alkalosis is a fairly common finding in several CNS disorders. Patients with cerebral hemorrhage or edema caused by trauma, cerebrovascular accident (CVA), or a ruptured aneurysm may develop chronic respiratory alkalosis. This alkalosis is the result of direct stimulation of the respiratory center, as is the case with CNS tumors or infections, such as encephalitis. The presence of sustained respiratory alkalosis in patients with CNS diseases is a grave prognostic sign.

Pregnancy

Chronic respiratory alkalosis is a normal finding during pregnancy because of the stimulating effect progesterone has on the respiratory center. The normal $Paco_2$ is 30 mm Hg during pregnancy, with the expected renal compensation decreasing the serum HCO_3^- concentration to approximately 19 mEq/L. Remember that with chronic respiratory alkalosis, the serum HCO_3^- concentration decreases 5.0 mEq/L for every 10 mm Hg decrease in $Paco_2$.

High-altitude residence

Living at high altitudes for prolonged periods can result in chronic respiratory alkalosis, with a $Paco_2$ less than 35 mm Hg. The hyperventilation is induced by hypoxemia, and although hypocapnia is present, the pH is normal because of renal compensation. Upon return to a lower altitude ($<$6,500 feet), the hypoxemia is corrected and the $Paco_2$ returns to normal. Only rarely do complications arise from prolonged high-altitude residence. These complications, which are referred to as *chronic mountain sickness,* include excessive polycythemia, thrombosis, cyanosis, and heart failure.

Adaptive Physiologic Responses

Chemical buffering

With chronic respiratory alkalosis, intracellular and extracellular buffering is limited. Because of the sustained nature of the acid-base imbalance, renal compensation plays the largest role.

Respiratory

The primary problem with chronic respiratory alkalosis is the regulation of CO_2, and thus respiratory compensation is not available.

Renal

Renal compensation occurs via a decrease in net acid excretion (diminished excretion of ammonia and titratable acids). Serum HCO_3^- concentration is then decreased because it is used up in buffering the retained hydrogen ions (H^+). (Remember that HCO_3^- is generated for each H^+ excreted. If H^+ excretion is decreased, less HCO_3^- will be generated, which assists in correcting the acid-base imbalance.) This process is clinically evident 6 hours after hypocapnia first occurs. Over the first 24 hours, the decrease in acid excretion by the kidneys accounts for more than 50% of the decrease in serum HCO_3^-. Increasing levels of urine HCO_3^- (bicarbonuria) are also found as renal compensation progresses. If the hypocapnia is sustained, a new *steady* state is achieved within 1.5 to 3 days. Renal compensation occurs as a direct response to the decreased CO_2 and is independent of systemic pH, Pao_2, or HCO_3^-. Maximal renal compensation will cause serum HCO_3^- to decrease 5 mEq/L for each 10 mm Hg decrease in $Paco_2$ and takes 7 to 9 days to reach maximal effect with the *normalization* of pH. Persistent or chronic respiratory alkalosis is the *only* acid-base disturbance in which the compensatory mechanism (the kidney) can return the pH to the normal 7.40.

Assessment

Individuals with chronic respiratory alkalosis are not likely to exhibit signs and symptoms related to hypocapnia because renal compensation will have returned the pH to normal (Table 16-2).

Central nervous system (CNS)/neuromuscular manifestations

When there is cortical damage, the patient will have Cheyne-Stokes respirations (periodic breathing with moments of apnea) caused by a hyperreflexic response of the respiratory center to CO_2. Cerebral lesions (e.g., cerebral hemorrhage) often result in Biot's respirations (periods of apnea interspersed with a few deep breaths in a totally irregular pattern), which cause hypocapnia and respiratory alkalosis.

Cardiovascular effects

Because of the chronic hypoxia, polycythemia may occur in chronic high-altitude residence, some cyanotic heart diseases and, in rare instances, restrictive pulmonary diseases. When polycythemia is present, the risk of thrombosis increases. Although uncommon, cardiac failure can occur in chronic mountain sickness because of the excessive polycythemia, which can lead to thrombosis.

Table 16-2 Quick assessment guidelines for individuals with chronic respiratory alkalosis

Signs and Symptoms	Physical Assessment
Usually asymptomatic, but often have an increased respiratory rate and depth	Tachypnea, hyperventilation

History and Risk Factors

- *Cerebral disease:* Tumor, encephalitis
- *Chronic hepatic insufficiency*
- *Pregnancy*
- *Chronic hypoxemia:* Adaptation to high altitude; cyanotic heart disease; lung disease resulting in decreased compliance (e.g., fibrosis)

Pulmonary effects

Dyspnea during exertion is often present in patients with restrictive pulmonary disorders (chest deformities or pulmonary fibrosis). Often, individuals with long-term anxiety are unaware of overbreathing but do acknowledge frequent deep sighing. See Table 16-2 for quick assessment guidelines for individuals with chronic respiratory alkalosis.

Diagnostic Studies

Arterial blood gas (ABG) values

$Paco_2$ will be less than 35 mm Hg. The pH will be normal or very near normal, depending on the length of time the patient has been hypocapneic. (Remember that maximal renal compensation may take up to 7 to 9 days.) The Pao_2 may be decreased if the chronic respiratory alkalosis is induced by chronic hypoxemia (as in restrictive pulmonary disease or chronic mountain sickness).

Bicarbonate (HCO_3^-)

A large decrease in serum HCO_3^- is seen with chronic respiratory alkalosis. It will decrease by 5 mEq/L for every 10 mm Hg decrease in $Paco_2$. It is unusual to see an HCO_3^- less than 12 mEq/L in pure chronic respiratory alkalosis. A HCO_3^- lower than 12 mEq/L indicates a concurrent metabolic acidosis.

Serum electrolytes

Serum sodium and potassium will be normal, and the serum chloride will be increased slightly but to a lesser degree than the decrease in HCO_3^-, resulting in a small elevation in the anion gap. There will be a 3 mEq/L increase in the anion gap when the $Paco_2$ has decreased to 20 mm Hg or below. If there are severe derangements in the electrolytes, the presence of another acid-base disturbance is likely. Hypophosphatemia (phosphorus as low as 0.5 mg/dl) may be seen because alkalosis results in the intracellular shift of phosphorus (see Chapter 11).

Complete blood cell count (CBC)

Polycythemia may be present in any condition that causes chronic hypoxemia, such as high-altitude residency, cyanotic heart disease, or some chronic pulmonary disorders. Polycythemia that develops in response to chronic hypoxemia is referred to as secondary polycythemia. The hematocrit will be increased in males to greater than 54% and in females to greater than 49%.

Medical Management

Frequently, treatment is unnecessary because of the chronicity of the acid-base disturbance and the renal compensation that occurs.

Oxygen therapy

If hypoxemia is present and identified as a causative factor in the respiratory alkalosis, oxygen therapy may be used to maintain the Pa_{O_2} at ≥ 60 mm Hg.

Nursing Diagnoses and Interventions

Nursing diagnoses and interventions are specific to the underlying pathophysiologic process.

Case Study 16-1

L.D. is a moderately obese 57-year-old woman who had an abdominal hysterectomy 2 days ago. She has had an uncomplicated postoperative course and has been up walking in her room twice. She is resting in bed when she suddenly develops left chest pain that increases with respirations. She describes a feeling of breathlessness and appears very anxious.

Her vital signs 1 hour prior to the onset of the symptoms were as follows:

RR 14 breaths/min, HR 80 bpm, BP 134/82 mm Hg, and temperature 37.1° C (98.8° F).

Her current vital signs are as follows:

RR 28 breaths/min and shallow, HR 124 bpm and irregular, BP 140/86 mm Hg, and temperature 37° C (98.6° F).

Blood gases are assessed with the following results:

L.D.'s Blood Gas Values	**Normal (*perfect*) Values**
pH 7.49	7.40
$Paco_2$ 29 mm Hg	40 mm Hg
Pao_2 64 mm Hg	range: 80-95 mm Hg
HCO_3^- 22 mEq/L	24 mEq/L
O_2 saturation 87%	range: 95%-99%

1. Question: Based on the ABG values, which acid-base disturbance is present?

Answer: The pH is increased, which reflects an alkalosis. The $Paco_2$ is decreased to 29 mm Hg, which corresponds to the increased pH. The serum HCO_3^- is slightly decreased, which would correspond to an acidosis, not to the alkalosis and increased pH that are present. The acid-base disturbance is respiratory alkalosis.

2. Question: Has any compensation occurred? If so, what compensatory mechanism is responsible?

Answer: The serum HCO_3^- has decreased by approximately 2 mEq/L (from 24 to 22). The $Paco_2$ has decreased by approximately 11 mm Hg and the expected acute compensatory response would be a decrease in serum HCO_3^- by 2 mEq/L for each 10 mm Hg decrease in $Paco_2$, which is what has occurred. The pH is still abnormal and only a small decrease in HCO_3^- is seen, so the compensatory mechanism is the release of nonbicarbonate buffers and tissue buffers that prevent life-threatening dramatic increases in pH in acute situations. If renal compensation had occurred, the HCO_3^- would have decreased by 4 to 5 mEq/L for each 10 mm Hg decrease in $Paco_2$ and a near-normal or normal pH would be present. Therefore the patient's clinical presentation and ABG findings suggest an acute event causing the alkalosis. Her physician suspects a pulmonary embolus.

L.D.'s Serum Electrolyte Values	**Normal Values**
Na^+ 138 mEq/L	137-147 mEq/L
K^+ 4.0 mEq/L	3.5-5.0 mEq/L
Cl^- 105 mEq/L	95-108 mEq/L
HCO_3^- 22 mEq/L	22-26 mEq/L

3. Question: Do the venous electrolytes reflect any other acid-base disturbance?

Answer: No, the components Na^+, K^+, and Cl^- are all within normal limits. The HCO_3^- has decreased by 2 mEq/L. The anion gap, $Na^+ - (Cl^- + HCO_3^-)$, is 11, which is within normal limits (12 + or − 2). Based on available data, no concurrent acid-base disturbance accompanies the acute respiratory alkalosis.

4. Question: Do you think the Pao_2 of 64 mm Hg is responsible for the hyperventilation? What other stimulus may be contributing to the hyperventilation?

Answer: The Pao_2 is probably not the stimulus for the hyperventilation. Usually, the Pao_2 must decrease to 50 to 60 mm Hg before the chemoreceptors stimulate the respiratory center to increase respirations. The pulmonary receptors (sensitive to irritation, congestion, and stretch) in the lung, thorax, and small airways may be stimulating the respiratory center due to the presence of pulmonary emboli. The pain and anxiety L.D. is experiencing also may be contributing to her hyperventilation.

Case Study 16-2

N.L. is a 46-year-old man with a 20-year history of alcohol abuse. He has been admitted to hospitals numerous times over the past 3 or 4 years for treatment of jaundice, ascites, and other problems associated with hepatic dysfunction. A diagnosis of cirrhotic liver disease based on liver biopsy was made 2 years ago. N.L. admits he has continued to drink heavily. He is lethargic and confused (i.e., he is oriented to person only), and he has ascites, spider angiomata on his trunk, +1 ankle edema, and is icteric (jaundiced).

N.L.'s vital signs are as follows:

HR 76 bpm and regular, RR 24 breaths/min and unlabored, BP 112/70 mm Hg, and temperature 37.6° C (99.8° F).

N.L.'s ABG and Serum Electrolyte Values	Normal (*perfect*) Values
pH 7.46	7.40
$Paco_2$ 21 mm Hg	40 mm Hg
Pao_2 82 mm Hg	range: 80-95 mm Hg
O_2 saturation 94%	range: 95%-99%
HCO_3^- 14 mEq/L	24 mEq/L
Na^+ 138 mEq/L	137-147 mEq/L
Cl^- 108 mEq/L	95-108 mEq/L
K^+ 3.9 mEq/L	3.5-5.0 mEq/L
HCO_3^-* 16 mEq/L	26 mEq/L

*Measured with serum electrolytes. Note the normal serum HCO_3^- value is 2 mEq/L higher than the normal calculated HCO_3^- seen with the ABG analysis (see the section on CO_2 content in Chapter 14 for a review of these differences).

1. **Question:** Based on the history and ABG values, which acid-base disturbance is present?

   Answer: The pH is increased, which corresponds to an alkalosis. The $Paco_2$ is decreased by 19 mm Hg, which correlates with the increase in pH. Both of the values reflect an alkalosis. The serum HCO_3^- is quite low (8 mEq/L below normal). (Remember, the venous plasma content may be 1 to 3 mEq/L higher than the calculated HCO_3^- from the ABGs.) Based on the available information, a respiratory alkalosis is present. Metabolic acidosis might have been considered, based on the decrease in $Paco_2$ and HCO_3^-, with the reasoning that the decreased $Paco_2$ was a compensation for the decreased HCO_3^-. This reasoning is not valid, however, because the pH reflects alkalosis, not acidosis, and the body does not *overcompensate* to correct acid-base disturbances. N.L.'s history and ABG values indicate a respiratory alkalosis only.

2. **Question:** Has any compensation for the respiratory alkalosis occurred? Is this an acute or a chronic acid-base disturbance? Why?

   Answer: The kidneys have compensated for the decreased $Paco_2$ by decreasing the serum HCO_3^- concentration from the normal 24 to 16 mEq/L. If this were an acute respiratory alkalosis, you would anticipate the serum HCO_3^- concentration to decrease by 2 mEq/L for every 10 mm Hg decrease in $Paco_2$. Since the $Paco_2$ decreased by 19 mm Hg, a total decrease of 4 mEq/L of HCO_3^- would be expected, but the HCO_3^- is 16 mEq/L. In chronic respiratory alkalosis, maximal renal compensation occurs within several days, decreasing the HCO_3^- 4 to 5 mEq/L for each 10 mm Hg decrease in $Paco_2$. N.L.'s HCO_3^- did decrease 8 mEq/L in response to renal compensation. He has long-standing hepatic dysfunction and the kidneys have had ample time to respond. Based on this information it can be determined that he has chronic respiratory alkalosis.

3. **Question:** Are the electrolytes within normal range? What is the anion gap? Does it indicate a second acid-base disturbance?

   Answer: The electrolytes are normal, with the exception of a decreased HCO_3^- concentration, which has been discussed. The Cl^- is elevated to the upper limit of normal, which corresponds to the presence of a chronic respiratory alkalosis. The anion gap is 14, which is slightly elevated but still within normal range (12 + or − 2). The elevation in anion gap (increased by approximately 3 mEq/L) is expected in chronic respiratory alkalosis when the $Paco_2$ is decreased to approximately 20 mm Hg. The presence of a second acid-base disturbance cannot be detected, based on the ABG and electrolyte values.

4. **Question:** What mechanism(s) is(are) responsible for stimulating the chronic respiratory alkalosis?

Answer: The hyperventilation is not caused by hypoxemia since the patient's Pao_2 is 85 mm Hg. Hyperventilation is not related to the decrease in oxygen tension, even in individuals who are experiencing hypoxemia, until the Pao_2 is below 50 to 60 mm Hg. The hyperventilation and respiratory alkalosis associated with hepatic failure is believed to be related to the accumulation of toxins, but no specific causative agent has been identified. Although conclusively no one toxin has been tied to hypocapnia, the blood ammonia level tends to correspond best. When respiratory alkalosis with a $Paco_2$ less than 25 mm Hg occurs in individuals with liver disease, it is a very poor prognostic sign. Patients with this degree of hypocapnia, regardless of the cause of their liver disease, usually do not survive for longer than several months.

References

Anderson L and Heinrich W: Alkalemia-associated morbidity and mortality in medical and surgical patients, South Med J 80(6):729-733, 1987.

Behrman R and others: Textbook of pediatrics, ed 13, Philadelphia, 1987, WB Saunders Co.

Braunwald E and Isselbach K, editors: Harrison's principles of internal medicine, ed 11, New York, 1987, McGraw-Hill Book Co.

Dreisback R and Robertson W: Handbook of poisoning, ed 12, San Mateo, Calif, 1987, Appleton & Lange.

Easterday U: Acid-base balance. In Horne M and Swearingen PL: Pocket guide to fluids and electrolytes, St Louis, 1989, The CV Mosby Co.

Easterday U and Howard C: Respiratory dysfunctions. In Swearingen PL and others: Manual of critical care: applying nursing diagnoses to adult critical illness, St Louis, 1988, The CV Mosby Co.

Goldberger E: A primer of water, electrolyte, and acid-base syndromes, ed 7, Philadelphia, 1986, Lea & Febiger.

Larson E and others: Acute mountain sickness and acetazolamide: clinical efficacy and effect on ventilation, JAMA 248(3):328-332, 1982.

Maxwell M, Kleeman C, and Narins R: Clinical disorders of fluid and electrolyte metabolism, New York, 1987, McGraw-Hill Book Co.

Montgomery A and others: Effects of dexamethasone on the incidence of acute mountain sickness at two intermediate altitudes, JAMA 261(5):734-736, 1989.

Murray J and Nadel J: Textbook of respiratory medicine, Philadelphia, 1988, WB Saunders Co.

Narins R and Emmet M: Simple and mixed acid-base disorders: a practical approach, Medicine 59(3):161-185, 1980.

Rose B: Clinical physiology of acid-base and electrolyte disorders, ed 3, New York, 1989, McGraw-Hill Book Co.

Shapiro B and others: Clinical application of blood gases, ed 4, Chicago, 1989, Year Book Medical Publishers, Inc.

Valtin H and Gennari F: Acid-base disorders, Boston, 1987, Little, Brown & Co, Inc.

Chapter

17

Metabolic Acidosis

Metabolic acidosis occurs when there is a primary decrease in serum bicarbonate (HCO_3^-) to <24 mEq/L, with a pH of <7.40. The $Paco_2$, which will be <40 mm Hg, decreases to compensate for the decreased HCO_3^-. The decrease in serum HCO_3^- is caused by one or a combination of the following mechanisms: (1) loss of base (HCO_3^-) from the body, (2) excess fixed acid (H^+) production, and (3) inability of the kidneys to excrete the daily fixed acid load. Each of these mechanisms causes an imbalance in the normal acid-base ratio. (Remember, to maintain a pH of 7.40, the ratio must be 1 acid [H^+] for every 20 bases [HCO_3^-].)

Loss of base usually occurs through loss of bicarbonate-rich fluid (e.g., diarrhea, pancreatic and biliary drainage) from the gastrointestinal (GI) tract. HCO_3^- also can be lost in the urine because of renal tubular dysfunction (e.g., renal tubular acidosis) or due to medications (e.g., carbonic anhydrase inhibitors such as acetazolamide [Diamox]). When base is lost, the 1:20 acid to base ratio is altered, and the pH decreases to below 7.40.

Excess acid production or ingestion alters the acid-base ratio by increasing the number of acids in proportion to the bases. The bicarbonate concentration decreases as the excess acid load is buffered, causing the pH to decrease below 7.40. The excess acid production refers to fixed acid (H^+) production rather than to CO_2 (which would cause a respiratory acidosis). The most common causes of increased acid production are diabetic ketoacidosis (DKA) and lactic acidosis. Ingestion of certain toxins (e.g., methanol) or toxic doses of some medications (e.g., salicylates) also can produce metabolic acidosis by increasing the H^+ load.

When the kidneys are unable to eliminate the normal daily acid load, acid (H^+) accumulates, HCO_3^- stores become depleted, and metabolic acidosis develops. When the daily acid load is not excreted, the ratio of acids to bases gradually increases (acids increase, bases decrease). This commonly occurs in acute renal failure, renal insufficiency, and chronic renal failure.

Potential Causes of Metabolic Acidosis

Loss of HCO_3^-	Excess Acid Production/ Ingestion	Inability of the Kidneys to Excrete Acid (H^+) Load
Gastrointestinal 1. Diarrhea 2. Biliary and pancreatic drainage 3. Urethral sigmoidostomy or ileostomy 4. Cholestyramine	**Ketoacidosis** 1. Diabetic 2. Alcohol induced	**Renal Insufficiency; Acute or Chronic Renal Failure**
	Lactic acidosis	**Renal Tubular Acidosis Type (RTA) I**
	Massive Rhabdomyolysis	**Hypoaldosteronism**
Renal 1. Renal tubular acidosis (RTA) type II 2. Carbonic anhydrase inhibitors (acetazalomide)	**Ingestion** 1. Salicylates 2. Hyperalimentation fluids if acetate or lactate is not added	**Potassium-sparing Diuretics**
		Posthypercapneic Metabolic Acidosis after Correction of Chronic Respiratory Alkalosis

Many of the disorders resulting from the three mechanisms discussed above are acute and resolve with appropriate treatment. Some, however, are chronic, resulting in the gradual alteration in the acid-base environment. As shown in the accompanying box, causes of metabolic acidosis are numerous.

Anion Gap

Determining the presence of an anion gap is useful in identifying the underlying pathophysiologic process responsible for the acid-base disorder. The concept of anion gap has been defined and discussed in Chapters 1, 5, and 13. Its role in determining differential diagnosis in metabolic acidosis will be discussed further in this chapter. The anion gap is the difference or *gap* between the serum concentration of the major cation (Na^+) and the normally measured anions (Cl^- and HCO_3^-): $Na^+ - (Cl^- + HCO_3^-)$. The normal serum electrolyte values are Na^+ 140 mEq/L, Cl^- 104 mEq/L, and HCO_3^- 24 mEq/L, resulting in a normal anion gap of 12 (+ or − 2).

$$\begin{array}{l} Na^+ \quad Cl^- \; HCO_3^- \\ 140 - (104 + 24) = 12 \end{array}$$

Normal anion gap metabolic acidosis

Metabolic acidosis can be present with a normal anion gap. For example, when hydrochloric acid (HCl) is added to body fluids, it is buffered by HCO_3^-. When HCO_3^- becomes *used up* from buffering the H^+, Cl^-, which had been paired with the H^+, takes the place of the HCO_3^- in the serum to maintain the anion balance (electroneutrality). The total number of anions remains constant—only the specific numbers of HCO_3^- and Cl^- change (the HCO_3^- decreases and the Cl^- increases) so the anion gap remains normal. For example:

Normal serum

$$\begin{array}{ccc} Na^+ & HCO_3^- & Cl^- \\ 140 & - (24 + & 104) = 12 \end{array}$$

After adding 10 mEq/HCl⁻

$$\begin{array}{ccc} Na^+ & HCO_3^- & Cl^- \\ 140 & - (14 + & 114) = 12 \end{array}$$

In the above example, the HCO_3^- decreases by 1 mEq/L for each 1 mEq of HCl^- added. As the HCO_3^- neutralizes the H^+, a Cl^- is freed, leading to an increase in Cl^- by 1 mEq for each 1 mEq decrease in HCO_3^-. Since the total number of anions has not changed, the anion gap remains normal, even though there is a significant metabolic acidosis (HCO_3^- concentration is 14). This type of metabolic acidosis is called a normal anion gap metabolic acidosis or a hyperchloremic metabolic acidosis.

Increased anion gap metabolic acidosis

In the above discussion, HCO_3^- was used to buffer the HCl. HCO_3^- also buffers acids where the H^+ is joined with anions (A^-) other than Cl^-, for example, lactic acid. For the purposes of this discussion, this acid will be referred to as H^+ (hydrogen ion) + A^- (anion other than Cl^-), or HA. As HCO_3^- neutralizes the H^+, an A^- is freed. So again, for every HCO_3^- that is used up, an anion (A^-) is added to the serum. Anions other than Cl^- and HCO_3^- are usually not measured. The serum chloride concentration is not changed, and the freed anions (A^-) are not measured. So as the HCO_3^- concentration decreases, there is a greater gap between the cations and the measured anions (the anion gap). For example:

Normal serum

$$\begin{array}{ccc} Na^+ & HCO_3^- & Cl^- \\ 140 & - (24 + & 104) = 12 \end{array}$$

After adding 10 mEq lactic acid ($H^+ + A^- = HA$)

$$\begin{array}{ccc} Na^+ & HCO_3^- & Cl^- \\ 140 & - (14 + & 104) = 22 \end{array}$$

The anion gap, which is increased, is 22. Although the anions added to the body fluids are not measured specifically with the electrolyte profile, their presence is identified when the anion gap is calculated. The decrease in HCO_3^- concentration corresponds to the increase in unmeasured anions. Once a high anion gap acidosis

Causes of Normal and Increased Anion Gap Acidosis

Normal Anion Gap Acidosis (hyperchloremic acidosis)

GI loss of HCO_3^-
Renal loss of HCO_3^-
Inability of the kidneys to excrete H^+

1. Early stages of renal insufficiency/renal failure
2. Renal tubular acidosis (types I and IV)
3. Medications/infusion therapy:
   a. spironolactone
   b. amiloride (a diuretic)
   c. triamterene
   d. total parenteral nutrition (TPN)
   e. ammonium chloride
   f. amphotericin B
4. Chemical:
   a. toluene

Post chronic hypocapnea correction

Increased Anion Gap Acidosis

Ketoacidosis

1. Diabetic
2. Alcohol related
3. Fasting/starvation

Lactic acidosis
Renal insufficiency/failure (acute or chronic)
Ingestion of acids*

1. Salicylates
2. Methanol
3. Ethylene glycol

Massive rhabdomyolysis

*Ingestion of most of these toxins also will cause increased production of lactate, resulting in a severe mixed metabolic acidosis.

is identified, specific laboratory tests can be done to determine the specific anion involved in the acidosis (e.g., lactate or ketones). The involved anion is *not* responsible for the acidosis (the excess H^+ is), but the involved anion identifies the source of the increased acid production. See the accompanying box for causes of normal and increased anion gap acidosis.

Altered anion gap with normal pH

Anion gap also can be increased without acidemia being present. This occurs in many situations, such as dehydration, alkalemia, chronic hypocapnia, administration of carbenicillin, and administration of lactated Ringer's solution or IV sodium citrate (both solutions contain unmeasured anions). Although it is not clinically significant in terms of acid-base balance, a decreased anion gap can occur in individuals with hypoalbuminemia and when multiple myeloma is present. The mechanism for decreased anion gap in these instances is unclear.

ACUTE METABOLIC ACIDOSIS

Acute metabolic acidosis can cause many serious alterations in cellular and organ functioning and may pose a serious threat to life. Immediate chemical buffering and respiratory compensation occur with acute metabolic acidosis and usually prevent rapid, life-threatening alterations in pH. But these adaptive mechanisms can be overwhelmed quickly by the rapid accumulation of acids (lactic acidosis) or by the rapid loss of HCO_3^- (diarrhea). Because of the acute onset of the acid-base disturbance, the ability of the kidneys to intervene is limited. (Remember that maximal renal compensation can take hours to days to occur.)

Acute metabolic acidosis is a common acid-base disturbance requiring prompt identification and treatment. Applying the concept of anion gap is useful in understanding the underlying process leading to the acute metabolic acidosis. More than one cause of acute metabolic acidosis may be present, leading to severe alterations in pH. Common causes of mixed acute metabolic acidosis will be discussed in Chapter 19.

History and Risk Factors

The etiologies of acute metabolic acidosis can be categorized into two groups: those that exist with a normal anion gap and those that result in increased anion gap. In general terms, normal anion gap acidosis is the result of the loss of HCO_3^- (i.e., in stool, pancreatic or biliary drainage, or urine) or by the inability of the kidneys to excrete the daily load of H^+ (e.g., with renal insufficiency). Increased anion gap acidosis is generally the result of the accumulation of organic acids other than H^+, such as ketones, lactate, and metabolites from toxins.

Normal anion gap acute metabolic acidosis (hyperchloremic acidosis)

Gastrointestinal loss of HCO_3^-

A GI loss of HCO_3^- (diarrhea) is the most common cause of hyperchloremic acidosis. HCO_3^- is the major anion in the ileum and colon, and a small amount of HCO_3^- is lost in the stool normally. In diarrhea, absorption of HCO_3^- is impaired because of the decrease in transit time, and thus a larger amount is lost in the stool, causing metabolic acidosis. Although absorption of HCO_3^- is decreased during diarrhea, the bowel continues to absorb Cl^- and return it to the circulation. This results in a stool free of Cl^- and an increased serum Cl^- concentration. Organic acids are also present in intestinal fluids and are metabolized to HCO_3^- in the bowel under normal circumstances. When diarrhea occurs, these organic acids (potential HCO_3^-)

are lost to a greater degree, also contributing to the development of metabolic acidosis. Very severe acidosis and electrolyte disorders can occur with frequent and voluminous stools. In fact, diarrhea resulting in severe metabolic acidosis is a leading cause of death in infants and small children worldwide. Laxative abuse also can cause excess loss of HCO_3^- through the stool.

Large amounts of HCO^- also can be lost through biliary and pancreatic secretions, which are rich in HCO_3^- and low in Cl^-. Pancreatic secretions have a greater concentration of HCO_3^- and are more likely to be implicated in the development of metabolic acidosis. The loss of HCO_3^- leads to an excess of H^+ (again, the 20:1 ratio of bases to acids has been disturbed). Normally functioning kidneys usually can eliminate the excess H^+ as it accumulates, but if renal insufficiency or failure exists, the individual is at high risk for developing acute metabolic acidosis. The last cause of GI loss of HCO_3^- is use of the drug cholestyramine (Questran), which is used to treat hypercholesterolemia. This drug acts as an anion-exchange resin—the Cl^- present in the drug is exchanged for HCO_3^- in the GI tract. This causes a decrease in HCO_3^- concentration and can result in metabolic acidosis.

Renal loss of HCO_3^-

In renal insufficiency/failure, as the glomerular filtration rate (GFR) falls below 20 ml/min, bicarbonate reabsorption is decreased, resulting in bicarbonaturia and metabolic acidosis. (As renal insufficiency progresses to end-stage renal failure and organic acids no longer are excreted, a high anion gap metabolic acidosis will develop.) Carbonic anhydrase inhibitors such as acetazolamide (Diamox) used to treat glaucoma or seizure disorders can cause a mild metabolic acidosis because they decrease reabsorption of HCO_3^- in the proximal tubule. Another carbonic anhydrase inhibitor is mafenide acetate (Sulfamylon), a bacteriostatic ointment used to treat burns, which also can cause a mild metabolic acidosis if large enough quantities are absorbed into the systemic circulation.

Decreased renal excretion of H^+

Renal insufficiency from diabetic nephropathy, for example, causes a decrease in H^+ excretion because ammonia (NH_3) synthesis is impaired. (Remember, greater quantities of H^+ can be excreted when they combine with NH_3 to form NH_4.) If less NH_3 is available to combine with H^+, fewer H^+ can be excreted, and the concentration of H^+ increases. As mentioned earlier, the anion gap does not increase in renal insufficiency/failure until ***unmeasured anions*** (the organic acids such as sulfuric and phosphoric acid) accumulate due to the very low GFR.

Certain drugs and therapies also can decrease the kidneys' ability to excrete H^+. The potassium-sparing diuretics spironolactone (Aldactone), amiloride (Midamor), and triamterene (Dyrenium) can cause metabolic acidosis because they decrease H^+ and K^+ secretion in the distal tubule. These drugs are not usually given to individuals with renal failure. Amphotericin B can cause a decrease in H^+ excretion due to alteration in membrane permeability for H^+. The administration of large quantities of isotonic saline can cause a dilutional (or expansion) metabolic acidosis because the ratio of acids to bases is altered as the extracellular water volume increases. Finally, metabolic acidosis can occur when paint thinner, model glue, or transmission fluid is sniffed. The metabolic acidosis is the result of toluene, a component present in each of the above. Toluene can cause metabolic acidosis in one of two ways: (1) decreasing H^+ secretion in the collecting tubules (a type I renal tubular acidosis) resulting in a normal anion gap acidosis or (2) by adding organic acids to the body fluids resulting in a high anion gap metabolic acidosis.

TPN solutions, depending on the amino acids used, can cause metabolic acidosis because H^+ is generated as the amino acids (especially arginine and lysine) are used. Many TPN solutions now contain buffers (acetate or lactate) to decrease the incidence of metabolic acidosis. A more indirect cause of metabolic acidosis from TPN results when *starved* patients are fed and their serum phosphorus decreases as adenosine triphosphate (ATP) is formed in the cells. With less phosphate available, there is a decrease in phosphate buffering of H^+ in the urine, so less H^+ is excreted. (Remember that phosphate is required for titratable acid excretion.)

Post chronic hypocapnia

Post chronic hypocapnia-induced metabolic acidosis occurs when sustained hypocapnia (decreased $Paco_2$) is resolved quickly and the decrease in HCO_3^- concentration from renal compensation of the chronic respiratory alkalosis remains. Until the kidneys adjust to the newly corrected $Paco_2$, a metabolic acidosis is present. When individuals who have lived at high altitudes for a prolonged period return to a normal altitude, they may experience a transient metabolic acidosis as their kidneys adjust (increase) the serum HCO_3^- concentration to normal. Information regarding a recent past history of living at a high altitude is a significant component of a nursing history.

High anion gap metabolic acidosis

Basically, this type of metabolic acidosis is caused by the addition of organic acids to the body. Normally, the kidneys can adjust to the daily variation in the fixed acid load (e.g., the increases in protein consumption). High anion gap acidosis occurs when the organic acids are added to the body or produced by the body more rapidly than they can be metabolized by the body or excreted by the kidneys. The accumulation of these *unmeasureable anions* increases the anion gap and gives rise to the term *high anion gap acidosis.* It is important to emphasize that the presence of these anions does not cause the metabolic acidosis, instead they signal the source of the increased acid production.

Lactic acidosis

Lactic acidosis can result from any process that causes increased lactate and H^+ production or decreased lactate metabolism by the liver. Lactic acid is generated by the metabolism of glucose and alanine, an amino acid. Lactic acid is then buffered by

HCO_3^- to form lactate. Normally, the liver converts lactate to carbon dioxide and water, which is either excreted or used to form HCO_3^-. When the liver is unable to metabolize the lactate that has been formed, lactate accumulates and a high anion gap metabolic acidosis can occur. The inability of the liver to metabolize lactate can be the result of a liver disease, such as cirrhosis.

Hypoperfusion leading to hypoxia also can be responsible for the accumulation of lactate. Hypoxia can result from numerous processes, such as shock, severe anemia, cardiac or respiratory arrest, and any process that leads to vascular collapse. The hypoxia causes lactate to accumulate more rapidly than it can be metabolized by the body, with a resulting high anion gap metabolic acidosis. Hypoperfusion can occur without a measurable decrease in blood pressure, and hypoxemia may or may not be present. Lactic acidosis can occur any time tissue demands for oxygen exceed the available oxygen supply. When metabolic demands are greatly increased, for example, with a grand mal seizure, vigorous exercise, or prolonged shivering secondary to hypothermia, large quantities of lactate accumulate rapidly. In these instances, the lactic acidosis is usually of short duration and resolves when the muscle activity decreases and oxygen delivery is adequate to meet cellular metabolic demands. The acidosis resolves because the body metabolizes the lactate to CO_2 and H_2O and then to HCO_3^-. This new HCO_3^- replaces that which was used during the acidosis, and the acid-base status returns to normal. (A similar process occurs in DKA when ketoanions are metabolized to HCO_3^- after appropriate treatment.)

Lactate can accumulate and may lead to lactic acidosis in the absence of hypoperfusion. Most commonly, this occurs with DKA, respiratory or metabolic alkalosis, severe bacterial infections, leukemia, thiamine and ethanol ingestion, and certain enzyme deficiencies. In these disorders, the cause of the lactate accumulation is not clearly understood.

The normal arterial lactate concentration is 0.5 to 1.6 mEq/L. When the arterial lactate level exceeds 7 mEq/L with acidemia (pH <7.40), lactic acidosis is present. Occasionally, a transient increase in lactate level will be noted as the patient responds to treatment, perfusion improves, and lactate enters the systemic circulation from the tissues. In general, lactic acidosis is associated with a high mortality rate.

Ketoacidosis

Ketoacidosis can be triggered by alcohol ingestion and starvation, but the most common cause is diabetes mellitus. DKA usually occurs in individuals with insulin-dependent diabetes mellitus. Hyperglycemia is present, frequently exceeding 400 mg/dl. The degree of ketoacidosis is determined by the ABG analysis and serum ketone values, as well as by the presence of ketonuria. The severity of the DKA may be underestimated if the ketone level results are relied on too heavily, because nitroprusside, the reagent used to detect the presence of ketones, is only sensitive to acetoacetic acid. The other ketoacid, β-hydroxybutyric acid, is not measured and it makes up a large part of the acid load in DKA. Both ketoacids dissociate into ketoanions and H^+, so they both contribute to the development of the acidosis. If hypoxia is present with the DKA, the ratio of β-hydroxybutyric acid to acetoacetic acid increases, but since it is not measured, the increase goes undetected. This is important because as the hypoxia caused by dehydration is corrected with fluid administration, the β-hydroxybutyric acid is converted to acetacetic acid, which *is* measured. This will cause a sudden increase in the *serum ketone* value, which may, mistakenly, lead to the conclusion that the ketosis is worsening. When insulin is given as part of the treatment, the generation of ketoacids stops. The ketoacids present in the body fluids are metabolized to CO_2 and H_2O and then to HCO_3^-. So in theory, the base stores could be restored at least in part by the circulating ketoacids. But often enough ketones are lost in the urine because of the osmotic diuresis that occurs with DKA, so that regeneration of HCO_3^- through ketoacids is inadequate or the acidosis is so severe (<7.20) that exogenous HCO_3^- must be administered. Sodium bicarbonate therapy is usually reserved for patients with severe metabolic acidosis. Many mixed acid-base disturbances may be present in individuals with DKA. See Chapter 19 for a discussion of mixed disorders. (For further discussion of treatment for DKA, see Case Study 3-1.)

Alcohol-related ketoacidosis usually occurs in chronic alcoholics who suddenly stop drinking for a day or more. Because the alcohol has provided the carbohydrates necessary for energy, cessation or reduction of alcohol intake causes the body to break down protein and fat for energy, resulting in ketoacidosis. Glycogen stores in the liver are not available for use because they have been depleted due to a chronically poor diet. These patients often present with severe vomiting, which causes volume depletion, which in turn impairs the kidneys' ability to excrete the ketoanions that are produced. The volume deficit also can lead to hypoperfusion, causing lactic acidosis, which further intensifies the metabolic acidosis. The effects of the ethanol cause a large proportion of ketoacids to be in the form of β-hydroxybutyrate, which will not be detected by the usual ketone reagent, nitroprusside. Serum ketone values may be negative, even though ketosis is present. Treatment of this disorder usually consists of hydration and glucose administration. Insulin is rarely necessary and can be very dangerous because individuals with alcohol-induced ketoacidosis tend to have low blood sugar.

Fasting is another cause of ketoacidosis. When there is a lack of carbohydrate intake, the blood sugar is lowered, which removes the stimulation for production of insulin and stimulates glucagon release. Ketoacid accumulation is usually small because the presence of ketones in the blood causes secretion of insulin, which will limit any further ketone production. Ketoacidosis from fasting is usually not severe and is self-limiting.

Renal failure

The high anion gap found in acute and chronic renal failure results from the inability of the kidneys to excrete the normal acid load, mainly the diet-generated acids (sulfuric, phosphoric, and other organic acids). Since H^+ is not excreted, HCO_3^- is not generated. Remember, in normally functioning kidneys, for every H^+ excreted, an

HCO_3^- is generated. However, when renal failure is present, HCO_3^- is used to buffer the accumulating H^+, but it is not replaced. The increase in H^+ and decrease in HCO_3^- results in the metabolic acidosis. The specific pathology underlying the development of metabolic acidosis in acute and chronic renal failure may vary, but the outcome is similar: high anion gap metabolic acidosis. In chronic renal failure (CRF), the HCO_3^- may decrease gradually to 15 mEq/L or lower as nephron mass and function diminish, causing an asymptomatic, gradual metabolic acidosis. On the other hand, individuals with acute renal failure may experience the same decrease in HCO_3^- over a period of hours to days and will be symptomatic.

Ingestion of acids

The metabolites and byproducts of some substances add organic acids to body fluids. Ingesting these toxins will also cause an increase in lactic acid production, resulting in a simultaneous lactic acidosis, which compounds the metabolic acidosis.

Ingestion of salicylates can cause varied acid-base disturbances. The two that are seen most commonly are metabolic acidosis and respiratory alkalosis. The metabolic acidosis is caused by the accumulation of lactate resulting from the effect salicylic acid has on cellular metabolism. Other organic acids, such as ketoanions, also contribute to the metabolic acidosis. Salicylic acid itself only contributes to the metabolic acidosis minimally. Adults and children respond differently to salicylate overdose. Adults tend to develop mixed acid-base disturbances: metabolic acidosis and respiratory alkalosis, with the respiratory alkalosis usually the predominant acid-base disorder (see Chapter 16 for more information). Children, especially those younger than 6 years of age, develop a more profound metabolic acidosis, often with an anion gap >20 mEq/L and ketonuria. The profound metabolic acidosis with children is the result of the accumulation of organic acids, the loss of Na^+ and K^+ in the urine as organic acids are excreted, and ketosis from the dehydration and starvation. Levels of salicylate high enough to cause signs of intoxication vary, but serum salicylate levels greater than 30 mg/dl are usually considered above the therapeutic level. Hypokalemia, hypouricemia, ketonuria, and either hyperglycemia or hypoglycemia may be present, along with a prolonged prothrombin time.

Methanol (methyl alcohol, wood alcohol) is a toxin used as a commercial denaturant and organic solvent. Ingestion of small quantities (15 to 30 ml) can result in death. Methanol is converted to formaldehyde and formic acid. These metabolites are responsible for the toxic effects of methanol. Individuals who have ingested methanol usually present with headaches, vomiting, and optic changes, such as blurred vision or even blindness.

Ethylene glycol is present in antifreeze and occasionally is consumed intentionally by alcoholics who do not have access to ethanol alcohol. The initial signs and symptoms of ethylene glycol ingestion resemble those seen with acute ethanol alcohol intoxication but progress to vomiting, deeper coma, and death. Ethylene glycol is metabolized into several end products, but glycolic acid is the acid most responsible for the increased anion gap and the metabolic acidosis. Permanent blindness and the development of acute renal failure can result from ingestion.

Massive rhabdomyolysis

Rhabdomyolysis is most often caused by severe, traumatic, and crushing injuries but can occur with severe influenza, strenuous exercise, heat stroke, seizures, hypokalemia, hypophosphatemia, and muscle ischemia. When muscle cells are damaged, myoglobin, a muscle protein, is released along with H^+ and other organic acids. The resultant metabolic acidosis is believed to be caused by the release of these substances from the damaged muscle cells. Rhabdomyolysis is also associated with the development of acute renal failure (myoglobin clogs the tubules), which can increase the degree of metabolic acidosis that occurs with this disorder.

Physiologic Responses

Chemical buffering

In the presence of metabolic acidosis, chemical buffering by intracellular and extracellular buffers occurs instantaneously. As mentioned in Chapter 13, HCO_3^- is the most important buffer in the extracellular fluid (ECF), and it buffers the fixed acids (H^+) that are added to body fluids. Intracellular buffers (phosphates and proteins) and bone carbonate buffer approximately 50% to 60% of the acid load. (Intracellular and extracellular buffering is discussed further in Chapter 13.) The entry of H^+ into the cells to be buffered causes K^+ to move out of the cells, and this frequently causes hyperkalemia. (For more details, see Chapter 9 and 13.) The immediate action of the chemical buffers prevents the pH from decreasing to life-threatening range, thereby providing the respiratory system time with which to compensate for the metabolic acidosis.

Respiratory

Alveolar hyperventilation occurs immediately as the decrease in pH is sensed by the respiratory center. The $Paco_2$ will decrease proportionately to the decrease in serum bicarbonate concentration (i.e., the ventilatory response depends on the degree of metabolic acidosis present). The $Paco_2$ will decrease approximately 1.2 mm Hg for every 1 mEq/L decrease in HCO_3^-. These guidelines for respiratory compensation only apply after the metabolic acidosis has been present for 12 to 24 hours because a *steady state* must be established before determinations can be made about the appropriateness of the compensatory respiratory response. If the $Paco_2$ is greater or less than expected, based on the serum bicarbonate concentration obtained at the same time, then a mixed acid-base disturbance is present. For example, in a metabolic acidosis with a $Paco_2$ that is lower than expected, a respiratory alkalosis is also present (see Chapter 16).

The maximum respiratory compensation for metabolic acidosis will cause the $Paco_2$ to decrease to 10 to 15 mm Hg. This maximum compensation may take several hours to a day to achieve. The delay in achieving maximal respiratory compensation is the result of the lag in acidification of the central nervous system (CNS). Remember that HCO_3^- crosses the blood-brain barrier slowly. Changes in serum HCO_3^- concentration are not fully appreciated as quickly as changes in $Paco_2$. As metabolic acidosis continues, the CNS also will experience the decrease in bicarbonate concentration. Coincidentally, when a patient is treated with HCO_3^-, the serum concentration increases before that of the CNS. Again, because of this lag (slow crossing of the blood-brain barrier), the respiratory center does not sense the correction of the serum HCO_3^-, and the respiratory compensation (hyperventilation) continues. Respiratory alkalosis may occur after the correction of metabolic acidosis and may persist for up to 36 hours after the metabolic acidosis has been corrected.

Lactic acidosis causes a greater degree of hypocapnia (decreased $Paco_2$) than do other causes of metabolic acidosis, even when the systemic pH is the same. This is believed to occur because increased concentrations of lactic acid are also present in the cerebrospinal fluid (CSF), and they exert a more direct effect on the respiratory center.

Regardless of the cause of the metabolic acidosis, the decrease in $Paco_2$ helps minimize the decrease in pH, but respiratory compensation cannot return the pH to normal. Alveolar hyperventilation is more a result of the increase in the depth of ventilation than an increase in the respiratory rate. In severe acidosis, when the pH is 7.00 or less, the tidal volume can increase 5 to 6 times greater than normal (from 5 to 6 L/min to 30 L/min). The deep respirations seen with metabolic acidosis are called **Kussmaul respirations** and are very obvious on physical assessment. If the patient has an underlying or concurrent respiratory dysfunction, the respiratory compensation will be diminished. Individuals who experience a respiratory acidosis in addition to a metabolic acidosis are at risk for developing severe acidosis because the respiratory compensation is no longer available and pH can decrease rapidly as carbon dioxide and fixed acids accumulate.

Individuals with chronic lung disease have a very limited ability to compensate during metabolic acidosis. The degree of compensation possible depends on the extent of the disease process and the respiratory reserve available. In these patients the work of breathing is so great that metabolic demands increase along with the generation of increased carbon dioxide, leading to a diminished ability to increase carbon dioxide elimination in the presence of metabolic acidosis. Respiratory muscle fatigue also occurs, which further limits the individual's ability to eliminate increased quantities of carbon dioxide.

Renal

When faced with a metabolic acidosis, the kidneys are able to increase the excretion of H^+ by increasing the excretion of ammonium, NH_4 (see Chapter 13). The renal compensation takes days to reach its maximum effect. When faced with a rapidly accumulating fixed acid load, metabolic acidosis develops before the kidneys are able to compensate. In some disorders, dysfunctional kidneys (e.g., with acute renal failure or renal tubular acidosis) are responsible for the development of metabolic acidosis, and thus they are unable to assist in the alleviation of the metabolic acidosis.

Assessment

The signs and symptoms seen with metabolic acidosis depend on the underlying cause. There are numerous causes of metabolic acidosis, so the clinical picture can vary considerably. Often, the signs and symptoms are more closely related to the underlying process than to the presence of the metabolic acidosis itself. Not only does the underlying process affect the clinical picture, but the degree of acidosis and the speed with which it occurs also influence the signs and symptoms. Rapidly developing, severe metabolic acidosis will cause more signs and symptoms. With mild metabolic acidosis, the patient may be asymptomatic. In general, when the serum HCO_3^- is 15 to 18 mEq or less or the pH is below 7.20, most patients will be symptomatic. See Table 17-1 for quick assessment guidelines for individuals with acute metabolic acidosis.

Pulmonary effects

The most prominent physical sign noted with acute, severe metabolic acidosis is the increase in the depth of respirations. Some degree of tachypnea (increased respiratory rate) also may be noted, but the deep, labored respirations (Kussmaul respirations) are the most striking respiratory feature. The increase in depth of respirations may not be noted in mild metabolic acidosis, as small increases in tidal volume and rate allow the lungs to eliminate the necessary amount of CO_2 to compensate for mild decreases in serum HCO_3^-.

Cardiovascular effects

Metabolic acidosis stimulates the sympathetic nervous system, causing the release of epinephrine. Epinephrine causes the heart rate to increase (tachycardia) until the pH falls below 7.0. After the pH falls below 7.0, the myocardium is no longer sensitive to the effects of epinephrine and the heart rate will decrease (bradycardia). Acidosis not only causes changes in the heart rate but also directly depresses the myocardium and impairs myocardial contraction (a negative inotropic effect). Initially, the two opposing effects of acidosis on the heart—increased heart rate secondary to the stimulation by epinephrine and the heart's decreased ability to contract—tend to balance each other out to maintain cardiac output. Cardiac output is equal to heart rate times stroke volume (contractility), so if heart rate is increased and stroke volume is decreased, cardiac output will remain unchanged. But the cardiac output will decrease progressively as the acidemia worsens since bradycardia will occur and stroke volume will be impaired further. As the cardiac output decreases, tissue perfusion worsens and lactic acid production increases. The concurrent lactic acidosis causes a more severe acidosis, which further affects myocardial function.

The heart is less vulnerable to the negative effects of acidosis if the pH is greater than 7.20, and it is therefore extremely important to keep the pH greater than this level, especially in the presence of cardiac arrest or cardiogenic shock.

Table 17-1 Quick assessment guidelines for individuals with metabolic acidosis

Signs and Symptoms	Physical Assessment
Findings vary, depending on the underlying disease states and severity of acid-base disturbance. There may be changes in LOC that range from fatigue and confusion to stupor and coma	Decreased BP; tachypnea leading to alveolar hyperventilation (Kussmaul respirations); cold and clammy skin with mild to moderate acidosis; flushed, warm, and dry skin with severe acidosis; dysrhythmias; and shock state

History and Risk Factors

- *Renal disease:*Acute renal failure
- *Ketoacidosis:* Diabetes mellitus, alcoholism, starvation
- *Lactic acidosis:* Respiratory or circulatory failure, drugs and toxins, hereditary disorders, septic shock. It can be associated with other disease states, such as leukemia, pancreatitis, bacterial infection, and uncontrolled diabetes mellitus
- *Poisonings and drug toxicity:* Salicylates, methanol, ethylene glycol, ammonium chloride
- *Loss of alkali:* Draining wounds (e.g., pancreatic fistulas), diarrhea, ureterostomy

Ventricular dysrhythmias, especially fibrillation, frequently develop with metabolic acidosis, compromising cardiac output even further. These rhythm disturbances may be related to electrolyte disturbances (i.e., alterations in potassium, magnesium, and phosphorus) that are often present with metabolic acidosis.

In addition to the myocardial changes, acidosis also causes changes in the peripheral vasculature. In mild to moderate acidosis, vasoconstriction occurs secondary to sympathetic stimulation. The vascular constriction increases the amount of blood returning to an exhausted heart, but with worsening acidosis, the vessels are no longer responsive to sympathetic stimulation and vasodilatation occurs. The vasodilatation seen with worsening acidosis causes the skin to be flushed, warm, and dry, even though the individual is hypotensive.

Central nervous system (CNS) effects

Individuals with acute metabolic acidosis often complain of a dull headache, weakness, and general malaise. Confusion, stupor, and coma also may be present. The changes in sensorium are more closely related to the underlying cause of the metabolic acidosis than to the acidosis itself. Changes in levels of consciousness are especially common after ingestion of certain toxins (e.g., methyl alcohol and ethylene glycol). Neurologic impairment is generally less prominent in metabolic acidosis than in respiratory acidosis. (Remember that carbon dioxide crosses the blood-brain barrier quickly and directly effects cerebral blood flow and functioning.)

Gastrointestinal effects

Complaints of nausea, vomiting, abdominal distention, and abdominal pain are frequent occurrences among individuals with DKA. It is not clear if these findings are related specifically to the metabolic acidosis or to the ketosis.

Diagnostic Studies

Arterial blood gas (ABG) values

The pH will be less than 7.40, and the $Paco_2$ will be less than normal (40 mm Hg) when the metabolic acidosis is diagnosed. The decrease in $Paco_2$ is proportional to the decrease in serum HCO_3^-. The $Paco_2$ will be decreased by 1.2 mm Hg for every 1 mEq/L decrease in HCO_3^-. $Paco_2$ rarely decreases below 10 to 15 mm Hg as compensation in pure metabolic acidosis. The Pao_2 may be normal or decreased, depending on the underlying pathologic process. Lactic acidosis leading to metabolic acidosis often occurs when hypoxemia (Pao_2 less than 60 mm Hg) or hypoxia (inadequate tissue oxygenation) is present. The acidosis will cause a shift to the left in the oxygen-hemoglobin dissociation curve (see Figure 14-1) causing oxygen to be less tightly bound to hemoglobin and assisting oxygenation at the tissue level.

Bicarbonate (HCO_3^-)

In acute metabolic acidosis, the HCO_3^- will be less than 24 mEq/L. When an elevated anion gap metabolic acidosis is present, the increase in the anion gap will be approximately equal to the decrease in HCO_3^- from normal (i.e., if the anion gap increases from 12 to 26, a difference of 14, the HCO_3^- can be expected to decrease by 14, from 24 to 10 mEq/L). Remember that electroneutrality must be maintained: as the HCO_3^- decreases, another anion takes its place in a 1:1 ratio. In normal anion gap acidosis, the Cl^- would increase to compensate for the decrease in HCO_3^-, in a 1:1 ratio. In high anion gap acidosis, unmeasured anions (lactate, ketoanions) increase to compensate for the decrease in HCO_3^-, again in a 1:1 ratio.

Serum electrolytes

Usually, potassium is elevated ($>$4.0 mEq/L) with metabolic acidosis as H^+ moves into the cells to be buffered and K^+ moves out of the cells to maintain electroneutrality. Hyperkalemia related to acidosis occurs most commonly in renal failure. A normal or low K^+ in the presence of acidosis usually signals low body stores of K^+. Sodium may be low, normal, or high, depending on the underlying cause. Sodium may be low in DKA, severe diarrhea, and nephritis, while it will be normal with fasting or hyperchloremic acidosis, and it can be high if dehydration is present. Chloride will be elevated in normal anion gap metabolic acidosis (hyperchloremic metabolic acidosis), for example, with diarrhea or pancreatic drainage. Chloride can be normal or low with elevated anion gap acidosis (e.g., with DKA). Hypophosphatemia may occur in some disorders (e.g., DKA, alcohol-induced ketosis, and diarrhea) that result in metabolic acidosis, although acidosis is not directly implicated in this electrolyte imbalance.

Urine pH

A very acidic urine (pH 4.5 to 5.2) will be excreted when renal function is normal. If renal dysfunction (e.g., renal tubular acidosis) is present, the kidneys are unable to increase the excretion of H^+ and the urine pH will not be as acidic.

Electrocardiogram (EKG)

Hyperkalemia is usually present with metabolic acidosis. Signs of hyperkalemia on the EKG include prolonged PR interval, widened QRS, and peaked or tented T waves. (For more information, see Chapter 9.) Severe acidosis also can cause ST deviations that mimic those seen with myocardial injury (i.e., ST segment elevation).

Medical Management

Identification and treatment of the underlying disorder is the only effective therapy for acute metabolic acidosis.

Alkalinizing agents

Sodium bicarbonate ($NaHCO_3$) is indicated in most clinical situations when arterial pH is <7.2. Correcting the pH to 7.20 decreases the risk of cardiac dysfunction. Acidosis depresses myocardial contractility and increases the risk of dysrhythmias, especially ventricular fibrillation. It is unnecessary and dangerous to attempt an immediate correction of the pH to >7.20-7.25 because doing so would cause a shift to the right in the oxygen-hemoglobin dissocation curve, resulting in decreased oxygen delivery at the tissue level. The usual mode of delivery of $NaHCO_3$ is IV drip: 100 to 150 mEq $NaHCO_3$ in 1,000 ml D_5W. In cases of severe metabolic acidosis, rapid administration is extremely important because any further reduction in bicarbonate will cause a dramatic, life-threatening drop in pH. In severe metabolic acidosis (cardiac arrest), $NaHCO_3$ is usually given IV push, 1 mEq/kg as the initial dose, with subsequent doses of 50 to 75 ml of an 8.4% solution, depending on serial pH results. $NaHCO_3$ must be given *cautiously,* regardless of how it is administered, because overtreating with $NaHCO_3$ can result in a metabolic alkalosis that can be severe and dangerous. Fluid overload resulting in pulmonary edema can occur when $NaHCO_3$ is given IV because it is hypertonic and causes fluid to shift into the intravascular space.

Potassium replacement

Usually, hyperkalemia is present but a potassium deficit can occur as well, with DKA and diarrhea, for example. If a potassium deficit exists (K^+ <3.5), it should be corrected before $NaHCO_3$ is administered because correction of the acidosis may worsen the hypokalemia (the K^+ would move from the extracellular space back into the cellular space, causing a further reduction in the serum K^+ level). (See Chapter 9 for more information on hypokalemia.)

Mechanical ventilation

Mechanical support of ventilation may be necessary to ensure that respiratory compensation is adequate, especially for semicomatose or comatose patients. If mechanical ventilation is used, the respiratory rate and tidal volume need to be adjusted to allow for maximal elimination of carbon dioxide. The respiratory compensation can lessen the cardiac effects of the metabolic acidosis. If the patient has been treated with $NaHCO_3$, increased carbon dioxide will be generated. Adequate ventilation must be ensured to eliminate the additional carbon dioxide.

Treatment of the underlying disorder

Diabetic ketoacidosis (DKA)

Usually, insulin administration and fluid replacement will stop the production of ketones in the presence of DKA and correct the metabolic acidosis. Once insulin is given, ketoacidosis is usually corrected as β-hydroxybutyrate (an anion present in ketosis) is metabolized into HCO_3^- by the body. Usually it is only when the pH is below 7.10 (HCO_3^- 6 to 8 mEq/L) that exogenous $NaHCO_3$ is required (in very small amounts) to raise the pH to greater than 7.20.

Alcoholism-related ketoacidosis

$NaHCO_3$ therapy is not usually required. Administration of saline and glucose usually will reverse the acidosis.

Diarrhea

Usually, diarrhea occurs in association with other fluid and electrolyte disturbances (i.e., hypokalemia, fluid deficit, hyperchloremia) and correction of the acidosis requires treatment of the underlying process as well as the appropriate fluid and electrolyte replacement.

Acute renal failure

Hemodialysis or peritoneal dialysis may be necessary to maintain an adequate level of serum bicarbonate in the patient with acute renal failure (ARF). As functioning nephrons decrease, H^+ is not secreted and bicarbonate concentration decreases. $NaHCO_3$ is usually not appropriate because of the risk of intravascular fluid overload, which is often already present with ARF.

Poisoning and drug toxicity

Treatment of poisoning or drug toxicity depends on the specific drug ingested or infused. Hemodialysis or peritoneal dialysis may be necessary.

Lactic acidosis

When the underlying disorder is treated successfully, the acidosis usually resolves. As tissue perfusion is restored, the body can metabolize the lactate into HCO_3^-, thus correcting the acidosis. If the underlying problem causing the lactic acidosis cannot be successfully treated, the use of $NaHCO_3$ is only transiently helpful, and its use is controversial (Narins and Cohen, 1987). The mortality rate associated with lactic acidosis is high.

Nursing Diagnoses and Interventions

Nursing diagnoses and interventions are specific to the pathophysiologic process. In addition, see Chapter 15, the section on acute respiratory acidosis for

Potential sensory/perceptual alterations related to sensorium changes, p. 287;

Potential alteration in oral mucous membrane related to abnormal breathing pattern, p. 288;

Sleep pattern disturbance related to frequent treatments and procedures, p. 288, and

Potential for ineffective family coping related to stress reaction, p. 288.

CHRONIC METABOLIC ACIDOSIS

Chronic metabolic acidosis is caused by any sustained pathologic process that results in the kidneys' inability to excrete adequate amounts of acid (H^+), for example, chronic renal failure (CRF) and renal tubular acidosis (RTA). Chronic metabolic acidosis also can occur when there is a long-term, 3 to 5 day loss of alkaline fluid from the body, such as with pancreatic or biliary drainage or when renal HCO_3^- reabsorption has been impaired (e.g., with use of carbonic anhydrase inhibitors). (See the box on p. 322 for potential causes of metabolic acidosis.) In most instances in which chronic metabolic acidosis develops, the process is gradual, with slowly decreasing HCO_3^- and pH. The patients are often asymptomatic until the serum HCO_3^- is 15 mEq/L or less.

History and Risk Factors

Chronic renal insufficiency/failure

Chronic renal insufficiency or failure is the most common cause of chronic metabolic acidosis. Numerous pathologic events are responsible for the development of CRF, for example, glomerulonephritis, diabetic nephropathy, nephrosclerosis, and vascular changes, among others. Regardless of the specific cause of the CRF, the sequence of events leading to chronic metabolic acidosis is clear. With CRF, the functional unit of the kidney, the nephron, decreases in number, which decreases the GFR and renal acid excretion. As more acid is generated than can be excreted, H^+ accumulates and the serum bicarbonate concentration decreases. Remember that fixed acid (H^+) is generated daily from metabolic processes as well as dietary sources. H^+ will accumulate if the amount excreted is not equal to the amount added to the body. In addition to the decreasing nephron mass, excretion of H^+ is impaired because of the kidneys' inability to manufacture ammonia, which is the main buffer for excretion of urinary H^+. With renal insufficiency, the anion gap is normal because dietary acids (i.e., sulfuric and phosphoric) are still excreted. As the GFR decreases to 50% or below, dietary acids are no longer excreted, and thus H^+ accumulates. Serum bicarbonate concentration decreases because it is used to buffer the excess H^+. Since H^+ excretion is decreased, generation of new HCO_3^- is also decreased. Generation of HCO_3^- is dependent on excretion of H^+. Thus the ***unmeasured anions*** (fixed and organic acids) that replace the decreased HCO_3^- to maintain electroneutrality cause the anion gap to increase. Numerous fluid and electrolyte imbalances are seen with CRF. (See Chapters 6, 9, and 10.)

Renal tubular acidosis (RTA)

There are four types of RTA, each with one or more specific causes and each causing specific derangements in nephron functioning. Basically, in all types of RTA the kidneys have a reduced ability to excrete H^+, although their ability to excrete dietary organic acids is much less affected. Because organic acids are excreted, the anion gap does not increase. A hyperchloremic metabolic acidosis with a normal anion gap occurs because large quantities of chloride are reabsorbed to make up for the decrease in HCO_3^-, thus preserving electroneutrality but increasing serum chloride.

Carbonic anhydrase inhibitors

Carbonic anhydrase inhibitors can cause chronic metabolic acidosis with sustained use. For a discussion refer to the earlier section on acute metabolic acidosis in this chapter.

Loss of alkali

Loss of alkali from extrarenal sources can lead to chronic metabolic acidosis if the loss is sustained (e.g., with diarrhea, pancreatic or biliary drainage). (Refer to the earlier section on acute metabolic acidosis for details.)

Physiologic Responses

Chemical buffering

In chronic metabolic acidosis, the intracellular and extracellular buffering is limited. Because of the sustained nature of the acid-base disturbance, chemical buffers achieve a new, steady state with a lower HCO_3^- and a lower pH. The bone buffers, calcium carbonate and calcium phosphate, play a major role in buffering a chronic acid load. The calcium buffers are released from the bone in the presence of a chronic metabolic acidosis to neutralize the excess H^+. In CRF the prolonged use of bone buffers can cause osteomalacia because calcium is *leeched* out of the bone. Other factors also play a role in the development of bone disease in CRF (refer to Chapters 10 and 11 for details).

Respiratory

The net effect of respiratory compensation in metabolic acidosis is lost after several days of sustained acidosis. Although the $Paco_2$ remains low because of hyperventilation, the serum HCO_3^- decreases as the renal HCO_3^- reabsorption is lowered in response to the decrease in $Paco_2$. (Remember that $Paco_2$ concentration plays a direct role in renal reabsorption of HCO_3^-. Decreased levels of $Paco_2$ cause HCO_3^- to be excreted in the urine.) The serum HCO_3^- concentration is reduced even further, so that although the $Paco_2$ has decreased from normal, the pH in chronic metabolic acidosis is the same as that in uncompensated metabolic acidosis. Respiratory compensation is very effective, preventing life-threatening acidosis during acute metabolic acidosis, but its effectiveness is lost when the metabolic acidosis lasts longer than a few days.

Renal

When the kidneys are not responsible for the development and maintenance of chronic metabolic acidosis (e.g., loss of HCO_3^- in sustained diarrhea or pancreatic drainage), they respond to the acidosis by increasing their acid excretion. Even diseased kidneys (i.e., in renal insufficiency) are stimulated to increase acid excretion by the presence of acidosis, thereby assisting in stabilizing the pH to a limited extent.

Assessment

Most individuals with chronic metabolic acidosis will be asymptomatic because of the gradual progression of the disorder. Fatigue, malaise, and anorexia may be present, but they are more closely associated with the underlying disease process than with the acidosis.

Diagnostic Studies

Arterial blood gas (ABG) values

The $Paco_2$ will be decreased by 1.2 mm Hg for every 1 mEq/L decrease in serum HCO_3^-. An interesting finding is that the $Paco_2$ will almost always correspond to the last two digits of the pH (e.g., an individual with a serum HCO_3^- of 12 mEq/L and a pH of 7.26 will have a $Paco_2$ of about 26 mm Hg). This fact can be applied only when the chronic metabolic acidosis is in a steady state and maximal respiratory compensation has occurred. The pH will be moderately decreased in the chronic metabolic acidosis from stable CRF. If individuals with CRF are suddenly exposed to a large acid load (exogenous or endogenous), they are at risk for developing a severe acidosis because alkali stores are already depleted and renal compensation is absent. The pH may be lower in chronic metabolic acidosis, caused by some types of RTA, and with chronic loss of bicarbonate-rich body fluids, such as with diarrhea or biliary or pancreatic drainage. The $Paco_2$ value will depend on the individual's underlying state of health. Hypoxia is not a usual finding in chronic metabolic acidosis.

Bicarbonate (HCO_3^-)

Serum bicarbonate will decrease to less than 22 mEq/L, usually stabilizing at 18 to 21 mEq/L, and rarely falling below 10 mEq/L with CRF. With CRF a serum HCO_3^- below 15 mEq/L often signals the need to initiate treatment, such as sodium bicarbonate ($NaHCO_3$) or dialysis, depending on the clinical picture.

Serum electrolytes

Usually, the ability to regulate potassium is not impaired until the late stages of CRF. If the patient is oliguric (less than 400 ml urine produced per day), hyperkalemia can develop because of the inability of the kidneys to excrete the necessary amount of potassium to maintain balance. Hyperkalemia also may occur if the patient is anuric or when acidosis develops. (Remember that with acidosis, H^+ moves into the cell and potassium moves out of the cell.) Hypokalemia may be present in some types of RTA secondary to wasting of potassium in the urine.

Chloride

Usually chloride is normal with CRF, but if a hyperchloremic metabolic acidosis is present secondary to loss of alkali (e.g., with RTA, diarrhea, or carbonic anhydrase inhibitor therapy), the chloride may be increased. In CRF the anion gap is rarely increased beyond 23 mEq/L. If the anion gap is greater than this, an additional superimposed organic acidosis may be present.

Phosphorus

Phosphorus is not directly altered secondary to chronic metabolic acidosis, but it will be increased if CRF is present (see Chapter 11 for more information about phosphorus disturbances).

Calcium

Calcium levels are usually low in CRF. Hypocalcemia, depending on its severity, may require treatment before treatment for chronic metabolic acidosis is initiated. Rapidly increasing the pH can cause a decrease in ionized calcium and thereby induce tetany (see Chapter 10 for a discussion of calcium).

Urine pH

The ability to produce an acidic urine depends on the underlying pathology. The urinary pH in CRF can be as low as 4.5, but since ammonia synthesis is decreased, only about half of the H^+ load produced daily can be excreted in the urine. In RTA the urinary pH is elevated because of urinary HCO_3^- wasting.

Electrocardiogram (EKG)

No specific EKG changes are noted with chronic metabolic acidosis. In the presence of hyperkalemia, EKG changes will be noted. For more information, see the earlier section on acute metabolic acidosis.

Medical Management

Correction of the chronic metabolic acidosis is usually unnecessary in adults. Children with metabolic acidosis require correction of pH to normal, with a HCO_3^- in the range of 22 to 24 mEq/L, so that normal growth will not be affected, especially in the presence of RTA.

Alkalizing agents

When serum bicarbonate levels drop below 15 mEq/L, treatment with oral alkali is often initiated, using sodium bicarbonate or sodium citrate—Shohl's solution. The dosage is increased gradually until serum HCO_3^- reaches 18 to 20 mm Hg. The gradual correction of the acidosis is important so that tetany does not develop from hypocalcemia, induced by rapid alkalinization of plasma. With severe acidosis (HCO_3^- < 10 mEq/L and pH < 7.20), use of parenteral $NaHCO_3$ may be necessary (refer to the earlier section on acute metabolic acidosis for detail). Be alert to the possibility of pulmonary edema if oliguria is present and $NaHCO_3$ is administered parenterally. Severe acidosis is usually not the result of CRF alone, but rather it occurs secondary to the addition of a sudden acid load.

Hemodialysis or peritoneal dialysis

When the chronic metabolic acidosis is caused by CRF, dialysis may be indicated if the metabolic acidosis cannot be controlled by oral medications.

Avoidance of certain medications

Some medications should be avoided in CRF because of their ability to increase metabolic acidosis. The antibiotic tetracycline, for example, causes an increased acidosis secondary to its catabolic effects.

Treatment of renal tubular acidosis (RTA)

RTA treatment depends on the specific type of RTA present. Usually, oral bicarbonate is administered (1 to 15 mEq/kg) daily. The HCO_3^- is either administered as sodium bicarbonate or as sodium citrate (citrate is converted to HCO_3^- by the body).

Nursing Diagnoses and Interventions

Potential alteration in nutrition: Less than body requirements related to decreased intake secondary to fatigue, dietary restrictions, and metallic taste in the mouth (caused by the uremic state from CRF)

Desired outcome: Patient's weight remains stable.

1. Provide foods that correspond to patient's prescribed diet and preference.
2. Offer small meals and snacks at frequent intervals.
3. Offer oral care often to minimize the metallic taste. Brushing teeth before meals and using mouthwash frequently throughout the day are helpful.
4. Provide hard candy to keep the oral mucosa moist, diminish metallic taste, and supply calories.
5. Monitor hemoglobin and hematocrit levels to determine if anemia may be contributing to patient's fatigue.
6. Provide periods of uninterrupted rest by clustering necessary treatments and procedures. If possible, avoid performing unpleasant or uncomfortable treatments 1 hour before and after mealtimes.
7. Encourage family members to eat with patient or be present at mealtimes to provide social interaction.

NOTE: Other nursing diagnoses and interventions are specific to the underlying pathophysiologic process.

Case Study 17-1

M.J. is a 65-year-old woman with advanced ovarian cancer. A tumor was surgically removed but evidence of microscopic disease remains. M.J. received whole-abdomen irradiation. Subsequently, after 2½ weeks of treatment, she experienced progressive diarrhea, which has intensified over the last 2 to 3 days. The diarrhea is refractory to outpatient medical management, necessitating hospital admission.

M.J. exhibits the following indicators of volume depletion: dry mucous membranes and poor skin turgor. Her chest is clear, and her respirations are deep but not labored. Except for ice chips, she has had no oral intake for 3 days because of nausea. Her vital signs are as follows:

BP 100/50 mm Hg (supine) and 70/40 mm Hg sitting
HR 130 bpm
RR 24 breaths/min and deep
T 37.5° C (99.5° F).

M.J.'s ABG Values	Normal (*perfect*) Values
pH 7.32	7.40
$Paco_2$ 29 mm Hg	40 mm Hg
Pao_2 90 mm Hg	95 mm Hg
Oxygen saturation 97%	99%

M.J.'s Serum Electrolytes (mEq/L)	Normal Ranges (mEq/L)
Na^+ 150	137-147
Cl^- 115	95-108
K^+ 3.3	3.5-5.0
HCO_3^- 15	22-26 (24 mEq/L perfect value)

1. **Question:** Which acid-base disturbance is present? Has compensation for the disturbance occurred?

   Answer: A metabolic acidosis is present. The pH is less than 7.40. The serum HCO_3^- is decreased from 24 mEq/L to 15 mEq/L. The $Paco_2$ has decreased to 30 mm Hg as respiratory compensation for the metabolic acidosis.

2. **Question:** Is the degree of compensation appropriate for the disorder?

   Answer: The $Paco_2$ should decrease by 1.2 mm Hg for every 1 mEq/L decrease in HCO_3^-. The HCO_3^- decreased by 9 mEq, so the corresponding respiratory compensation should be $1.2 \times 9 = 11.7$. The respiratory compensation is appropriate for the disturbance (see Table 14-3).

3. **Question:** What is the mechanism for developing metabolic acidosis with diarrhea?

   Answer: The stool contains a moderate amount of HCO_3^-, which is lost with diarrhea. Organic acids present in the bowel are also lost and these are potential sources of HCO_3^- (these organic acids are usually metabolized to HCO_3^- by the body). The volume depletion that frequently accompanies diarrhea also contributes to the development of metabolic acidosis.

4. **Question:** What is the role of volume depletion in the development of metabolic acidosis?

   Answer: Volume depletion can cause hypoperfusion of tissues as the cardiac output falls, with resultant generation of lactic acid. The increase in lactic acid production and concentration causes an increase in H^+

concentration, and thus metabolic acidosis. Acidemia itself has negative effects on the heart, decreasing cardiac output further and leading to hypoperfusion of tissues and hypoxia with increased generation of lactic acid. While the pH is not markedly decreased at this time, the degree of hypovolemia present is significant. Rehydration is critical in preventing a further decrease in pH.

5. **Question:** What changes in M.J.'s hematocrit would you anticipate, based on her hypovolemia?

   Answer: The hematocrit would increase because of the volume depletion. This is referred to as hemoconcentration. The hypernatremia present is also the result of hypovolemia.

References

Astiz M and others: Relationship of oxygen delivery and mixed venous oxygenation to lactic acidosis in patients with sepsis and acute myocardial infarction. Crit Care Med 16(7):565-568, 1988.

Battle D: Renal tubular acidosis. In Kurtzman N and Battle D: Med Clin North Am 67(4):859-878, 1983.

Du Bose T: Clinical approach to patients with acid-base disorders. In Kurtzman N and Battle D: Med Clin North Am 67(4):799-814, 1983.

Easterday U: Acid-base balance. In Horne M and Swearingen PL: Pocket guide to fluids and electrolytes, St Louis, 1989, The CV Mosby Co.

Easterday U and Howard C: Respiratory dysfunctions. In Swearingen PL and others: Manual of critical care: applying nursing diagnoses to adult critical illness, St Louis, 1988, The CV Mosby Co.

Fells P: Ketoacidosis. In Kurtzman N and Battle D: Med Clin North Am 67(4):831-843, 1983.

Frommer P: Lactic acidosis. In Kurtzman N and Battle D: Med Clin North Am 67(4):815-836, 1983.

Goldberger E: A primer of water, electrolyte, and acid-base syndromes, ed 7, Philadelphia, 1986, Lea & Febiger.

Hricik D and Kassirer J: Understanding and using the anion gap. Consultant 23(7):130-143, 1983.

Jeffers L: The effects of acidosis on cardiovascular function. J Assoc Nurse Anesthetists 23(2):148-150, 1986.

Koko J and Tannen R: Fluids and electrolytes, Philadelphia, 1986, WB Saunders Co.

Laski M: Normal regulation of acid-base balance: renal and pulmonary response and other extrarenal buffering mechanisms: In Kurtzman N and Battle D: Med Clin North Am 67(4):771-780, 1983.

Maxwell M and others: Clinical disorders of fluid and electrolyte metabolism, ed 4, New York, 1987, McGraw-Hill Book Co.

Mehta P and Kloner R: Effects of acid-base disturbances in cardiovascular function. In Geheb M and Carlson R: Crit Care Clin 5(4):747-753, 1987.

Middlebaugh R and Middlebaugh D: Current considerations in respiratory and acid-base management during cardiopulmonary resuscitation, Crit Care Nurse Quar 10(4):25-33, 1988.

Narins R and Cohen J: Bicarbonate therapy for organic acidosis: the case for its continued use. Ann Intern Med 106(4):615-617, 1987.

Narins R and Emmet M: Simple and mixed acid-base disorders: a practical approach. Medicine 59(3):161-185, 1980.

Pfister S and Bullas J: Arterial blood gas evaluation: metabolic acidemia. Crit Care Nurse, 8(8):14-19, 1988.

Pfister S and Bullas J: Arterial blood gas evaluation: metabolic acidemia. Crit Care Nurse 9(1):70-72, 1989.

Riley L and others: Metabolic acid-base disorders. In Geheb M and Carlson R: Crit Care Clin (5)4:699-724, 1987.

Ryder R: Lactic acidosis: high-dose or low-dose bicarbonate therapy? Diabetes Care 7(1):99-100, 1984.

Sabatini S: The acidosis of chronic renal failure. In Kurtzman N and Battle D: Med Clin North Am 67(4):845-858, 1983.

Schrier R: Renal and electrolyte disorders, ed 3, Boston, 1986, Little, Brown & Co, Inc.

Schrier R and Gottschalk C: Diseases of the kidney, Boston, 1988, Little, Brown & Co, Inc.

Vanatta J and Fogelman M: Fluid balance: a clinical manual, Chicago, 1982, Year Book Medical Publishers, Inc.

Weil M and others: Difference in acid-base state between venous and arterial blood during cardiopulmonary resuscitation, N Engl J Med 15(3):153-156, 1986.

Weldy NJ: Body fluids and electrolytes: a programmed presentation, ed 5, St Louis, 1988, The CV Mosby Co.

Chapter

18

Metabolic Alkalosis

Metabolic alkalosis occurs when there is an excess serum bicarbonate (HCO_3^-) concentration, either from a gain of HCO_3^- or a loss of hydrogen ions (H^+). The end result is an imbalance in the acid-base ratio. The proportion of base to acid is increased (i.e., there are more than 20 HCO_3^- for every 1 H^+). Metabolic alkalosis results from an elevated serum HCO_3^- (>24 mEq/L), an increased pH (>7.40), and a $Paco_2$ >40 mm Hg. The $Paco_2$ increases to compensate for the increase in HCO_3^-. An increase in the acid (CO_2) assists in neutralizing the increase in the base (HCO_3^-).

Metabolic alkalosis is a commonly occurring acid-base disturbance (see the accompanying box for a list of potential causes). Severe alkalosis (pH >7.60) is associated with high morbidity and mortality rates and usually is the result of a mixed disorder—both a metabolic and respiratory alkalosis—but it can occur from pure metabolic alkalosis alone. There are numerous causes of acute and chronic metabolic alkalosis. The most common cause of acute metabolic alkalosis in hospitalized patients is loss of gastric secretions via vomiting or NG suction. It also can be seen with acute, aggressive diuretic therapy. Chronic metabolic alkalosis is commonly associated with the long-term use of diuretics and mineralocorticoid excess.

Metabolic alkalosis is unique among the acid-base disturbances because factors that cause or generate the initial rise in serum HCO_3^- may not be responsible for the maintenance of the disorder. In fact, if the *cause* of the disorder is eliminated, acid-base balance may not be restored until the other factors maintaining it have been corrected. For example, even if diuretics are discontinued, the metabolic alkalosis will continue until it is treated appropriately. The distinction between the initiating or generating events of metabolic alkalosis and the factors that maintain it are important and will be discussed next.

Potential Causes of Metabolic Alkalosis

H^+ Loss

1. Gastrointestinal
   - Loss of gastric secretions (NG suctioning or vomiting)*
   - Villous adenoma
   - Congenital chloridorrhea
2. Renal
   - Diuretics (especially *loop* and thiazides*)
   - Mineralocorticoid excess*
   - Post chronic hypocapnia
   - Carbenicillin/penicillin derivation
   - Hypercalcemia/hypoparathyroidism
3. H^+ shift into the cells
   - Hypokalemia*
   - Carbohydrate refeeding after starvation

HCO_3^- Retention

1. Administration of bicarbonate or bicarbonate precursors
2. Massive blood transfusion
3. Milk-alkali syndrome

Contraction Alkalosis

Diuretics
Cystic fibrosis

*Most common causes

Generation of Metabolic Alkalosis

The loss of H^+ or gain of HCO_3^- must occur to generate or initiate a rise in serum HCO_3^- concentration. The increase in serum HCO_3^- concentration can be the result of (1) loss of H^+ from the GI tract, (2) increase in H^+ excretion by the kidneys in amounts greater than necessary to eliminate the normal acid load, (3) administration of HCO_3^- or precursors to HCO_3^- (citrate, lactate) at a rate faster than can be excreted by the kidneys, or (4) loss of fluid that contains more Cl^- than HCO_3^- so that a greater HCO_3^- concentration is present in a smaller amount of fluid *(contraction alkalosis)*. Even in the presence of these factors, the kidneys have the ability to excrete a tremendous amount of HCO_3^-, so metabolic alkalosis is ***maintained*** only when renal bicarbonate excretion is impaired.

Maintenance of Metabolic Alkalosis

The following are the three most common reasons for maintenance of metabolic alkalosis and are related to the impaired excretion of HCO_3^-: volume depletion, hypokalemia, and mineralocorticoid excess.

Volume depletion

A decrease in the effective circulating volume (ECV) promotes the continuation of metabolic alkalosis because Na^+ reabsorption is increased with hypovolemia. When Na^+ is reabsorbed, it must be accompanied by Cl^- or the excretion of H^+ or K^+ to maintain electroneutrality. Because of the difference between the amount of Na^+ (140 mEq) and the amount of Cl^- (115 mEq) present in each liter of glomerular filtrate, not all the Na^+ can be accompanied by a Cl^-, and this results in increased excretion of H^+ or K^+. In metabolic alkalosis, the serum chloride concentration is already decreased, further limiting its ability to accompany Na^+ as it is reabsorbed. Hence, as Na^+ is reabsorbed to restore a normal volume, an equal number of H^+ must be excreted. (K^+ is also lost in exchange for some Na^+, resulting in hypokalemia.) The increase in H^+ secretion causes an increase in HCO_3^- reabsorption and generation, maintaining metabolic alkalosis. Remember that for every mEq of H^+ excreted, a mEq of HCO_3^- is reabsorbed or regenerated. Hypovolemia *maintains* metabolic alkalosis because it causes increased renal reabsorption of HCO_3^-. Hypovolemia only *causes* metabolic alkalosis when the fluid lost contains more Cl^- than HCO_3^-, for example, with use of diuretics or with cystic fibrosis.

Hypokalemia

Hypokalemia also plays a role in maintaining metabolic alkalosis but a smaller role than that of decreased ECV. Hypokalemia stimulates ammonia production, thus increasing the excretion of H^+. Hypokalemia also causes a decrease in the reabsorption of Cl^-, causing more H^+ to be excreted in exchange for the reabsorption of Na^+. (Thus HCO_3^- excretion is decreased and metabolic alkalosis is maintained.)

Mineralocorticoid excess

Mineralocorticoid excess from any cause results in the increased release of aldosterone, which increases the excretion of H^+ and K^+, resulting in maintenance of the metabolic alkalosis. Again, the excretion of HCO_3^- is impaired until appropriate treatment is initiated. Remember, if the metabolic alkalosis is to be corrected, both the event generating the acid-base disturbance and the factors maintaining it must be treated.

Causes of Metabolic Alkalosis

Based on the fact that a decreased chloride concentration can maintain a metabolic alkalosis, the causes of metabolic alkalosis are frequently divided into two groups, based on their ability to be corrected by saline (NaCl) administration: saline-responsive metabolic alkalosis and saline-resistant metabolic alkalosis.

Saline-responsive metabolic alkalosis

Causes of metabolic alkalosis that respond favorably to the administration of saline are categorized as saline-responsive and include vomiting, NG suction, diuretic use, and post chronic hypercapnia. All of the saline-responsive causes of metabolic alkalosis involve hypovolemia and a loss of Cl^- or a decrease in serum chloride concentration. Therefore supplying the necessary Cl^- along with volume in the form of saline can resolve the metabolic alkalosis.

Saline-resistant metabolic alkalosis

Those disturbances causing metabolic alkalosis that do not respond to saline administration are categorized as saline-resistant and include severe hypokalemia, renal failure, some edematous states, and mineralocorticoid excess. These disorders are *not* maintained because of a Cl^- deficit and hypovolemia, and therefore supplying Cl^- does not resolve the disorder. In fact volume expansion, not depletion, is usually present. See the accompanying box for a list of saline-responsive and saline-resistant disorders.

ACUTE METABOLIC ALKALOSIS

The major causes of acute metabolic alkalosis are loss of gastric acid from vomiting or NG suction and the use of diuretics. These disorders not only cause a loss of H^+ (thus an increase in HCO_3^-) but also a decrease in ECV (i.e., volume contraction) and a loss of KCl. Causes of metabolic alkalosis can be categorized as follows: H^+ loss, HCO_3^- retention, contraction alkalosis, and H^+ shift into the cells. Causes of acute metabolic alkalosis and the events initiating it will be discussed along with the factors that maintain the disorder (see the accompanying box for potential causes of acute metabolic alkalosis).

History and Risk Factors

Gastrointestinal (GI) H^+ loss

Metabolic alkalosis occurs when gastric secretions rich in HCl are lost either by vomiting or NG suction. In addition to the HCl that is lost, small quantities of KCl are also removed. For each mEq of H^+ secreted into the stomach, a mEq of HCO_3^- is generated and added to the serum. Normally, this increase in serum HCO_3^- concentration is brief because as the digestive process continues, the H^+ from the gastric secretions are eventually reabsorbed. The reabsorbed H^+ enter the blood and neutralize the HCO_3^- that had been added previously. However, if the H^+ in the gastric secretions are removed via vomiting or NG suction, they are not available to buffer the increased serum HCO_3^-, and a metabolic alkalosis occurs. Normally the kidneys can excrete excess HCO_3^- and return the pH to normal, but the accompanying volume depletion and hypokalemia increase the likelihood of

Potential Causes of Metabolic Alkalosis Based on Their Response to Saline

Saline-responsive Alkalosis	Saline-resistant Alkalosis
Diuretics	Severe hypokalemia
Vomiting	Renal failure
NG suctioning	Mineralocorticoid excess
Post chronic hypercapnia	Edematous states

Potential Causes of Acute Metabolic Alkalosis

H^+ Loss

1. Gastrointestinal
   - Vomiting
   - NG suction
2. Renal
   - Diuretics
   - Post chronic hypercapnia
   - High doses of carbenicillin/penicillin derivatives
   - Mg^{2+} deficiency

HCO_3^- Retention

1. Acute administration of $NaHCO_3$ or HCO_3^- precursors (citrate, acetate)
2. Massive blood transfusion

H^+ Shift into the Cells

Hypokalemia

Contraction Alkalosis

1. Diuretics
2. Vomiting

maintaining the metabolic alkalosis. Metabolic alkalosis secondary to GI loss of H^+ is responsive to treatment that replaces the lost volume and Cl^- (saline-response metabolic alkalosis).

Renal H^+ loss

Diuretics

Metabolic alkalosis is commonly generated by the use of diuretics that inhibit the reabsorption of NaCl (e.g., furosemide, thiazides). In fact, use of diuretics is probably the most common cause of metabolic alkalosis. Water is lost with the NaCl, leading to an acute decrease in effective circulating volume (ECV). Because HCO_3^- has been reabsorbed as the Cl^-, Na^+, and water are excreted, serum bicarbonate concentration remains constant even though the ECV is decreased, resulting in *contraction alkalosis.* The decrease in ECV stimulates aldosterone release, which leads to H^+ and K^+ excretion. These mechanisms combine to generate and maintain the alkalosis. Thus the acute effects of diuretics on acid-base balance are basically related to their ability to alter the proportions of HCO_3^- to ECV. Normally the metabolic alkalosis caused by diuretics is milder than that caused by gastric suction losses. Patients especially at risk for the development of metabolic alkalosis during acute diuretic administration are those who are treated with a potent diuretic (furosemide) and those who have been on a sodium-restricted diet, for instance, individuals with congestive heart failure (CHF). The rapid loss of NaCl in the urine can produce a more severe alkalosis than is normally seen. Metabolic alkalosis generated by diuretic administration is responsive to treatment with chloride and volume expansion.

Post chronic hypercapnia

When chronic hypercapnia (increased $Paco_2$) is corrected rapidly, for example with mechanical ventilation, a metabolic alkalosis can occur. Remember that in chronic respiratory acidosis the serum bicarbonate concentration increases as the kidneys compensate for the elevated $Paco_2$ by increasing H^+ excretion. When chronically elevated $Paco_2$ decreases rapidly, the serum bicarbonate concentration remains elevated, causing metabolic alkalosis. The metabolic alkalosis results in an acute increase in pH, causing severe consequences, including neurologic disturbances and death. A chronically elevated $Paco_2$ never should be decreased rapidly. This is applicable, for example, for the individual with chronic obstructive pulmonary disease (COPD) who has a chronically elevated $Paco_2$ and who experiences respiratory arrest requiring emergency intubation and manual ventilation. Overly vigorous manual ventilation given before mechanical ventilation can be instituted may result in a rapid decrease in $Paco_2$ and a severe metabolic alkalosis. In addition to an increase in serum HCO_3^- with chronic respiratory acidosis, there is a loss of Cl^- in the urine, leading to hypochloremia and volume depletion. The exact mechanism for the loss of Cl^- is not known, but the metabolic alkalosis will persist until the hypochloremia is corrected. If adequate Cl^- is available, the post chronic hypercapnia metabolic alkalosis will be short term.

Use of carbenicillin/penicillin

High doses of sodium carbenicillin or other penicillin derivatives can generate a metabolic alkalosis. This disorder occurs because these drugs contain a large load of Na^+ and are excreted as a nonreabsorbable anion that increases excretion of the available cations K^+ and H^+, which can lead to metabolic alkalosis.

Magnesium deficiency

Magnesium deficiency can contribute to the development of metabolic alkalosis and hypokalemia via release of aldosterone and renin, resulting in increased H^+ excretion.

HCO_3^- retention

Acute administration of $NaHCO_3$

The kidneys have a remarkable ability to regulate bicarbonate concentration and are usually able to excrete excess HCO_3^-. This ability to excrete excess HCO_3^- can be overwhelmed when sodium bicarbonate ($NaHCO_3$) or its precursor is administered rapidly in large doses. The most common example of acute metabolic alkalosis from administration of $NaHCO_3$ occurs during severe metabolic acidosis (e.g., with cardiopulmonary arrest or diabetic ketoacidosis) if $NaHCO_3$ is not used judiciously. This occurs because as the acidosis resolves with treatment, the involved anions, lactate and ketones (β-hydroxybutyrate), are metabolized to HCO_3^- (see Chapter 17 for further discussion). This *new* HCO_3^- replaces that which was *lost* during the acidosis, so there is no net decrease in HCO_3^- concentration (except for the ketones excreted in the urine) and acid-base status should return to normal. If the acidosis has been treated with exogenous $NaHCO_3$ and overcorrection occurs, metabolic alkalosis can result as the HCO_3^- metabolized from lactate and ketones is added to the serum. Exogenous $NaHCO_3$ is often necessary in severe metabolic acidosis but must be administered judiciously.

Massive blood transfusion

Massive blood transfusion (usually more than 8 units given rapidly) also can cause a metabolic alkalosis because banked blood contains *precursors* to HCO_3^- (i.e., acetate, citrate), which are metabolized rapidly into HCO_3^- in the body. Citrate is part of the compound used to anticoagulate banked blood. Some volume expanders (i.e., human plasma protein fractions, such as Protenate and Plasmatein), also can cause metabolic alkalosis because these solutions contain both citrate and acetate.

Physiologic Responses

Chemical buffering

Chemical buffering of the excess base occurs immediately as H^+ are released from tissue stores. About a third of the excess HCO_3^- added to the extracellular fluid (ECF) is buffered. Almost all of the buffering is the result of H^+ shifting from the intracellular fluid (ICF) to the ECF.

Respiratory response

The respiratory system also compensates for metabolic alkalosis. The alkalemia inhibits both central and peripheral chemoreceptors, causing alveolar hypoventilation. This leads to the increase in $Paco_2$ (this is a compensatory response to the metabolic alkalosis and is not considered a respiratory acidosis). Maximal respiratory compensation may take 24 hours to achieve, but when this steady state has been reached, the $Paco_2$ should increase 0.7 mm Hg for every 1.0 mEq/L increase in serum bicarbonate concentration. Alveolar hypoventilation is limited because of the hypoxemia that can occur. Remember that hypoxemia stimulates respirations. In the past it was believed that the $Paco_2$ levels would not exceed 55 mm Hg as a compensatory mechanism for metabolic alkalosis. While this is generally true, severe hypercapnia ($Paco_2$ 60 to 65 mm Hg) has been reported as a compensatory mechanism for metabolic alkalosis. In patients with hypoxemia or underlying respiratory alkalosis, this compensatory response may be impaired. Individuals with underlying lung disease or resting hypoxemia may reach a maximum respiratory compensation before the $Paco_2$ reaches 55 mm Hg. If hypoxemia is limiting the degree of alveolar hypoventilation, administration of oxygen may diminish the hypoxemic stimulus to increase ventilation because it increases the Pao_2. Thus higher levels of $Paco_2$ can be retained, decreasing the effects the excess HCO_3^- will have on the pH.

Renal response

The kidneys have a tremendous ability to excrete large quantities of HCO_3^-, and thus restore acid-base equilibrium after the generation of metabolic alkalosis. This will occur unless factors are present that maintain the alkalosis or renal insufficiency or failure is present.

Assessment

The signs and symptoms of acute metabolic alkalosis are related more closely to the volume depletion and hypokalemia that are often present than they are to the metabolic alkalosis itself. See Table 18-1 for quick assessment guidelines for individuals with acute metabolic alkalosis.

Central nervous system/neuromuscular manifestations

The symptoms of neurologic involvement frequently seen in respiratory alkalosis (e.g., carpopedal spasm, paresthesias, light-headedness) occur less frequently with metabolic alkalosis. However, mental confusion, obtundation, and a predisposition to seizures can occur. The discrepancy in symptoms between respiratory and metabolic alkalosis most likely occurs because HCO_3^- crosses the blood-brain barrier much more slowly than CO_2, so the change in cerebral spinal fluid (CSF) pH is not as great with metabolic alkalosis. The rapid development of alkalosis in an individual with a low serum Ca^{2+} can result in tetany because of the decrease in ionized calcium when alkalosis is present.

Cardiovascular effects

Atrial tachycardias may be present. Severe alkalemia (pH >7.60) can cause serious ventricular cardiac rhythm disturbances in patients with heart disease, but these disturbances also can occur in individuals with healthy hearts. An alkalemia this severe is usually caused by a mixed respiratory and metabolic alkalosis. It is rare for a simple metabolic alkalosis to be responsible for alkalemic-induced dysrhythmias. (The hypokalemia that is usually present with alkalosis plays a more direct role in inducing cardiac rhythm disturbances.)

Table 18-1 Quick assessment guidelines for individuals with acute metabolic alkalosis

Signs and Symptoms	EKG Findings
Neuromuscular irritability (i.e., hyperreactive reflexes, tetany) secondary to the accompanying hypokalemia. Severe alkalosis can result in signs of neuromuscular excitability as well as apathy, confusion, and stupor	Numerous types of atrial-ventricular dysrhythmias as a result of the cardiac irritability occurring with hypokalemia; atrial or sinus tachycardia; prolonged QT interval (T-P phenomenon)

History and Risk Factors

- *Clinical circumstances associated with volume/chloride depletion:* Vomiting or gastric drainage
- *Rapid correction of chronic hypercapnia*
- *Excessive alkali intake:* May be iatrogenic from overcorrection of metabolic acidosis (frequently seen during CPR). Excessive ingestion of sodium bicarbonate (e.g., Alka-Seltzer) or calcium carbonate (e.g., Tums, Rolaids)

Gastrointestinal manifestations

Anorexia, nausea, and vomiting may be present, but they are more typical of chronic metabolic alkalosis caused by excessive alkali intake.

Pulmonary effects

The compensatory retention of CO_2 secondary to alveolar hypoventilation may result in decreased respiratory rate and depth. Periods of apnea also may be present, but are self-limiting because ventilation is stimulated by the hypoxemia that develops. In individuals with chronic lung diseases, hypoxemia can be aggravated by the alveolar hypoventilation.

Diagnostic Studies

Arterial blood gas (ABG) values

The pH will be greater than 7.40. It may be mildly ($\leq$7.50) or severely ($>$7.50) elevated, depending on the underlying cause and the degree of respiratory compensation. The serum bicarbonate concentration will be greater than 24 mEq/L. In severe metabolic alkalosis the HCO_3^- can be equal to or greater than 40 mEq/L. The $Paco_2$ will be elevated in proportion to the increase in serum HCO_3^-. There will be an increase in $Paco_2$ of 0.7 mm Hg for every 1 mEq/L increase in HCO_3^-. Although there is much individual variation in the respiratory response to metabolic alkalosis, patients who have a $Paco_2$ below normal have a concurrent respiratory alkalosis because the $Paco_2$ should be increasing, not decreasing. For example, this may occur in an individual who has hypoxemia secondary to pneumonia resulting in respiratory alkalosis. Often it is difficult to determine if the increased $Paco_2$ is a compensatory response to metabolic alkalosis or if it is a primary respiratory acidosis and the increased HCO_3^- is a renal compensatory response. In simple acid-base disturbances the answer to this question rests with the pH. The direction of the pH (decreased with acidosis; increased with alkalosis) will determine the primary cause of the acid-base disturbance. In a simple acid-base disturbance the pH will not return to normal, so if the pH is >7.40 with an increased HCO_3^- and $Paco_2$, a metabolic alkalosis exists and the increased $Paco_2$ is a compensatory response. If the pH is <7.40 with an increased HCO_3^- and $Paco_2$, a respiratory acidosis exists with a renal compensation.

The patient's clinical picture and history give the nurse extremely important information about the underlying cause of the acid-base disturbance, and a thorough nursing assessment with patient history must be obtained before accurate evaluation and interpretation of lab results can occur. The Pao_2 may be decreased, depending on the degree of alveolar hypoventilation present. The Pao_2 will not decrease below 60 to 70 mm Hg as a respiratory compensation because ventilation is strongly stimulated when the Pao_2 reaches this level.

Serum electrolytes

The degree of change seen in serum electrolytes varies, depending on the cause of the metabolic alkalosis. In general, there will be a decrease in sodium with a proportionately larger decrease in chloride (<95 mEq/L). The potassium also will be decreased (<4.0 mEq/L), sometimes to dangerously low levels (<2.0 mEq/L).

Hypercalcemia, hypocalcemia, and hypophosphatemia can occur, depending on the underlying cause and other concurrent disorders. Renal excretion of H^+ and reabsorption of HCO_3^- increase in the presence of elevated calcium levels. With hypercalcemia, phosphorus values usually will be decreased. Hypomagnesemia is often present in disturbances that lead to metabolic alkalosis (e.g., NG suction, diuretic use, hyperaldosteronism). Magnesium levels may be <1.5 mEq/L.

Urinalysis

If the patient's history does not provide adequate data, urinary chloride levels may provide valuable information when attempting to identify the cause of metabolic alkalosis. The urine chloride concentration should be less than 15 mEq/L when hypovolemia and hypochloremia are present (e.g., with vomiting or after diuresis following use of diuretics). This occurs because the kidneys are conserving chloride maximally. The urine chloride concentration will be well over 20 mEq/L in individuals with excess retained HCO_3^-. Urine chloride assessment may not be accurate in patients who have tubular defects and are unable to maximally conserve chloride (e.g., with renal insufficiency or severe hypokalemia [<2.0 mEq/L]). Urine chloride values are not useful if diuretics have been taken in the previous 12 hours because of diuretic-induced chloruresis. Urine pH should be more alkalotic than usual. If the urine is acidic, the acid-base disturbance may be progressing to a chronic metabolic alkalosis.

Electrocardiogram (EKG)

An EKG may show atrial or sinus tachycardia (an increased heart rate). There may be a prolonged Q-T interval. In fact, the T wave approaches and may merge with the P wave that follows (referred to as the T-P phenomenon). If hypokalemia is associated with the metabolic alkalosis, serious ventricular dysrhythmias may be present, especially if the patient is receiving digitalis preparations (see Chapter 9 for more detail).

Medical Management

The type of medical management will depend on the underlying disorder. Mild or moderate metabolic alkalosis usually does not require specific therapeutic interventions.

Saline infusion

In individuals with saline-responsive metabolic alkalosis (e.g., caused by gastric loss or diuretics), oral or IV administration of water and NaCl, (i.e., isotonic [0.9% NS] or half isotonic [0.45% NS]), will expand the ECV and correct the alkalosis.

Potassium chloride (KCl)

When hypokalemia is present, the potassium is administered as KCl because the K^+ and Cl^- losses can then be replaced simultaneously. In saline-responsive metabolic alkalosis, the hypokalemia is not the causative agent in the development or maintenance of the disorder—volume and Cl^- depletion are. The metabolic alkalosis can be corrected without correcting the hypokalemia, although treatment of the hypokalemia is usually indicated as well. Treatment of saline-resistant metabolic alkalosis (i.e., mineralocorticoid excess) will be discussed in the section on chronic metabolic alkalosis later in this chapter. See Chapter 9 for the nursing implications of KCl administration. Posthypercapneic alkalosis also requires Cl^- and K^+ replacement and KCl is usually the drug of choice.

Histamine H_2-receptor antagonists

Histamine H_2-receptor antagonists (cimetidine, ranitidine, and famotidine) reduce production of gastric HCl, and thus they are useful in preventing or decreasing the metabolic alkalosis that can occur with NG suctioning. Patients requiring NG suctioning or those with intractable vomiting may benefit from the administration of these agents.

Carbonic anhydrase inhibitors

Acetazolamide (Diamox) is especially useful for correcting metabolic alkalosis for patients who cannot tolerate rapid volume expansion, for example, individuals with CHF. It can be administered orally or intravenously. Acetazolamide causes a large increase in the renal excretion of HCO_3^- and K^+. Because this drug increases K^+ excretion, it may be necessary to provide potassium supplementation before giving this drug.

Acidifying agents

Severe metabolic alkalosis (pH >7.60 and HCO_3^- >40 to 45 mEq/L) may require treatment with acidifying agents, such as dilute hydrochloric acid, ammonium chloride, or arginine hydrochloride. Because of their serious side effects, these medications are not used frequently. Ammonium chloride can cause ammonia intoxication, while arginine hydrochloride can cause hyperkalemia.

Nursing Diagnoses and Interventions

Nursing diagnoses and interventions are specific to the underlying pathophysiologic process. The following may apply.

Decreased cardiac output related to electrical alterations (dysrhythmias) secondary to metabolic alkalosis induced by gastric suctioning or potassium-wasting diuretics

Desired outcome: EKG reveals a normal tracing; pH is ≤ 7.45.

1. Monitor laboratory values, especially pH and serum HCO_3^-, to determine patient's response to therapy. Notify MD if there are significant changes or lack of response to treatment.
2. Monitor EKG for the presence of dysrhythmias. Assess apical and radial pulses simultaneously when evaluating cardiac rate and rhythm to detect pulse deficit. Notify MD of any changes in cardiac rate and rhythm.
3. Monitor potassium levels, especially in patients receiving digitalis preparations. Hypokalemia sensitizes patients to the cardiotoxic effect of digitalis. (Recall that hypokalemia frequently coexists with metabolic alkalosis.) Notify MD if potassium levels drop to <3.5 mEq/L.
4. Weigh patient daily to determine fluid volume status.

5. Use isotonic saline solutions to irrigate NG tubes. Water is not recommended for irrigation because it can cause electrolyte washout.
6. If the patient is permitted to have ice chips, administer limited quantities to avoid washing electrolytes from the patient's stomach. Total volume of ice consumed over a shift frequently is underestimated. Determine the volume of a specific number of ice cubes or the quantity of crushed ice by melting and measuring. Establish the volume of fluid the patient may consume each shift. Document the volume consumed in ml or ounces, not by the number of cubes.
7. Measure and document the amount of fluid removed by suction.
8. Administer histamine H_2-receptor antagonists (e.g., cimetidine, rantitidine, or famotidine) as prescribed to block hydrochloride secretion and volume secretion by the stomach, thus lessening systemic metabolic alkalosis.

For other nursing diagnoses and interventions, see Chapter 6, *Disorders of Fluid Balance,* for **Fluid volume deficit** related to decreased circulating volume secondary to abnormal losses of body fluids or reduced intake, p. 109, and **Altered cerebral, renal, and peripheral tissue perfusion** related to hypovolemia, p. 110.

CHRONIC METABOLIC ALKALOSIS

Unresolved acute metabolic alkalosis can result in chronic metabolic alkalosis. The most common cause of chronic metabolic alkalosis is long-term diuretic therapy with thiazides or furosemide. The alkalosis is usually mild to moderate, with a pH >7.40, HCO_3^- >24 mEq/L, and $Paco_2$ >40 mm Hg, with an increase of 0.7 mm Hg for each 1.0 mEq/L increase in HCO_3^- (up to a HCO_3^- of 40 mEq/L). See the accompanying box on p. 360 for potential causes of chronic metabolic alkalosis.

History and Risk Factors

H^+ loss

Gastrointestinal (GI) losses

Long-term vomiting or NG suction, as well as external drainage from gastrostomy tubes or external fistulas, can cause chronic metabolic alkalosis. Chronic metabolic alkalosis can be caused by a rare congenital disorder, chloridorrhea. The alkalosis is a result of a defect in intestinal Cl^- reabsorption and the excretion of HCO_3^-. (The fecal pH is very low and the fecal chloride concentration very high.) Villous adenomas of the colon also can cause chronic metabolic alkalosis. These lesions are rare, but in some cases a metabolic alkalosis develops because of the development of a diarrhea rich in Cl^-, K^+, Na^+, and protein.

Renal losses

Any cause of mineralocorticoid excess can cause a metabolic alkalosis because the excess secretion of aldosterone causes an increase in H^+ and K^+ excretion.

Potential Causes of Chronic Metabolic Alkalosis

H^+ Loss

1. Gastrointestinal
   - Chronic vomiting/NG suction
   - External drainage of gastric fluids (gastrostomy tubes, external fistulas)
   - Villous adenoma
   - Congenital chloridorrhea
2. Renal
   - Mineralocorticoid excess: Cushing's syndrome, primary aldosteronism, ACTH secretion tumors, excess licorice intake
   - Diuretics: Potassium-wasting diuretics

H^+ Shift into the Cells

1. Hypokalemia
2. Carbohydrate refeeding after starvation

HCO_3^- Retention

1. Administration of HCO_3^- or HCO_3^- precursor
2. Milk-alkali syndrome

Cystic Fibrosis

Mineralocorticoid excess can be endogenous from primary hyperaldosteronism or Cushing's syndrome, either of which will cause minimal to moderate metabolic alkalosis. When the mineralocorticoid excess is caused by adrenal hyperfunction associated with tumors or secretory ACTH (i.e., bronchogenic cancer), the alkalosis may be more severe. Exogenous mineralocorticoid excess also can result in metabolic alkalosis from H^+ and K^+ loss, for example, with excess licorice consumption or licorice-containing chewing tobacco (licorice contains a steroid that has mineralocorticoid properties). Some over-the-counter nasal sprays contain a synthetic mineralocorticoid that causes alterations in H^+ and K^+ excretion, leading to a metabolic alkalosis. A thorough nursing history and assessment can help identify patient habits that may play a role in the development of chronic metabolic alkalosis from these causes.

Chronic administration of potassium-wasting diuretics may result in chronic metabolic alkalosis due to the combination of contraction alkalosis, hypovolemia-induced secretion of aldosterone, and increased delivery of Na^+ to the distal tubule with an increased excretion of H^+ and K^+. Hypokalemia is also present, but it plays a less significant role in the generation and maintenance of the chronic metabolic alkalosis. Thiazide diuretics can cause a massive depletion of K^+ stores with loss of up to 1,000 mEq, which is one third of the total body K^+. A serum K^+ of <2.0 can result.

H^+ shift into the cells

Potassium depletion

Profound potassium depletion can occur with chronic metabolic alkalosis. Although hypokalemia contributes to the development and generation of chronic metabolic alkalosis, it alone does not cause metabolic alkalosis. The role of hypokalemia in chronic metabolic alkalosis is the same as that in acute metabolic alkalosis.

Carbohydrate refeeding after starvation

Chronic metabolic alkalosis can occur when refeeding patients with carbohydrates after a prolonged fast. The exact mechanism of the intracellular shift of H^+ is unknown, but the shift of H^+ is believed to be responsible for the generation of this disorder.

HCO_3^- retention

HCO_3^- retention as a cause of chronic metabolic alkalosis is usually related to chronic milk ingestion and chronic use of antacids containing calcium carbonate. Over time the kidneys are affected by the large amounts of calcium, and renal insufficiency occurs. The kidneys are then no longer able to excrete the absorbed HCO_3^- in amounts equal to the intake. This is referred to as the *milk-alkali syndrome,* and a mild to moderate alkalosis will be seen. Continued administration of any absorbable alkali can lead to persistent increases in serum HCO_3^- and to chronic metabolic alkalosis without renal dysfunction. The kidneys will return the acid-base status to normal when the administration of alkali ceases.

Cystic fibrosis

Individuals with cystic fibrosis have a much higher concentration of NaCl in their sweat than other people. Cystic fibrosis can lead to metabolic alkalosis as a large amount of chloride is lost through sweat without a similar loss of HCO_3^-. The low-sodium diet that is usually prescribed for these individuals, combined with the loss of chloride, generates and maintains the alkalosis.

Physiologic Responses

In general, the physiologic response to chronic metabolic alkalosis is the same as that for acute metabolic alkalosis.

Assessment

Patients with chronic metabolic alkalosis are usually asymptomatic. The effects of the hypokalemia that often accompany chronic metabolic alkalosis may be all that will be noted. See Table 18-2 for quick assessment guidelines for individuals with chronic metabolic alkalosis.

Central nervous system/neuromuscular manifestations

With profound alkalosis and severe potassium depletion, the patient may experience weakness, neuromuscular irritability (tetany or hyperreactive reflexes), and a decrease in GI tract motility, resulting in ileus.

Cardiovascular effects

No specific cardiovascular effects are noted in most cases of chronic metabolic alkalosis since this disorder is usually not profound. If cardiac disturbances occur, the accompanying hypokalemia is usually the cause. See Chapter 9 for the effects of hypokalemia on cardiac function.

Pulmonary effects

Pulmonary effects with chronic metabolic alkalosis are the same as those seen with acute metabolic alkalosis.

Diagnostic Studies

The diagnostic studies performed and the expected results are the same as those seen with acute metabolic alkalosis.

Medical Management

The goal is to correct the underlying acid-base disorder via the following interventions:

Fluid management

If volume depletion exists, normal saline infusions are given. Many causes of chronic metabolic alkalosis are not responsive to saline (saline-resistant metabolic alkalosis).

Potassium replacement

Saline-resistant metabolic alkalosis usually responds to treatment with potassium replacement. If a chloride deficit is also present, potassium chloride is the drug of choice. If a chloride deficit does not exist, other potassium salts are acceptable.

Table 18-2 **Quick assessment guidelines for individuals with chronic metabolic alkalosis**

Signs and Symptoms

Patient may be asymptomatic. With severe potassium depletion and profound alkalosis, patient may experience weakness, neuromuscular irritability, and decrease in GI tract motility, which can result in ileus

EKG Findings

Frequent PVCs or U waves with hypokalemia and alkalosis

History and Risk Factors

- *Chronic vomiting or chronic GI losses through gastric suction*
- *Diuretic use:* Thiazide diuretics cause a loss of chloride, potassium, and hydrogen ions. Massive depletion of potassium stores with loss of up to 1,000 mEq, which is ⅓ of total body potassium, may occur, causing profound hypokalemia ($K \le 2.0$ mEq/L)
- *Hyperadrenocorticism:* Cushing's syndrome, primary aldosteronism. This is not a chloride deficit but a chronic loss of potassium, which can lead to total body depletion of potassium with profound hypokalemia ($K \le 2.0$ mEq/L)
- *Milk-alkali syndrome:* An infrequent cause of metabolic alkalosis. Hypercalcemic nephropathy and alkalosis develop secondary to excessive intake of absorbable alkali

IV potassium

If the patient is on a cardiac monitor, up to 20 mEq/hr of potassium chloride may be given for serious hypokalemia. Concentrated doses of KCl (>40 mEq/L) require administration through a central venous line because of blood vessel irritation.

Oral potassium

Oral potassium tastes *very* unpleasant. Most patients can only tolerate drinking approximately 15 mEq per glass, with a maximum daily dose of 60 to 80 mEq. Slow-release potassium tablets are an *acceptable* form of KCl in the treatment of hypokalemia associated with metabolic alkalosis, but they can cause bowel ulcerations and stenosis. All forms of KCl may be irritating to the gastric mucosa, with the exception of enteric-coated tablets, which may cause bowel irritation and ulceration instead.

Dietary potassium

A normal diet contains 3 g or 75 mEq of potassium, but not in the form of potassium chloride. Dietary supplementation of potassium is not effective if a concurrent chloride deficit is also present.

Potassium-sparing diuretics

Potassium-sparing diuretics, such as spironolactone (Aldactone), may be added to the treatment if potassium-wasting diuretics are the cause of hypokalemia and metabolic alkalosis.

Carbonic anhydrase inhibitors

Acetazolamide (Diamox) causes a large increase in the renal excretion of HCO_3^- and K^+. Because K^+ excretion is increased, potassium supplementation may be indicated before the administration of acetazolamide to avoid drug-induced hypokalemia or further reduction in an already low potassium level. This drug is especially useful in individuals with metabolic alkalosis who cannot tolerate rapid volume expansion (e.g., patients with CHF) to correct the alkalosis.

Nursing Diagnoses and Interventions

Knowledge deficit: Necessary precautions for taking potassium-wasting diuretics

Desired outcome: Patient verbalizes knowledge about thiazide diuretics and the necessary precautions that must be made.

1. Provide patient and significant others with the following information about the prescribed diuretic: name, purpose, dosage, precautions, and potential side effects.
2. Stress the importance of taking only the prescribed dose, as higher concentrations of the medication increase the risk of hypokalemia and alkalosis.
3. Explain that diets high in sodium increase the risk of alkalosis and hypokalemia, necessitating restrictions of sodium as prescribed.
4. If potassium chloride supplements are prescribed, teach patient the following:
   - Oral potassium has an unpleasant taste and is most palatable when mixed with orange juice or tomato juice.
   - Slow-release tablets should not be chewed.
   - Both forms of potassium can irritate the stomach and should be taken with meals.
   - Although many foods contain potassium, they should not be used as a substitute for the potassium chloride supplements prescribed by the patient's physician.

For other nursing diagnoses and interventions, see the section on disorders of fluid balance in Chapter 6 for **Fluid volume deficit** related to abnormal losses of body fluids or reduced intake, p. 109, and related to hypovolemia, p. 110. Also see the section on acute metabolic alkalosis earlier in this chapter for **Decreased cardiac output** related to electrical alterations (dysrhythmias) secondary to metabolic alkalosis induced by gastric suctioning or potassium-wasting diuretics, p. 358.

Case Study 18-1

K.R. is a 44-year-old woman who has arrived in the emergency room with a 3-day history of nausea, malaise, headache, and fever of 38.3° C (101° F). She has been vomiting for the last 12 hours and has not been able to eat or drink anything for at least 24 hours. Her mucous membranes are dry, and she has poor skin turgor. VS are as follows:

BP 94/62 mm Hg (supine); 80/60 mm Hg (sitting)
HR 118 bpm and irregular
RR 8 breaths/min, unlabored, and shallow.

K.R.'s ABG and Electrolyte Values	Normal (*perfect*) Values
pH 7.51	7.40
$Paco_2$ 49 mm Hg	40 mm Hg
Pao_2 90 mm Hg	range: 80-95 mm Hg
Oxygen saturation 97%	range: 95%-99%
Na^+ 130 mEq/L	range: 137-147 mEq/L
Cl^- 95 mEq/L	range: 95-108 mEq/L
K^+ 3.0 mEq/L	range: 3.5-5.0 mEq/L
HCO_3^- 38 mEq/L	24 mEq/L

1. **Question:** Based on the laboratory results, does the patient have an acid-base disturbance? If yes, is the disorder of a metabolic or respiratory origin?

   Answer: An acid-base disturbance is present, based on a pH of 7.51, an increased $Paco_2$, and an increased serum HCO_3^-. The acid-base disturbance is an alkalosis (pH $>$7.40). To determine the cause of the alkalosis, both the $Paco_2$ and the serum HCO_3^- must be analyzed. The $Paco_2$ is elevated, but since this is the acid component of acid-base balance, the pH would be decreased if the primary disturbance were respiratory in origin. Bicarbonate, the base, *is* elevated, which corresponds with the increase in pH seen in this patient. The increased HCO_3^- could be explained as a compensatory response to the increased $Paco_2$, but the pH indicates the presence of alkalosis rather than acidosis. Based on the laboratory values, the disorder is metabolic alkalosis.

2. **Question:** Does the patient's history and clinical picture support the conclusion drawn from the laboratory values?

   Answer: Yes. The patient has a history of persistent vomiting, which leads to the loss of gastric secretions rich in HCl. The loss of the H^+ leads to an excess of HCO_3^-, resulting in a metabolic alkalosis. The patient also appears dehydrated, and the Cl^- and K^+ level are low. These findings are consistent with metabolic alkalosis.

3. **Question:** How do you explain the elevated $Paco_2$?

   Answer: The elevated $Paco_2$ represents the body's attempt to compensate for the increased serum HCO_3^- (alkalosis) by retaining acids (CO_2) in order to lessen the alterations in pH.

4. **Question:** Is the increased $Paco_2$ appropriate, based on the HCO_3^- value?

   Answer: Yes. The $Paco_2$ should increase 0.7 mm Hg for every 1 mEq/L rise in HCO_3^-. The HCO_3^- increased from 24 mEq/L to 38 mEq/L, a difference of 14 mEq/L. The $Paco_2$ should then increase 9 mm Hg ($0.7 \times 14 = 9$), from 40 mm Hg to 49 mm Hg, which it has.

5. **Question:** What physical finding is consistent with the respiratory compensation that has led to the increase in $Paco_2$?

   Answer: The decreased respiratory rate (8) and the shallow respirations, causing alveolar hypoventilation. The Pao_2 is on the low end of normal and may be decreased from the patient's normal because of the alveolar hypoventilation. Severe hypoxemia (Pao_2 <60 mm Hg) does not occur with respiratory compensation because ventilation is stimulated when the Pao_2 reaches approximately 50 to 60 mm Hg.

6. **Question:** Given more time, would you expect respiratory compensation to correct the pH to normal?

   Answer: No. Respiratory compensation cannot correct the pH to normal in metabolic alkalosis. Respiratory compensation can prevent the pH from increasing to dangerously high levels that are incompatible with life. If the $Paco_2$ had not increased from 40 mm Hg to 49 mm Hg, the pH would have been 7.60, a much more dangerous increase than that which the patient experienced.

7. **Question:** Which factor(s) are responsible for generating and maintaining the metabolic alkalosis?

   Answer: The metabolic alkalosis was generated by loss of HCl and volume (in the form of gastric secretions) as a result of the persistent vomiting. The loss of H^+ causes an alkalosis by altering the ratio of acids to bases present (normal ratio is 1:20). With loss of acids, the proportion of bases increases. The Cl^- and volume loss from vomiting causes a *contraction alkalosis* because a greater amount of Cl^- than HCO_3^- is present in the lost fluid. This causes an alkalosis because the same amount of HCO_3^- is present in less ECV (i.e., the volume is *contracted* around the same serum HCO_3^- concentration). The alkalosis is maintained by the presence of a decreased ECV. Hypovolemia stimulates the release of aldosterone, which in turn increases the reabsorption of Na^+ and excretion of H^+ and K^+. Thus HCO_3^- continues to be reabsorbed and generated, maintaining the metabolic alkalosis. The hypokalemia also contributes to the maintenance of the metabolic alkalosis because it causes a shift of H^+ into the cells, and if K^+ is not available in the tubules for exchange with Na^+, H^+ is exchanged instead.

8. Question: Based on the cause of the acid-base disturbance and the laboratory values, do you think the metabolic alkalosis will be responsive to treatment with saline (NaCl)?

Answer: Yes. The metabolic alkalosis is being maintained because of a decrease in ECV. If Cl^- and volume (in the form of saline) are administered, the hypovolemia will be corrected, and the stimulus for aldosterone release will be removed. With adequate volume and Cl^-, the excess HCO_3^- can be excreted, and K^+ and H^+ can be reabsorbed, allowing the kidneys to reestablish a normal acid-base environment.

Case Study 18-2

S.H. is a 73-year-old man admitted to the emergency room from an extended care facility. He is confused and has a history of CHF that has been treated with thiazides and digoxin for the last several years. He had a cerebrovascular accident (CVA) 2 months ago, which has impaired his swallowing ability. He has a gastrostomy tube in place, and there is a moderate amount of thin, nonpurulent drainage around the tube, with excoriation of the skin at the tube site. S.H. is on a sodium-restricted diet (2 g) and receives enteral feedings. The reason for his admission is his increasing confusion and lethargy over the last 2 days. In addition, his skin and mucous membranes are dry, and his tongue is furrowed. His VS are as follows:

BP 100/70 mm Hg (supine); 70/50 mm Hg (sitting)
HR 110 bpm and irregular
RR 12 breaths/min and shallow
T 36.5° C (97.8° F)

S.H.'s ABG Values	Normal (*perfect*) Values
pH 7.48	7.40
$Paco_2$ 47 mm Hg	40 mm Hg
Pao_2 82 mm Hg	range: 80-95 mm Hg
Oxygen saturation 94%	range: 95%-99%
HCO_3^- (calculated from ABGs) 34 mEq/L	24 mEq/L

1. **Question:** Based on the patient's laboratory findings, is there evidence of an acid-base disorder? If so, what is it?

   Answer: Yes. The pH reflects an alkalosis. The serum HCO_3^- is elevated from a normal of 24 mEq/L to 34 mEq/L. This increase in the base corresponds to the increased pH (alkalosis). The $Paco_2$ is increased, which signals a possible acidosis, but this does not correlate with the pH. The elevated $Paco_2$ is caused by the respiratory compensation, and the alkalosis must be metabolic in origin.

2. **Question:** Does the clinical picture correspond with the laboratory values?

   Answer: The patient appears hypovolemic, based on the physical assessment of dry skin and mucous membranes and furrowed tongue, as well as the orthostatic drop in BP. The respiratory pattern of relatively slow and shallow breaths with an increased $Paco_2$ is consistent with alveolar hypoventilation. The patient's history reveals thiazide diuretic administration for treatment of his CHF. Thiazides can cause a decrease in ECF, which corresponds to the physical findings and laboratory findings of fluid volume deficit. The metabolic alkalosis, however, is more severe than that which would be expected from diuretic administration alone.

3. **Question:** How do thiazide diuretics contribute to the development and maintenance of metabolic alkalosis?

   Answer: The thiazide diuretics (or any nonpotassium-sparing diuretics) generate a metabolic alkalosis initially by inducing a decrease in the ECV, the volume that is perfusing the tissues. The body responds to the decreased ECF by increasing the release of aldosterone, which leads to an increase in K^+ and H^+ excretion and increased reabsorption of Na^+. With thiazide therapy the reabsorption of Na^+ is diminished in the proximal tubules, so a greater quantity of Na^+ is delivered to the distal tubule, where it is exchanged for H^+ and K^+, resulting in maintenance of the metabolic alkalosis. Hypokalemia increases HCO_3^- reabsorption, which increases the metabolic alkalosis to an even greater degree.

4. **Question:** Based on the patient's physical assessment and history, what else may be contributing to his metabolic alkalosis?

   Answer: The patient has a moderate amount of drainage around the gastrostomy tube. If the drainage contains gastric secretions, the loss of HCl can cause or contribute to the metabolic alkalosis. The skin is excoriated around the tube site, and this may be the result of contact with acidic fluid. The loss of gastric HCl explains the patient's unusually severe metabolic alkalosis. His metabolic alkalosis is a mixed alkalosis that is caused by two separate metabolic disturbances: renal and GI loss of

H^+. The presence and severity of the gastric drainage is a significant finding that easily can be overlooked when attempting to discover the cause of the metabolic alkalosis. The presence of the drainage and an estimate of the amount must be documented and communicated to the patient's physician. The nurse's diagnosis of impaired skin integrity will necessitate a specific plan of care to meet the patient's needs.

5. **Question:** What medication might be helpful in decreasing the HCl production and thus the loss of acid through gastric secretions?

   Answer: Histamine H_2-receptor antagonists (i.e., cimetidine, ranitidine, and famotidine) are a class of drugs that decrease both the concentration of acid and the volume of gastric secretions. This medication would decrease the HCl loss experienced by the patient. Cimetidine has been implicated as a cause of increasing confusion among the elderly, so if this drug is prescribed, a careful evaluation of the patient's orientation is important.

References

Anderson L and Henrick W: Alkemia associated morbidity and mortality in medical and surgical patients, South Med J 80(6):729-733, 1987.

Arieff AI and DeFronzo RA: Fluid, electrolyte, and acid-base metabolism: diagnosis and management, Philadelphia, 1985, JB Lippincott Co.

Brenner B and Rector F: The kidney, Philadelphia, 1986, WB Saunders Co.

Easterday U: Acid-base balance. In Horne M and Swearingen PL: Pocket guide to fluids and electrolytes, St Louis, 1989, The CV Mosby Co.

Easterday U and Howard C: Respiratory dysfunctions. In Swearingen PL and others: Manual of critical care: applying nursing diagnoses to adult critical illness, St Louis, 1988, The CV Mosby Co.

Galla J and Luke R: Pathophysiology of metabolic alkalosis, Hosp Pract 22(10):123-146, 1987.

Gilman A and others: Goodman & Gilman's the pharmacological basis of therapeutics, ed 7, New York, 1985, Macmillan Publishing Co.

Goldberger E: A primer of water, electrolyte, and acid-base syndromes, ed 7, Philadelphia, 1986, Lea & Febiger.

Kinney M and others: AACN's clinical reference for critical care nursing, New York, 1988, McGraw-Hill Book Co.

Kraut J: Disorders of acid-base metabolism. In Bricker N and Kirshenbaum M: The kidney: diagnosis and management, New York, 1984, John Wiley & Sons, Inc.

Perkins C and Bralley H: Metabolic alkalosis, Nursing 83 13(1):57, 1983.

Schrier R: Renal and electrolyte disorders, ed 3, Boston, 1986, Little, Brown & Co.

Tilkian S and others: Clinical implications of laboratory tests, ed 4, St Louis, 1987, The CV Mosby Co.

Valtin H and Gennari F: Acid-base disorders: basic concepts and clinical management, Boston, 1987, Little, Brown & Co.

Chapter

19

Mixed Acid-Base Disorders

A mixed acid-base disturbance occurs when two or more simple acid-base disorders are present at the same time. An understanding of mixed disorders requires a review of the basic principles of simple acid-base disorders. Applying the information presented in the previous acid-base chapters (see Chapters 13-18) can alert nurses to the presence of the signs that indicate the occurrence of a mixed acid-base disturbance. Once a mixed disturbance has been identified, a plan of care can be formulated that meets specific patient needs.

The effect of a mixed acid-base disturbance on the pH depends on the specific disorders involved and their severity. If a mixed disorder involves two types of acidosis, a larger drop in pH can be expected than would occur if a mixed acidosis-alkalosis disorder were present. When two opposing disorders (i.e., an alkalosis and acidosis) occur together, the pH level will be determined by the predominant disorder. The change in pH will be less severe since the opposing disorders tend to balance each other to a certain extent. In some cases the pH will be normal in the presence of a mixed acid-base disturbance.

Mixed acid-base disorders can be the result of numerous combinations of disturbances, both respiratory and metabolic (Table 19-1). The only combination of disorders that cannot occur simultaneously is a mixed respiratory acidosis and respiratory alkalosis because it is impossible to have alveolar hypoventilation and alveolar hyperventilation at the same time. A mixed respiratory disorder can occur when an acute respiratory disturbance (alkalosis or acidosis) is superimposed on a chronic respiratory disturbance, either alkalosis or acidosis, e.g., severe pneumonia causing an acute respiratory acidosis superimposed on chronic obstructive pulmonary disease (COPD).

In addition to mixed acid-base disturbances involving two simple disorders, triple acid-base disturbances, involving three separate simple acid-base disorders, can occur.

Table 19-1 Mixed acid-base disorders

Types	Examples
Metabolic acidosis and metabolic acidosis	Renal failure: diarrhea
Acute respiratory acidosis and chronic respiratory acidosis	Pneumonia: emphysema
Metabolic acidosis and metabolic alkalosis	Diabetic ketoacidosis: vomiting
Metabolic acidosis and respiratory acidosis	Lactic acidosis: respiratory arrest
Metabolic acidosis and respiratory alkalosis	Ethylene glycol ingestion: pneumonia
Metabolic alkalosis and respiratory acidosis	NG suction: sedative overdose
Metabolic alkalosis and respiratory alkalosis	Diuretic use: hepatic failure

Table 19-2 Triple acid-base disorders

Types	Examples
Metabolic acidosis, metabolic alkalosis, and respiratory acidosis	Lactic acidosis, diuretic use, and severe pulmonary edema
Metabolic acidosis, metabolic alkalosis, and respiratory alkalosis	Alcohol excess, ketoacidosis, and hepatic failure
Metabolic acidosis, metabolic acidosis, and respiratory acidosis	DKA, lactic acidosis, and sedative overdose
Metabolic acidosis, metabolic acidosis, and respiratory alkalosis	Cholestyramine use, uremia, and CVA
Acute respiratory acidosis, chronic respiratory acidosis, and metabolic alkalosis	Severe pneumonia, COPD, post chronic hypercapnia
Acute respiratory acidosis, chronic respiratory acidosis, metabolic acidosis	Obstructive lung cancer, COPD, and draining small bowel fistula

An example of a triple disturbance is an individual with vomiting (metabolic alkalosis), diarrhea (metabolic acidosis), and COPD (respiratory acidosis). These triple disturbances also may be referred to as *mixed* acid-base disturbances (Table 19-2).

It is important to understand the normal changes and compensatory responses in $Paco_2$ and HCO_3^- that are expected with various simple acid-base disturbances. If the $Paco_2$ and HCO_3^- vary from what is anticipated, a mixed acid-base disturbance may be responsible (Table 19-3). A review of simple acid-base disturbances and compensatory responses, found in Table 14-2, also will be helpful.

Table 19-3 Review of basic evaluation criteria with simple acid-base disorders

High $Paco_2$/High HCO_3^-	pH	$Paco_2$	HCO_3^-
Metabolic alkalosis	↑	↑ compensatory change	↑ primary disorder
Respiratory acidosis	↓	↑ primary disorder	↑ compensatory change
Low $Paco_2$/Low HCO_3^-			
Metabolic acidosis	↓	↓ compensatory change	↓ primary disorder
Respiratory alkalosis	↑	↓ primary disorder	↓ compensatory change

The clinical picture and a thorough history are essential for accurate interpretation of acid-base status. Mixed acid-base disturbances can be confusing, but applying basic principles of acid-base balance in a step-by-step approach as well as the patient's signs and symptoms and medical history, can help provide an accurate interpretation of the patient's acid-base status.

Case Study 19-1

C.H. is a 15-year-old boy who, when found at home, was arousable but very lethargic with deep respirations. An empty bottle of aspirin was found in the bathroom. His parents state that he has been upset about being grounded because of poor grades. They report no prior incidence of drug overdose.

C.H.'s VS are as follows:

HR 84 bpm

BP 110/60 mm Hg

RR 30 breaths/min, deep, and labored (Kussmaul respirations).

Arterial and venous blood samples are obtained for blood gas and electrolyte analysis. Results are as follows:

C.H.'s ABG and Electrolyte Values	Normal (*perfect*) Values
pH 7.46	7.40
$Paco_2$ 14 mm Hg	40 mm Hg
Pao_2 96 mm Hg	range: 80-95 mm Hg
Oxygen saturation 98% (on room air)	range: 95%-99%
Na^+ 140 mEq/L	range: 137-147 mEq/L
Cl^- 106 mEq/L	range: 95-108 mEq/L
K^+ 4.5 mEq/L	range: 3.5-5.0 mEq/L
HCO_3^- 10 mEq/L	24 mEq/L

1. **Question:** Based on the laboratory values, is the patient experiencing an acid-base disorder? If so, what type (acidosis/alkalosis; respiratory/metabolic) does he have? Which of the following explanations for C.H.'s condition seems most reasonable? (1) the kidneys are compensating for a respiratory alkalosis by excreting HCO_3^-, (2) the decreased $Paco_2$ is respiratory compensation for the decreased HCO_3^- (metabolic acidosis), or (3) there is more than one acid-base disturbance present.

   Answer: According to his pH value (>7.40), C.H. has an alkalosis. The $Paco_2$ is decreased (alveolar hyperventilation), which corresponds to the rise in pH. A respiratory alkalosis is present, although the alkalosis is less severe than you would anticipate from such a large decrease in $Paco_2$. His HCO_3^- is also decreased.

The first explanation is not correct because the patient's history clearly indicates that this is an acute disturbance, and maximal renal compensation takes days to occur. Additionally, renal compensation does not cause the HCO_3^- to fall below 15 to 18 mEq/L. The second explanation cannot be correct because the pH indicates an alkalosis, and if the respiratory system were compensating for a metabolic *acidosis,* it would not be able to correct the pH to normal and certainly would not be able to overcorrect to an alkalosis. The third explanation seems most reasonable. Another acid-base disturbance must be present, along with the respiratory alkalosis.

2. **Question:** What other acid-base disturbance is likely, based on the history and laboratory values?

   Answer: The HCO_3^- is very low at 10 mEq/L, which would correlate with metabolic acidosis. Salicylate intoxication frequently involves the development of a metabolic acidosis, and the laboratory values and clinical picture do correspond to this.

3. **Question:** Is the metabolic acidosis associated with a normal or elevated anion gap?

   Answer: The anion gap is calculated by subtracting the total of Cl^- and HCO_3^- from the Na^+. In this situation, the following values are present:

$$\begin{matrix} Na^+ & & Cl^- \;\; HCO_3^- & \\ 40 & - & (106 + 10) & = 24 \end{matrix}$$

The anion gap is elevated at 24, indicating a metabolic acidosis that involves the addition of unmeasured anions. C.H. is experiencing a mixed acid-base disturbance: respiratory alkalosis and high anion-gap metabolic acidosis.

4. **Question:** How does the salicylate overdose generate this mixed acid-base disorder?

Answer: The respiratory alkalosis is caused by the direct effect salicylates have on the respiratory center. The stimulation of the respiratory center results in hyperventilation and hypocapnia. The metabolic acidosis results from the effect of salicylates on cellular functioning, which increases production of organic acids, and to a lesser extent, from the organic acids that are generated from the salicylates themselves. The elevated anion gap is caused by the addition of these unmeasured anions to the body fluids.

Case Study 19-2

D.C. is a 60-year-old man brought to the emergency room at 2:00 AM, complaining of severe shortness of breath (SOB). He states that his SOB has gotten progressively worse and that he cannot tolerate lying flat and must sleep on three pillows at home.

D.C. appears anxious; his skin is pale, cool, and clammy; and he has 2+ (on a 0-4+ scale) pitting edema of his lower extremities. Auscultation of breath sounds reveals crackles (rales) at the lung bases, extending a third of the way into the lung fields. The cardiac monitor shows sinus tachycardia with premature ventricular contractions (PVCs). D.C. denies the presence of nausea and vomiting or chest pain. An indwelling urinary catheter is inserted, revealing a urinary output of less than 30 ml/hr. His history includes an extensive myocardial infarction (MI) 2 years ago and intermittent exertional angina over the past 6 months.

His VS are as follows: RR 28 breaths/min, HR 128 bpm, and BP 98/50 mm Hg. ABG and electrolyte values are obtained with the following results:

D.C.'s ABG and Electrolyte Values	Normal (*perfect*) Values
pH 7.10	7.40
$Paco_2$ 50 mm Hg	40 mm Hg
Pao_2 49 mm Hg	range: 80-95 mm Hg
Oxygen saturation 83% (on room air)	range: 95%-99%
Na^+ 142 mEq/L	range: 137-147 mEq/L
Cl^- 106 mEq/L	range: 95-108 mEq/L
K^+ 5.1 mEq/L	range: 3.5-5.0 mEq/L
HCO_3^- 15 mEq/L	24 mEq/L

1. **Question:** Which acid-base disturbance is present, based on the clinical picture and the laboratory values?

   Answer: A severe acidosis is present. A pH of 7.10 is life-threatening because of the negative effects it has on numerous body systems (i.e., cardiac depression occurs with decreased contractility and a tendency to develop ventricular dysrhythmias). The severe acidosis is caused by a combination of respiratory acidosis ($Paco_2$ 50 mm Hg) and a metabolic acidosis (HCO_3^- 15 mEq/L). The severity is in part due to the unavailability of the normal compensatory mechanism. For example, with metabolic acidosis, respiratory hyperventilation usually occurs to decrease the $Paco_2$, preventing the pH from decreasing precipitously. With respiratory acidosis, renal compensation occurs over time and bicarbonate (HCO_3^-) concentration is increased. Neither of the compensatory mechanisms (respiratory or renal) is available to buffer the great increase in acids. Chemical buffers are inadequate to neutralize the quantities of acids added to the body fluids.

2. **Question:** Is the anion gap elevated? Based on the anion gap, what is a possible cause for the metabolic acidosis?

   Answer: The anion gap is 21 mEq/L, which is elevated significantly from normal (12 + or − 2). The increased anion gap helps narrow the possible causes of the metabolic acidosis. With D.C., an unmeasured anion is present, which means that an acid has been added to the body fluid. His Pao_2 is 49 mm Hg, which could cause hypoxia; his skin is pale, cool, and clammy, which also points to the presence of shock, leading to tissue hypoperfusion and hypoxia. Hypoxia results in anaerobic metabolism and the production of lactate, which can cause metabolic acidosis. Lactic acidosis is a very common cause of metabolic acidosis with an elevated anion gap.

3. **Question:** Based on the patient's acidemia, would you anticipate the K^+ to be low, normal, or elevated?

   Answer: The K^+ is elevated during acidosis because H^+ moves into the cells in response to the decrease in pH and K^+ moves out of the cells to maintain electroneutrality. If the K^+ is normal or decreased with severe acidosis, a possible depletion of K^+ stores is suspected. The patient's K^+ is 5.1 mEq/L. This is elevated, as is expected, but it is not at a dangerous level.

4. **Question:** What pathologic process is responsible for the development of the respiratory acidosis?

   Answer: The SOB and crackles and the patient's history point to the presence of pulmonary edema. The presence of pulmonary edema causes respiratory disturbances as fluid accumulates in the alveoli and gas exchange becomes impaired. The initial disturbance in gas exchange usually occurs with the exchange of oxygen, causing hypoxemia.

5. **Question:** Based on the patient's statement that his SOB has progressively worsened and the fact that impaired oxygen exchange resulting in hypoxemia can occur early in pulmonary edema, what acid-base disturbance might have been present earlier in the previous day?

Answer: A respiratory alkalosis probably was present earlier, as hypoxemia from the pulmonary edema stimulated the respiratory center and hyperventilation occurred. When the Pao_2 drops to 50 to 60 mm Hg, the respiratory center is stimulated and hyperventilation occurs. As the pulmonary edema increases, gas exchange is impaired further, leading to worsening of the hypoxemia, and finally to retention of CO_2 and respiratory acidosis. The increased respiratory effort seen in pulmonary edema also increases the production of CO_2 and adds to the increasing $Paco_2$. CO_2 is exchanged 20 times more readily than O_2, so when CO_2 retention occurs, the Pao_2 on room air is usually markedly decreased.

Because of the life-threatening acidosis, the patient is treated cautiously with sodium bicarbonate ($NaHCO_3$). In addition, furosemide (Lasix), oxygen therapy, and morphine sulfate (MSO_4) are used to treat his pulmonary edema. His pH following treatment is 7.30, his skin has become warm and dry, and his SOB has resolved, with only a few scattered crackles in his lung bases.

6. **Question:** The $NaHCO_3$ immediately causes an increase in serum HCO_3^- concentration but does not treat the underlying cause of the metabolic acidosis. How does the medical therapy correct the underlying acid-base disturbances—lactic acidosis and respiratory acidosis?

   Answer: The administration of $NaHCO_3$ rapidly increases the pH and removes the negative influence severe acidosis has on myocardial functioning, which had resulted in a decreased cardiac output. As the acidemia resolves, the cardiac output increases, improving tissue perfusion. Besides the effects of $NaHCO_3$, as medical management improves the pulmonary edema, oxygen exchange is improved at the lung and tissue levels. The treatment of the pulmonary edema and the correction of the acidosis combine to increase tissue perfusion by improving oxygenation. The increased tissue perfusion stops further production of lactic acid.

7. **Question:** Over the next 4 to 6 hours, the patient's condition continues to improve. His ABG values reflect a pH of 7.46 without further administration of $NaHCO_3$. To what do you attribute this?

   Answer: When perfusion is restored, lactic acid production stops and the lactic acid present is converted to lactate, which is metabolized to HCO_3^- by the liver (see discussion of metabolic acidosis, Chapter 17). Thus the lactic acidosis that had produced the decreased pH is converted to HCO_3^-, which assists the body in correcting the acidosis. If overzealous administration of $NaHCO_3$ occurs during acidosis, a metabolic alkalosis can develop as lactate is converted to HCO_3^- by the body.

Case Study 19-3

N.D. is a 56-year-old woman who has a 20-year history of alcohol abuse and has had numerous admissions for treatment of the complications of alcoholic liver disease. She is now being admitted for evaluation and treatment of hematemesis. She has not had any alcohol for 2 weeks.

ABG and electrolyte values were taken on admission and results are as follows:

N.D.'s ABG and Electrolyte Values	Normal (*perfect*) Values
pH 7.45	7.40
Pa_{CO_2} 30 mm Hg	40 mm Hg
Pa_{O_2} 93 mm Hg	range: 80-95 mm Hg
Oxygen saturation 98%	range: 95%-99%
Na^+ 138 mEq/L	range: 137-147 mEq/L
Cl^- 108 mEq/L	range: 95-108 mEq/L
K^+ 3.4 mEq/L	range: 3.5-5.0 mEq/L
HCO_3^- 20 mEq/L	24 mEq/L

1. **Question:** Which acid-base disorder is present, based on the laboratory results?

   Answer: The pH reflects an alkalemia (>7.40). This corresponds to the decreased Pa_{CO_2} (alveolar hyperventilation).

2. **Question:** Is the respiratory alkalosis acute or chronic? How can you tell?

   Answer: The respiratory alkalosis is chronic because renal compensation has occurred. The serum HCO_3^- has decreased from a normal of 24 mEq/L to 20 mEq/L to compensate for the decrease in CO_2 levels, thus diminishing the effects respiratory alkalosis will have on pH. Renal compensation takes several days to occur, therefore the respiratory alkalosis must be chronic.

3. **Question:** What is a possible cause for the chronic respiratory alkalosis?

   Answer: Liver disease causes stimulation of the respiratory center, generating chronic respiratory alkalosis. The respiratory center is stimulated by ammonia and other metabolic byproducts that accumulate in liver disease.

On the day after admission, the patient experiences massive upper GI bleeding from esophageal varices. She is given 10 units of blood over an 8-hour period. Her systolic BP is maintained above 80 mm Hg with fluids. Her latest serial ABG and electrolyte values are as follows:

pH 7.59
Pa_{CO_2} 28 mm Hg
Pa_{O_2} 89 mm Hg
Na^+ 142 mEq/L

Cl^- 94 mEq/L
K^+ 4.8 mEq/L
HCO_3^- 26 mEq/L

4. **Question:** What is the patient's acid-base status now?

   Answer: Her alkalemia is worse. The $Paco_2$ has decreased by 2 mm Hg and the HCO_3^- has increased by 6 mEq/L. An increase in HCO_3^- is not the anticipated response to a decrease in $Paco_2$. Another acid-base disturbance has occurred—metabolic alkalosis.

The increase in K^+ from 3.4 mEq/L to 4.8 mEq/L has occurred secondary to the GI bleeding. Potassium was released as the RBCs hemolyzed (broke down) as a result of the GI bleeding. Another cause for the rise in K^+ could be related to the transfusion of the 10 units of blood. Hemolysis of RBCs with release of K^+ occurs in stored blood, especially blood that has been stored for 10 days or longer.

5. **Question:** What could be responsible for the patient's development of metabolic alkalosis?

   Answer: She received 10 units of whole blood over an 8-hour period. Banked blood usually has a preservative containing citrate that helps prevent coagulation. After the blood is transfused, the citrate is metabolized by the body to HCO_3^-. The blood was transfused so rapidly that her kidneys were unable to excrete the alkali load as quickly as it was given and a metabolic alkalosis developed.

6. **Question:** The $Paco_2$ decreased from 30 mm Hg (its chronic steady state) to 28 mm Hg. Can you think of a reason for this decrease?

   Answer: The increase in alveolar hyperventilation could be the result of anxiety. Remember that anxiety can cause or contribute to respiratory alkalosis.

7. **Question:** The patient's systolic BP was maintained at 80 mm Hg, thus avoiding tissue hypoperfusion. If the BP decreased and hypoxia were to occur, what additional acid-base disorder might develop?

   Answer: Lactic acidosis occurs as a result of tissue hypoxia and can cause metabolic acidosis. (Respiratory acidosis cannot occur simultaneously with respiratory alkalosis.) If this were to happen, the patient would have a triple acid-base disorder: respiratory alkalosis, metabolic alkalosis, and metabolic acidosis. The metabolic acidosis would cause the pH to decrease from its current state. The degree of change in pH would depend on the severity of the lactic acidosis.

References

Anderson L and Henrich W: Alkalemia: associated morbidity and mortality in medical and surgical patients, South Med J 80(6):729-733, 1987.

Easterday U: Acid-base balance. In Horne M and Swearingen PL: Pocket guide to fluids and electrolytes, St Louis, 1989, The CV Mosby Co.

Elms J: Demystifying mixed acid-base disorders, part I: guidelines to recognition, Emerg Med 20(13):24-33, 1988.

Elms J: Demystifying mixed acid-base disorders, part II: causes for the acid-base disturbances. Emerg Med 20(16):135-139, 1988.

Flamenbaum N: Acid-base disturbances, Emerg Med 16(3):59-89, 1984.

Galla J and Luke R: Pathophysiology of metabolic alkalosis, Hosp Pract 22(10):123-146, 1987.

Jeffers L: The effects of acidosis on cardiovascular function, J Assoc Nurse Anesthetists 54(2):148-150, 1986.

Miller W: The ABCs of blood gases, Emerg Med 16(3):37-57, 1984.

Narins R and Emmet M: Simple and mixed acid-base disorders: a practical approach, Medicine 59(3):161-185, 1980.

Neff J and Tidwell S: Acid-base balance: a tool for rapid evaluation, J Emerg Nurs 10(6):322-324, 1984.

Rose B: Clinical physiology of acid-base and electrolyte disorders, ed 3, New York, 1989, McGraw-Hill Book Co.

Valtin H and Gennari F: Acid-base disorders: basic concepts and clinical management, Boston, 1987, Little, Brown & Co.

Zonszein J and Baylor P: Diabetes ketoacidosis with alkalemia: a review, West J Med 149(2):217-219, 1988.

Appendix

Potential Fluid, Electrolyte, and Acid-Base Disturbances Occurring with Specific Clinical Disorders

FLUID, ELECTROLYTE, AND ACID-BASE DISTURBANCES RELATED TO ENDOCRINOLOGIC DISORDERS

DIABETIC KETOACIDOSIS (DKA)

DKA occurs as a result of an absolute or relative insulin deficiency. Insulin deficiency leads to hyperglycemia because of decreased glucose utilization by the cells and increased hepatic glucose production. It also results in ketonemia (ketones in the blood) secondary to altered fat metabolism with increased production of ketones. Hyperglycemia and ketonemia, in turn, lead to a combination of potentially life-threatening fluid and electrolyte disturbances.

Both hyperglycemia and ketonemia contribute to hyperosmolality, resulting in movement of water out of the cells. The neurologic changes seen in DKA are largely the result of intracellular dehydration within the central nervous system (CNS). The presence of ketones and excess glucose in the urine causes an osmotic diuresis, with loss of water and electrolytes in the urine. This polyuria may lead to profound fluid volume deficit and shock and contributes to the development of hypokalemia, hypophosphatemia, and hypomagnesemia.

Despite the loss of substantial quantities of electrolytes in the urine, patients initially may have normal or elevated levels of the major intracellular electrolytes (e.g., potassium, magnesium, and phosphorus). This occurs in part because of increased tissue catabolism (breakdown), with the release of intracellular electrolytes. In addition, insulin deficiency results in a decreased movement of K^+ into the cells, and acidosis causes an increased movement of K^+ out of the cells. Metabolic acidosis occurs with ketosis secondary to the increased production of organic acids (ketoacids). As H^+ moves into the cells to be buffered, K^+ moves out. The situation may be complicated further by the development of lactic acidosis if dehydration leads to decreased tissue perfusion. Hyperventilation (Kussmaul breathing) occurs as the body attempts to reduce the acid load by increasing elimination of carbon dioxide.

Potential Fluid, Electrolyte, and Acid-Base Disturbances

1. **Fluid volume deficit** occurring with polyuria secondary to hyperglycemia and ketonemia. See Chapter 6.
2. **Hyponatremia** (initially) secondary to the osmotic shift of water out of the cells (pseudo-hyponatremia). See Chapter 8.
3. **Hypernatremia** occurring with dehydration secondary to osmotic diuresis with the loss of free water (i.e., water is lost in excess of electrolytes). See Chapter 8.
4. **Hyperkalemia** (initially) occurring with movement of potassium out of the cells secondary to tissue catabolism, acidosis, and insulin deficiency. See Chapter 9.
5. **Hyperphosphatemia** (initially) occurring with the movement of phosphorus out of the cells secondary to tissue catabolism. See Chapter 11.
6. **Hypokalemia** occurring with the loss of potassium secondary to osmotic diuresis and the movement of potassium back into the cells occurring with administration of insulin and correction of acidosis. See Chapter 9.
7. **Hypophosphatemia and hypomagnesemia** occurring with the loss of these electrolytes secondary to osmotic diuresis and repair of the tissue with treatment. See Chapters 11 and 12.
8. **Metabolic acidosis** occurring with the abnormal production of ketoacids and lactic acid. See Chapter 17.

HYPEROSMOLAR HYPERGLYCEMIC NONKETOTIC COMA (HHNC)

HHNC is a life-threatening emergency characterized by severe hyperglycemia (blood glucose levels exceed 600 mg/dl and may be a high as 2,000 mg/dl) with the absence of significant ketonemia. As with DKA, hyperglycemia causes an osmotic diuresis with loss of electrolytes, most notably potassium, sodium, and phosphate. The diuresis is hypotonic in relation to the electrolytes (i.e., water is lost in excess of sodium and other electrolytes). The combination of hypotonic fluid loss and hyperglycemia leads to serum hyperosmolality. In turn, increased serum osmolality causes a shift of water out of the cells. The net result is a loss of both ICF and ECF, with individuals losing up to 25% of their total body water. Neurologic deficits (i.e., slowed mentation, confusion, seizures, or coma) occur as a result of the altered CNS cell function.

As extracellular volume decreases, the blood becomes more viscous and its flow is impeded. Thromboemboli are common because of increased blood viscosity, enhanced platelet aggregation and adhesiveness, and patient immobility. Cardiac workload is increased and may lead to myocardial infarction. Renal blood flow is decreased, potentially resulting in renal impairment or failure. Cerebrovascular accident may result from thromboemboli or decreased cerebral perfusion. These severe complications, in addition to the initial precipitating disorder, contribute to a mortality rate in excess of 50%. The onset of HHNC is often insidious and classically occurs in the older, noninsulin-dependent diabetic who has a concomitant reduction in renal function.

Potential Fluid, Electrolyte, and Acid-Base Disturbances

1. **Fluid volume deficit** occurring with losses secondary to hyperglycemia-induced osmotic diuresis. See Chapter 6.
2. **Hypokalemia** occurring with increased urinary losses secondary to osmotic diuresis. See Chapter 9.
3. **Hypophosphatemia** occurring with increased urinary losses secondary to osmotic diuresis. See Chapter 11.
4. **Hypomagnesemia** occurring with increased urinary losses secondary to osmotic diuresis. See Chapter 12.
5. **Metabolic acidosis** occurring with retention of lactic acid secondary to hypovolemia with tissue hypoxia. See Chapter 17.

DIABETES INSIPIDUS (DI)

DI is caused by either a deficiency in the synthesis or release of antidiuretic hormone (ADH) from the posterior pituitary gland (neurogenic or central) or a decrease in kidney responsiveness to ADH (nephrogenic), resulting in decreased water resorption by the renal tubules. Regardless of the cause, the individual with DI excretes large volumes of extremely dilute urine. As a result of increased water loss, ECF volume decreases and serum sodium and osmolality rise. Severe extracellular and intracellular dehydration, hypotension, and hypovolemic shock can occur. Loss of plasma water leads to increased blood viscosity, with risk of thromboemboli. Decreased cerebral perfusion and hypernatremia produce neurologic symptoms ranging from confusion, restlessness, and irritability to seizures and coma. The severity and prognosis of DI varies with its cause. Onset can be sudden and dramatic with inflammatory disease or cerebral trauma, or it can be gradual with tumor or infiltrative disease.

Potential Fluid and Electrolyte Disturbances

1. **Fluid volume deficit** occurring with decreased water reabsorption by the renal tubule secondary to decreased ADH production or effectiveness. See Chapter 6.
2. **Hypernatremia** occurring with increased free water loss secondary to decreased water reabsorption from the renal tubule. See Chapter 8.

SYNDROME OF INAPPROPRIATE ANTIDIURETIC HORMONE (SIADH)

SIADH develops as the result of excessive release of ADH or ADH-like substances. In the presence of increased ADH, water that normally would be excreted is reabsorbed into the circulation, resulting in water retention and eventually, water intoxication. Serum sodium and serum osmolality levels decrease secondary to dilutional effects. ECF volume expansion increases glomerular filtration and decreases the release of aldosterone, both of which act to increase urinary excretion of sodium, further reducing the serum sodium level. As the serum osmolality decreases, an osmotic gradient is created, favoring an ICF shift. Increased ICF in the brain can result in cerebral edema with altered neurologic function. Without prompt treatment, water intoxication, cerebral edema, and severe hyponatremia may lead to death.

Potential Fluid and Electrolyte Disturbances

1. **Hyponatremia** secondary to excessive water retention and continued urinary sodium losses. See Chapter 8.
2. **Fluid volume excess** occurring with water retention secondary to cellular volume expansion. See Chapter 6. NOTE: ECF volume expansion is usually minimized because of the decreased stimulus to aldosterone and increased stimulus to atrial natriuretic factor.

ADDISONIAN CRISIS

Addisonian or adrenal crisis is the severe deficiency of adrenocortical hormones caused by atrophy of the adrenal gland secondary either to inadequate stimulation from the pituitary gland or destruction of the adrenal glands themselves. Adrenal gland destruction is known as *primary Addison's disease,* and it results in inadequate circulating levels of all of the adrenocortical hormones: glucocorticoids (cortisol is the major hormone), mineralocorticoids (primarily aldosterone), and androgens. When adrenal gland dysfunction occurs secondary to pituitary insufficiency, it is called *secondary Addison's disease,* and its usual result is a deficiency of the glucocorticoids and androgens.

Deficiency of glucocorticoids retards the mobilization of tissue protein and inhibits the liver's ability to store glycogen, causing hypoglycemia and muscle weakness. Wound healing is slowed, and patients become particularly susceptible to infection. There is a loss of vascular tone in the periphery, as well as decreased vascular response to the catecholamines epinephrine and norepinephrine. In primary Addison's disease decreased secretion of aldosterone causes severe sodium and water loss from the kidneys and increased reabsorption of potassium. In either condition, shock can develop within hours and immediate treatment must be instituted to prevent death.

In primary Addison's disease, shock is a result both of hypovolemia and loss of vascular tone, while in secondary Addison's disease, shock results from the absence or decrease in vascular tone and normal vascular response to stress, resulting in peripheral vasodilatation and pooling of blood.

Potential Fluid and Electrolyte Disturbances

1. **Fluid volume deficit** occurring with decreased reabsorption of sodium and water by the renal tubule secondary to the lack of aldosterone. See Chapter 6.
2. **Hyperkalemia** occurring with decreased secretion and excretion of potassium by the renal tubule secondary to the lack of aldosterone. See Chapter 9.
3. **Hyponatremia,** which may occur in chronic primary adrenocortical insufficiency secondary to hypovolemia-induced release of ADH with the retention of free water. See Chapter 8.

FLUID, ELECTROLYTE, AND ACID-BASE DISTURBANCES RELATED TO CARDIAC DISORDERS

CHF AND PULMONARY EDEMA

Congestive heart failure (CHF) develops when the heart is unable to maintain a cardiac output sufficient to meet the metabolic needs of the tissues. As the heart fails and the cardiac output drops, there is a reduction in the volume of blood perfusing the tissues. Decreased tissue perfusion stimulates the release of ADH, causing retention of water by the kidneys. Decreased renal perfusion results in a reduction in the load of sodium and water filtered by the kidneys and stimulates the release of renin. Elevated renin levels lead to increased angiotensin II, a potent vasoconstrictor that heightens systemic vascular resistance and the workload of the heart (afterload). Angiotensin II, in turn, leads to an increase in aldosterone. The combination of decreased glomerular filtration and increased aldosterone levels decreases renal excretion of sodium and water. Retention of sodium and water increases the vascular volume (preload). Since the diseased heart is unable to circulate this increased volume, the pressure within the venous circuit increases, and peripheral edema develops.

When the left heart is unable to pump the blood returning from the lungs into the systemic circulation, the hydraulic pressure within the pulmonary circulation increases. If the hydraulic pressure exceeds the pulmonary oncotic pressure, fluid leaks into the pulmonary interstitium. This results in pulmonary edema with impairment of oxygen exchange. Right ventricular failure usually occurs secondary to left ventricular failure (the right heart must work harder as the pressure in the pulmonary vasculature increases), but may occur independently in such conditions as cor pulmonale. When the right heart fails, blood backs up in the venous circuit, with subsequent congestion of blood in the body organs (e.g., liver and spleen) and edema.

Potential Fluid, Electrolyte, and Acid-Base Disturbances

1. **Fluid volume excess** as evidenced by peripheral and pulmonary edema secondary to increased secretion of aldosterone and ADH. See Chapter 6.
2. **Hyponatremia** occurring with increased secretion of ADH (ADH affects the retention of water only, while aldosterone causes retention of both sodium and water). See Chapter 8.
3. **Hypokalemia** occurring with increased urinary excretion of potassium secondary to increased secretion of aldosterone and the use of potassium-wasting diuretics (e.g., furosemide). Furosemide is commonly used in the treatment of acute pulmonary edema because of its potent and rapid diuretic action when administered IV and its direct vasodilatory effect, which reduces preload. See Chapters 7 and 9.
4. **Respiratory alkalosis** may occur in pulmonary edema secondary to hypoxia-induced hyperventilation. See Chapter 16.
5. **Metabolic acidosis** may develop in individuals in cardiogenic shock secondary to increased production of lactic acid by hypoxic tissues and decreased excretion of acids by the kidney. See Chapter 17.
6. **Respiratory acidosis** may develop in *severe* pulmonary edema secondary to retention of carbon dioxide. See Chapter 15.

CARDIOGENIC SHOCK

Shock is a state in which blood flow to peripheral tissue is inadequate for sustaining life. Usually cardiogenic shock is caused by a massive myocardial infarction (MI) that renders 40% or more of the myocardium dysfunctional secondary to necrosis or ischemia. As a result, cardiac output is reduced and all tissues suffer from inadequate perfusion. With decreased perfusion to the heart, coronary flow is reduced, impairing cardiac function, which further decreases cardiac output.

The first stage of shock is characterized by increased sympathetic discharge as the **baroreceptors** at the carotid sinus and aortic arch are stimulated by the drop in blood pressure. The release of epinephrine and norepinephrine is a compensatory mechanism that increases cardiac output by increasing the heart rate and contractility of the uninjured myocardium. Vasoconstriction, a mechanism that increases blood pressure, also occurs. The second or middle stage of shock is characterized by decreased perfusion to the brain, kidneys, and heart. Lactate and pyruvic acid accumulate in the tissues, and metabolic acidosis occurs secondary to anaerobic metabolism. In the late stage of shock, which is usually irreversible, compensatory mechanisms become ineffective, and multiple organ failure occurs.

Potential Fluid, Electrolyte, and Acid-Base Disturbances

1. **Fluid volume deficit** may be present due to prior diuretic therapy with potent diuretics (e.g., furosemide). See Chapter 6.
2. **Fluid volume excess** may be present or develop due to stimulation of the renin-angiotensin system, resulting in retention of sodium and water. Overly aggressive fluid therapy may contribute. See Chapter 6.
3. **Hyponatremia** may be present secondary to an increase in the release of ADH, resulting in retention of water. See Chapter 8.
4. **Metabolic acidosis** occurring with accumulation of lactate and pyruvic acid in the tissues secondary to decreased tissue perfusion. See Chapter 17.
5. If shock is prolonged, the individual may develop acute tubular necrosis (ATN), which can lead to multiple fluid and electrolyte disturbances (see the following discussion).

FLUID, ELECTROLYTE, AND ACID-BASE DISTURBANCES RELATED TO ACUTE OR CHRONIC RENAL FAILURE

Acute renal failure (ARF) is a sudden loss of renal function that may or may not be accompanied by oliguria. The kidney loses its ability to maintain biochemical homeostasis, causing retention of metabolic wastes and dramatic alterations in fluid and electrolyte and acid-base balance. Although the renal damage is usually reversible, ARF has a high mortality rate.

A decrease in renal function secondary to decreased renal perfusion but without kidney damage is called *prerenal failure.* A reduction in urine output because of obstruction of urine flow is called *postrenal failure.* Prolonged prerenal or postrenal problems can lead to actual renal kidney damage, termed *intrarenal failure.*

The most common cause of intrarenal failure is acute tubular necrosis (ATN), which is usually caused by prolonged ischemia or exposure to nephrotoxins (i.e., antibiotics, heavy metals, or radiographic contrast media). ATN also can occur after transfusion reactions, septic abortions, or crushing injuries. The clinical course of ATN can be divided into the following three phases: (1) oliguric (lasting approximately 7 to 21 days), (2) diuretic (lasting approximately 7 to 14 days), and (3) recovery (lasting approximately 3 to 12 months). The overall mortality rate for ATN is 40% to 60%. Causes of intrarenal failure other than ATN include acute poststreptococcal glomerulonephritis, malignant hypertension, and hepatorenal syndrome.

Chronic renal insufficiency is a progressive, irreversible loss of kidney function that can develop over days or years. Eventually it may progress to **chronic renal failure (CRF),** also known as end-stage renal disease (ESRD). The patient with ESRD requires dialysis or a kidney transplant to sustain life. Prior to ESRD, the individual with chronic renal insufficiency can lead a relatively normal life managed by diet and medications. This period can last from days to years, depending on the cause of renal disease and the individual's level of renal function at the time of diagnosis.

There are many causes of CRF. Some of the most common include glomerulonephritis, diabetes mellitus, hypertension, and polycystic kidney disease. For some individuals, the cause of CRF is unknown.

Potential Fluid, Electrolyte, and Acid-Base Disturbances

1. **Fluid volume excess** occurring with anuria or oliguria. See Chapter 6.
2. **Fluid volume deficit** during the diuretic phase of ATN occurring with excretion of large volumes of hypotonic urine, combined with existing fluid restriction. Hypovolemia also may occur in postrenal failure after release of the obstruction (postobstructive diuresis). Hypovolemia may be the precipitating event in prerenal failure. See Chapter 6.
3. **Hyponatremia** occurring with excessive consumption or administration of hypotonic fluids. See Chapter 8.
4. **Hyperkalemia** secondary to the kidneys' inability to excrete potassium and increased tissue catabolism with the release of intracellular potassium. See Chapter 9.
5. **Hypokalemia** during the diuretic phase of ATN, especially in the potassium-restricted patient. See Chapter 9.
6. **Hyperphosphatemia** secondary to the kidneys' inability to excrete phosphorus. See Chapter 11.
7. **Hypocalcemia** occurring with decreased levels of the metabolically active form of vitamin D; hyperphosphatemia (calcium and phosphorus have a reciprocal relationship: as one increases, the other tends to decrease); skeletal resistance to parathyroid hormone (the hormone released by the parathyroid gland in response to a low serum calcium level); and hypoalbuminemia. See Chapter 10.
8. **Hypermagnesemia** secondary to the kidneys' inability to excrete excess magnesium. See Chapter 12.
9. **Metabolic acidosis** secondary to the kidneys' inability to excrete the body's daily load of nonvolatile acid. See Chapter 17.

FLUID, ELECTROLYTE, AND ACID-BASE DISTURBANCES RELATED TO ACUTE PANCREATITIS

Acute pancreatitis is an inflammatory condition in which the pancreas is damaged by the digestive enzymes it produces. Although the exact mechanism involved in the development of acute pancreatitis is unclear, the presence of gallstones, hypercalcemia, and chronic alcohol abuse are commonly associated factors. A variety of complications may accompany acute pancreatitis, including peritonitis, hypovolemic shock, and respiratory failure. Chemical peritonitis may occur secondary to rupture of pancreatic ducts and loss of pancreatic enzymes into the peritoneum. Vasodilatation, increased capillary permeability, and a shift of fluid into the peripancreatic tissues result in a decrease in effective circulating volume (ECV). Hemorrhage from gastroduodenal ulceration or rupture of necrotic tissue, hypoalbuminemia, and loss of GI fluids secondary to vomiting and nasogastric suction also contribute to a reduction in ECV. If volume depletion is not detected and treated promptly, shock and acute renal failure may ensue. Mild to severe respiratory failure is common with acute pancreatitis and is attributed to the release of phospholipase A. This enzyme is believed to destroy alveolar surfactant, thereby decreasing lung compliance and impairing ventilatory capacity. Hypocalcemia is common and is attributed to multiple factors, including calcium binding in areas of fat necrosis within the pancreas, decreased secretion of parathyroid hormone (PTH), and increased release of glucagon. Hypoalbuminemia also contributes to a lowered serum calcium level.

Potential Fluid, Electrolyte, and Acid-Base Disturbances

1. **Fluid volume deficit** secondary to loss of fluid into the interstitium and retroperitoneum, vomiting, NG suction, hemorrhage, and diaphoresis. See Chapter 6.
2. **Hyponatremia** may develop secondary to loss of sodium-rich fluids, accompanied by hypovolemia-induced increase in ADH secretion. See Chapter 8.
3. **Hypocalcemia** occurring with calcium deposition in areas of fat necrosis, decreased secretion of parathyroid hormone (parathyroid hormone is the primary regulator of serum calcium levels), hypoalbuminemia, and acute renal failure (if it develops). See Chapter 10.
4. **Hypomagnesemia** secondary to abnormal GI losses and deposition of magnesium in areas of fat necrosis. See Chapter 12.
5. **Hypokalemia** may occur secondary to abnormal GI losses. See Chapter 9.
6. **Respiratory alkalosis** may develop in the patient with respiratory complications secondary to hypoxemia-induced hyperventilation or to pain and anxiety. See Chapter 16.
7. **Metabolic acidosis** if shock or renal failure develops. See Chapter 17.

FLUID, ELECTROLYTE, AND ACID-BASE DISTURBANCES RELATED TO CIRRHOSIS

Cirrhosis is a condition characterized by progressive destruction and fibrosis of the liver. Initially the liver enlarges, but as the disease progresses, the liver decreases in size secondary to contraction of scar tissue. Although the most common cause of cirrhosis in the United States is chronic alcoholism, it also may develop secondary to other causes, including viral hepatitis, medications, and hereditary metabolic disorders.

Clinical manifestations of cirrhosis develop as the result of mechanical and functional changes within the liver. Structural damage to the liver leads to hepatic venous outflow obstruction, portal venous hypertension, and shunting of blood. Bleeding esophageal and gastric varices may develop secondary to portal hypertension. Metabolic alterations include encephalopathy, coagulapathy, and hypoalbuminemia. Acute renal failure secondary to hepatorenal syndrome may develop in end-stage cirrhosis.

Ascites is a common finding in hepatic cirrhosis. It develops secondary to hepatic venous obstruction and retention of sodium and water by the kidneys, which together increase the hydraulic pressure in the liver sinusoids, favoring movement of fluid into the peritoneal space. The increased sodium and water retention that occurs with ascites is believed to be caused both by abnormal handling of sodium and water by the kidneys and compensatory retention of sodium and water secondary to a reduction in ECV that stimulates the renin-aldosterone system. Reduction in ECV is the result of decreased hepatic synthesis of albumin (albumin helps to hold the vascular volume in the vascular space), peripheral vasodilatation, and ascites formation itself. An increase in pressure within the peritoneal cavity caused by ascites results in increased femoral venous pressure. This, combined with hypoalbuminemia, leads to the development of peripheral edema.

Potential Fluid, Electrolyte, and Acid-Base Disturbances

1. **Fluid volume excess** occurring with abnormal retention of sodium and water by the kidneys. However, this excess volume (ascites) is an example of a third-space fluid shift, thus the individual may exhibit the signs and symptoms of **fluid volume deficit.** See Chapter 6.
2. **Hyponatremia** is common in cirrhotic patients with ascites and edema, especially in the terminal stage. Hyponatremia is dilutional and is the result of abnormal renal handling of water. See Chapter 8.
3. **Hypokalemia** is commonly seen in the cirrhotic patient with ascites and edema. Causes of hypokalemia in these individuals include poor dietary intake, administration of potassium-wasting diuretics, elevated aldosterone levels (aldosterone causes an increased urinary excretion of potassium), magnesium

depletion (hypomagnesemia may induce hypokalemia secondary to altered function of the sodium-potassium pump), and vomiting. NOTE: Hypokalemia may cause an increase in serum ammonia levels secondary to increased production of renal tubular cell ammonia and precipitate hepatic coma. See Chapter 9.

4. **Hyperkalemia** may occur when liver disease is complicated by renal failure or use of potassium-sparing diuretics (see Chapter 7 for a discussion of diuretic therapy). See Chapter 9.
5. **Hypomagnesemia** may occur in alcoholic cirrhosis secondary to poor oral intake, decreased GI absorption, abnormal GI losses, and increased urinary excretion. See Chapter 12.
6. **Hypocalcemia** may occur in alcoholic cirrhosis secondary to magnesium depletion (hypomagnesemia causes a reduction in the release and action of PTH) or poor oral intake. See Chapter 10.
7. **Hypophosphatemia** occurs with chronic alcoholism, especially during acute withdrawal, secondary to poor dietary intake, increased GI losses with vomiting and diarrhea, use of phosphorus-binding antacids, hyperventilation (respiratory alkalosis causes an intracellular shift of phosphorus), and increased urinary losses. See Chapter 11.
8. **Respiratory alkalosis** may occur in all types of liver disease. The respiratory center is stimulated directly by ammonia and other metabolic byproducts that accumulate in hepatic insufficiency. Additionally, hyperventilation may occur in response to hypoxia. Hypoxia develops secondary to alterations in pulmonary vasculature, ventilation-perfusion mismatch, and poor diaphragmatic excursion. Alkalosis increases the cellular uptake of ammonia and, combined with hypokalemia, may precipitate hepatic coma. See Chapter 16.
9. **Metabolic alkalosis** also may occur in the setting of liver failure owing to diuretic therapy (see Chapter 7 for a discussion of diuretic therapy). See Chapter 18.
10. **Metabolic acidosis** may develop in severe chronic liver disease secondary to the liver's inability to eliminate lactic acid, the presence of alcoholic- and starvation-induced ketoacidosis, renal failure with the retention of acids, and the loss of bicarbonate in diarrhea. See Chapter 17.

FLUID, ELECTROLYTE, AND ACID-BASE DISTURBANCES RELATED TO BURNS

Composed of two layers, the epidermis and dermis, the skin is a complex organ with multiple functions. It provides a vital barrier between the external and internal environments that helps prevent the loss of body water and electrolytes. Thermal injury to the skin can lead to dramatic alterations in fluid and electrolyte balance secondary to loss of this protective layer and altered capillary dynamics.

Burns are classified according to the extent of injury and the layer of the skin involved. Superficial partial-thickness injury (first degree burns) destroy the epidermis, while deep partial-thickness injury (second degree burns) extend into a portion of the dermis. A full-thickness injury (third degree burn) destroys all layers of the skin. Factors affecting burn severity include the extent (estimated using the "rule of nines"—refer to any medical surgical nursing text for a discussion), depth of the injury, age of the individual, preexisting medical conditions, location of the burn, and concomitant injury.

Potential Fluid, Electrolyte, and Acid-Base Disturbances

1. **Fluid volume deficit** secondary to increased capillary permeability in damaged tissue with loss of intravascular fluid and proteins into the interstitium and evaporative loss of fluid through the burn wound. Since water leaks more readily into the interstitium than plasma proteins do, the blood becomes hemoconcentrated and water shifts out of the cells; the result is cellular dehydration in undamaged tissue. The plasma-to-interstitial fluid shift occurs during the first 2 to 3 days. Later, there is a shift of fluid from the interstitium back into the plasma.
2. **Fluid volume excess** may develop at this time, especially if aggressive fluid replacement was necessary during the initial phase to maintain adequate intravascular volume. See Chapter 6.
3. **Hyponatremia** occurring with hypovolemia-induced increase in ADH. See Chapter 8.
4. **Hyperkalemia** secondary to release of potassium from damaged cells. This is most likely to occur when the burn injury is complicated by acute renal failure. Hypokalemia may develop during the recovery phase secondary to shift of potassium into new tissue and increased excretion of potassium in the urine. See Chapter 9.
5. **Hypocalcemia** may develop secondary to loss of ECF from the burn wound and shift of calcium to the wound. See Chapter 10.
6. **Hypophosphatemia** is commonly associated with burns and occurs several days post burn injury. The exact cause is unknown, but it may occur secondary to elevated calcitonin levels or respiratory alkalosis. See Chapter 11.
7. **Respiratory alkalosis** secondary to pain-induced hyperventilation. See Chapter 16.
8. **Metabolic acidosis** may occur secondary to release of acids from damaged tissue and production of lactic acid if hypovolemia has led to shock. See Chapter 17.

Appendix

B

Potential Fluid, Electrolyte, and Acid-Base Disturbances Occurring with Gastrointestinal Disorders

As stated in Chapter 4, the gastrointestinal (GI) tract plays an important role in maintaining fluid and electrolyte balance since in health it is the primary site of fluid and electrolyte gain. Approximately 1.5 to 2.0 L of fluid are gained each day through the consumption of fluids and solid food. Additionally, 6 to 8 L of fluid are secreted into and reabsorbed out of the GI tract daily, equaling approximately half of the extracellular fluid (ECF) volume. Despite the large volume of fluid that moves through it daily, in health the GI tract contributes minimally to normal daily fluid loss (approximately only 100 to 200 ml/day). In disease, however, the GI tract becomes the most common site of abnormal fluid and electrolyte loss, potentially resulting in profound fluid and electrolyte imbalance. The composition of GI secretions varies with their location within the GI tract. See Chapter 4 for additional information.

LOSS OF UPPER GI FLUIDS

Losses from the upper GI tract (above the pylorus) are essentially isotonic and rich in sodium, potassium, chloride, and hydrogen (see Table 4-1). Losses may result from such problems as vomiting and procedures such as gastric suction.

Potential Fluid, Electrolyte, and Acid-Base Disturbances with Loss of Upper GI Fluids

1. **Fluid volume deficit** occurring with abnormal fluid loss. See Chapter 6.
2. **Hyponatremia** occurring with loss of sodium-rich fluids with inadequate electrolyte replacement; or due to ADH and thirst-induced retention of water. See Chapter 8.

3. **Hypokalemia** occurring with loss of potassium-rich fluids with inadequate replacement. Increased production of aldosterone secondary to hypovolemia will contribute to the development of hypokalemia owing to increased renal losses. Remember that aldosterone causes both an increased retention of sodium and an increased excretion of potassium. See Chapter 9.
4. **Hypomagnesemia** occurring with *prolonged* loss of upper GI fluids, combined with inadequate replacement. Compared to lower GI fluids, upper GI fluids have a relatively low concentration of magnesium. See Chapter 12.
5. **Metabolic alkalosis** occurring with loss of fluids rich in chloride and hydrogen. The loss of hydrogen causes a relative increase in plasma bicarbonate levels, resulting in metabolic alkalosis. Hypovolemia and hypochloremia contribute to the development and perpetuation of metabolic alkalosis due to renal conservation of $NaHCO_3$. Normally the kidneys compensate for metabolic alkalosis by increasing the excretion of bicarbonate. But when metabolic alkalosis is combined with volume and chloride depletion, there is an increased stimulus to both sodium reabsorption and hydrogen excretion, which inhibits the excretion of sodium bicarbonate. See Chapter 18.

LOSS OF LOWER GI FLUIDS

Losses from the lower GI tract (below the pylorus) are isotonic and rich in sodium, potassium, and bicarbonate. Lower GI contents may be lost through diarrhea (see Table 4-4), intestinal fistulas, or intestinal resection (ostomies). Vomiting also may contribute to lower GI loss because duodenal fluid is often lost along with the gastric contents. Losses from both the upper and lower GI tract can occur with bowel obstruction (see the section on bowel obstruction). Acid-base balance is usually maintained when both gastric and duodenal contents are lost since a loss of both hydrogen and bicarbonate occurs.

Diarrhea

Diarrhea may be defined as *an increased loss of fluid and electrolytes via the stool.* The causes of diarrhea are divided into two main categories: (1) those causing *osmotic diarrhea* and (2) those causing *secretory diarrhea.* Osmotic diarrhea occurs when poorly absorbable solutes are present in the colon. According to the principle of osmosis, water will move across a membrane from a solution of lower solute concentration to a solution of higher solute concentration (assuming that the membrane is permeable to water but not to the solute). In this situation, the poorly absorbed solutes create an osmotic gradient for water to move from the ECF into the lumen of the bowel. Additionally, water that normally would be reabsorbed from the bowel remains within the lumen. This is the means by which sorbitol and lactulose induce diarrhea. Malabsorption or maldigestion of carbohydrates also will result in osmotic diarrhea because bacteria in the colon convert unabsorbed carbohydrates to organic acids. These organic acids create an osmotic gradient favoring the production of diarrhea. Osmotic diarrhea usually stops within 24 to 48 hours of fasting.

Both water and electrolytes are secreted and reabsorbed in the bowel. Under normal conditions there is net reabsorption, thus reducing the volume of the bowel contents. Secretory diarrhea develops when there is either increased secretion or decreased reabsorption of water and electrolytes. Unlike osmotic diarrhea, secretory diarrhea does not cease with fasting and is characterized by large losses of water and electrolytes. Bacterial infections cause secretory diarrhea by irritating the bowel mucosa and causing increased secretion of water and electrolytes. The diarrhea that occurs with inflammatory bowel diseases (ulcerative colitis and Crohn's disease) is believed to be the result of inflammation of the bowel wall that causes both an increase in the secretion and decrease in the reabsorption of water and electrolytes. The individual with inflammatory bowel disease may pass more than five stools per day, causing both physical and emotional debilitation.

Bowel Obstruction

The type and extent of the fluid and electrolyte disturbance that occurs with bowel obstruction will depend on the location of the obstruction and its duration. The longer the bowel is obstructed, the more profound the fluid and electrolyte imbalance. Lower GI loss occurs because of sequestering of fluid in the distended bowel. Several liters of fluid may collect in the intestinal lumen, leading to a dramatic increase in lumen pressure and eventual damage to the intestinal mucosa. Peritonitis may then occur if bacteria enter the peritoneal cavity through the damaged intestinal wall. The fluid sequestered in the bowel is inaccessible to the ECF, thereby creating a separate third space. Peritonitis also causes a shift of fluid into a temporarily inaccessible third space. Normally the peritoneum aids in the rapid transport of fluid from the peritoneal cavity to the circulation. When the peritoneum becomes damaged or inflamed, fluid and electrolytes collect in the peritoneal cavity. The loss of ECF into the bowel and peritoneal cavity causes a reduction in effective circulating volume, which stimulates thirst and release of ADH, resulting in the retention of water. These two factors may lead to the eventual development of *dilutional hyponatremia* (see Chapter 8 for further discussion). Upper GI loss can occur with bowel obstruction owing to the increased stimulus to vomit.

Potential Fluid, Electrolyte, and Acid-Base Disturbances with Loss of Lower GI Contents

1. **Fluid volume deficit** occurring with abnormal fluid loss and third-space fluid shifts. See Chapter 6.
2. **Hyponatremia** secondary to loss of sodium-rich fluids with inadequate electrolyte replacement; or due to thirst and ADH-induced retention of water. See Chapter 8.

3. **Hypokalemia** occurring with loss of potassium-rich fluids, combined with inadequate replacement. See Chapter 9.
4. **Hypocalcemia** occurring with loss of calcium-rich fluid (may occur with chronic loss of lower GI fluids as with ulcerative colitis). See Chapter 10.
5. **Hypomagnesemia** secondary to abnormal fluid loss. Typically, hypomagnesemia occurs only with prolonged loss of GI fluids, such as with ulcerative colitis, or with conditions that decrease intestinal absorption of magnesium, such as malabsorption syndromes (e.g., sprue and celiac disease) or surgical resection of the small bowel. See Chapter 12.
6. **Metabolic acidosis** occurring with loss of fluids rich in bicarbonate. See Chapter 17.

Appendix

C

Potential Fluid, Electrolyte, and Acid-Base Disturbances Occurring in the Surgical Patient

Disturbances in fluid and electrolyte balance are common in the surgical patient owing to a combination of factors that occur preoperatively, intraoperatively, and postoperatively.

PREOPERATIVE FACTORS

1. **Preexisting conditions** (e.g., diabetes mellitus, hypertension, liver disease, or renal insufficiency) that may be aggravated by the stress of surgery. See Appendix A.
2. **Diagnostic procedures** (e.g., arteriogram or intravenous pyelogram) that require administration of intravenous dyes, which may cause inappropriate urinary excretion of water and electrolytes owing to the osmotic diuretic effect. See Chapters 1 and 7 for a discussion of osmotic diuresis.
3. **Administration of medications** (e.g., steroids or diuretics) that may affect the excretion of water and electrolytes. See Chapter 7.
4. **Surgical preparations** (e.g., enemas or laxatives) that may act to increase fluid and electrolyte loss from the GI tract. See Chapter 4 and Appendix B.
5. **Medical management of preexisting conditions** (e.g., NG suction and gastric lavage). See Chapter 4.
6. **Preoperative fluid restriction** (i.e., *NPO after midnight*). During an average 6-hour period of fluid restriction, the healthy individual loses approximately 300 to 500 ml of fluid owing to normal fluid loss. Fluid loss may be increased greatly if the patient is experiencing abnormal fluid loss or fever. See Chapter 4.

INTRAOPERATIVE FACTORS

1. **Induction of anesthesia,** which may lead to the development of hypotension in the patient with preoperative hypovolemia owing to the loss of compensatory mechanisms, such as tachycardia or vasoconstriction.
2. **Abnormal blood loss** related to preoperative trauma or the surgical procedure itself.
3. **Abnormal loss of extracellular fluid (ECF) into a third space** (e.g., loss of ECF into the wall and lumen of the bowel during bowel surgery). This fluid is temporarily unavailable either to the intracellular fluid (ICF) or ECF, hence it is termed *third-space fluid.* Loss of ECF also occurs when intravascular volume is lost into a nonequilibrating space (e.g., bleeding into a fractured hip).
4. **Loss of fluid from the surgical wound.** This is usually of concern with large wounds and prolonged operative procedures.

   NOTE: All of the above factors relate to fluid volume. See Chapter 6.

POSTOPERATIVE FACTORS

1. **Stress of surgery and postoperative pain,** which leads to an increased release of antidiuretic hormone (ADH) by the posterior pituitary gland and an increased release of adrenocorticotropic hormone (ACTH) by the anterior pituitary gland. Increased ADH results in retention of water by the kidneys. Excessive production may lead to the development of hyponatremia (see Chapter 8). ACTH acts on the adrenal cortex to cause an increase in the release of aldosterone and hydrocortisone. Both aldosterone and hydrocortisone lead to increased retention of sodium and water and increased excretion of potassium by the kidneys. The combined effects of these hormones may result in postoperative fluid retention lasting approximately 48 to 72 hours. See Chapter 6.
2. **Increase in tissue catabolism** (breakdown) secondary to tissue trauma, which causes the patient to produce a greater than normal amount of water from oxidation.
3. **Reduction in effective circulating volume (ECV),** which stimulates production of ADH and aldosterone. Potential causes include bleeding, fluid loss from the surgical wound, abnormal sequestration of fluid (i.e., third-space shift), draining fistulas, NG suction, vomiting, and increased insensible fluid loss from fever. These factors also may contribute to the development of electrolyte and acid-base disturbances.
4. **Risk or presence of postoperative ileus,** which may restrict the patient's ability to take oral fluids and necessitate NG suction. See Chapter 4.
5. **Hyperkalemia,** which may occur immediately postoperatively, due to the release of intracellular potassium secondary to tissue trauma. See Chapter 9.
6. **Metabolic acidosis,** which may occur due to an abnormal production of lactic acid in the hypotensive patient who experiences tissue hypoxia. See Chapter 17.

7. **Respiratory acidosis,** which may develop due to inadequate ventilation secondary to respiratory depression from anesthesia or pain medication, splinting of the operative site, or restriction to bed, which increases the risk of atelectasis or pneumonia, especially in the patient with chronic obstructive pulmonary disease (COPD). See Chapter 15.

Appendix

D

Potential Fluid and Electrolyte Disturbances Occurring in the Patient at Nutritional Risk

Standard criteria are used to evaluate the patient's potential for nutritional risk. These include:

1. History of inadequate nutrient intake.
2. IV support (5% dextrose solutions) or NPO for >5 days.
3. Recent, unplanned weight loss (>10 pounds).
4. Underweight condition (≤80% of ideal body weight).
5. Overweight status (>20% above ideal body weight).
6. Age: Infancy, childhood, advanced.
7. Pregnancy (especially in an adolescent).
8. Drug or alcohol abuse.
9. Organ or system failure (e.g., ARDS, COPD, renal failure, diabetes mellitus, pancreatitis, neuromuscular dysfunction).
10. Trauma, surgery, or disease of the oral cavity or GI tract (e.g., fractured mandible, radical head and neck resection, malabsorption syndrome).

FLUID AND ELECTROLYTE DISTURBANCES ASSOCIATED WITH INADEQUATE NUTRITION OR NUTRITION THERAPY

1. **Fluid volume excess** occurring with overfeeding (excess volume of total parenteral nutrition [TPN] or enteral formula) or overhydration. See Chapter 6.
2. **Fluid volume deficit** can occur because of diarrhea from an inappropriate choice of enteral formula, too rapid an infusion rate, malabsorption, or enteral administration of hyperosmolar medications. Bolus administration of tube feedings or delayed gastric emptying may result in vomiting. Gastric emptying can

be facilitated with the selection of a formula lower in fat or feeding the patient distal to the pyloric valve. Gastric emptying can also be improved by positioning patients on their right sides and elevating the HOB to a 45-degree angle. See Chapter 6.

Hypovolemia also can be caused by diuresis and the mobilization of edema fluid, which occurs shortly after initiation of TPN in a malnourished patient who is hypoproteinemic and edematous. A change in formula to reduce the percentage of carbohydrate may be required to manage hyperosmolar diuresis caused by hyperglycemia.

3. **Hypernatremia** occurring with excessive sodium intake from an enteral or parenteral formula in combination with inadequate water intake. See Chapter 8.
4. **Hyponatremia** secondary to water overload (see Chapter 6) or inadequate sodium intake. See Chapter 8.
5. **Hyperkalemia** occurring with excessive enteral or parenteral potassium supplementation or increased tissue catabolism, especially in renal insufficiency. See Chapter 9.
6. **Hypokalemia** occurring with excessive losses from diarrhea or emesis, which can occur secondary to selection of a hyperosmolar enteral feeding or bolus method of administration. Hypokalemia also occurs secondary to potassium release and excretion during muscle breakdown in the catabolic patient with normal renal function who is utilizing muscle for energy. See Chapter 9.
7. **Hypercalcemia** occurring with excessive infusion of calcium or vitamin D. See Chapter 10.
8. **Hypocalcemia** secondary to hypoalbuminemia. Roughly, for every 1 g/dl decrease in serum albumin from 4 g/dl, serum calcium drops 0.8 mg/dl. For example, in a patient with a serum albumin of 3.0 and a serum calcium of 7.5, the corrected calcium value is 8.3 (7.5 + 0.8). An increase in total calories administered or administration of albumin in severe hypoalbuminemia can normalize serum albumin and therefore, serum calcium. See Chapter 10.
9. **Hyperphosphatemia** can occur due to an excessive intake of dietary phosphates combined with a condition that decreases excretion of phosphorus (e.g., excess dairy products in the renal patient). See Chapter 11.
10. **Hypophosphatemia** secondary to transcellular shifts caused by an infusion of glucose. This process is accentuated in patients who are malnourished. Hypophosphatemia is also caused by decreased intake or loss of phosphates in a patient who is starving, malabsorbing, or vomiting. See Chapter 11.
11. **Hypomagnesemia** occurring with increased requirements of magnesium for new tissue synthesis in patients receiving nutritional support. See Chapter 12.

Appendix

E

Effects of Age on Fluid, Electrolyte, and Acid-Base Balance

INFANTS AND CHILDREN

- Relative to their size, infants and children have a greater body surface area (both external and internal) than the adult, and thus have a greater potential for fluid loss via the skin and gastrointestinal tract.
- Infants and children have a higher percentage of total body water (TBW) than adults. The greater percentage of the infant's body water is extracellular. As cellular growth occurs, more fluid becomes intracellular.
- Infants have a decreased ability to concentrate their urine, while at the same time they have an increased solute load to excrete because of their increased caloric need. These two factors result in a relatively greater obligatory fluid loss, meaning that they must produce a relatively larger volume of urine in order to excrete their daily load of metabolic wastes.
- The daily I&O for infants (e.g., 650 ml) is equal to approximately half the volume of their extracellular fluid (ECF) (e.g., 1,300 ml), as compared to adults, whose daily I&O (e.g., 2,500 ml) is approximately one sixth of their ECF (e.g., 15.0 L). Thus infants can lose a volume equal to their ECF in 2 days, while it takes an adult 6 days to do the same.
- Because of an infant's small size and decreased ability to excrete excess fluid, IV fluid administration necessitates caution via use of monitored pumps.
- Infants are less able to compensate for acidosis because of their decreased ability to acidify urine.
- Infants are at increased risk for developing hypernatremia since they are unable to verbalize thirst. Remember that thirst is the body's primary defense against symptomatic hypernatremia.

- Children have an increased incidence and intensity of fever, upper respiratory infections, and gastroenteritis, which can lead to abnormal fluid and electrolyte loss. See Appendix B for a discussion of the various fluid and electrolyte imbalances that occur with loss of upper and lower gastrointestinal contents.
- Infants and small children are prone to fluid volume deficit caused by a combination of the factors listed previously. Unfortunately, some of the common indicators of fluid volume deficit are less reliable in the infant or small child. Infants are unable to verbalize thirst, although their cries may become increasingly high-pitched. Skin turgor also may be a less reliable sign. Skin turgor may appear normal in the obese infant because of increased subcutaneous fat, or it may appear abnormal in the adequately hydrated but undernourished infant. Irritability is an early indicator of hypovolemia. Sunken fontanels, a traditional indicator of dehydration, does not occur until there has been moderate to severe fluid loss.

THE OLDER ADULT

- Weight (body fat) tends to increase with advancing age, thus the percentage of TBW decreases. Recall that fat cells contain little water. The percentage of TBW increases in the emaciated individual who has lost significant body fat.
- Renal function decreases with advancing age. Glomerular filtration rate drops; thus the older adult is less likely to compensate for an increased metabolic load. There is also a reduction in the ability to concentrate urine, resulting in greater obligatory water losses. The older adult must produce a larger volume of urine to excrete the same amount of metabolic waste as the younger adult.
- In the older adult, the kidneys are less able to compensate for an acid load, owing to decreased formation of ammonia. (Ammonia produced by the renal tubular cell diffuses into the lumen of the tubule and combines with hydrogen to form ammonium, which is then excreted in the urine . In this way, ammonia acts as a urinary buffer, allowing increased excretion of hydrogen ions). Normally, ammonia production increases in the presence of an acid load.
- Decreased respiratory function also reduces the older adult's ability to compensate for acid-base imbalance. The older adult is also more likely to develop hypoxemia.
- There is a reduction in the secretion of HCl by the stomach, which may affect the individual's ability to tolerate certain foods. The older adult is especially prone to constipation because of decreased gastrointestinal tract motility. Limited fluid intake, a restricted diet, and a reduced level of physical activity may contribute to the development of constipation. Excessive or inappropriate use of laxatives may lead to problems with diarrhea.
- As the skin ages, there is a reduction in insensible and sensible water loss secondary to decreased skin hydration and decreased functioning of the sweat glands. Thus the skin is less efficient in cooling the body, and the skin tends to be dry. In addition, skin turgor is a less reliable indicator of fluid status due to decreased skin elasticity.

- Older adults are at increased risk for developing hypernatremia since they have a less sensitive thirst center and may have problems with obtaining fluids (e.g., impaired mobility) or expressing their desire for fluids (e.g., the individual who has suffered a stroke). Thirst is the body's primary defense against symptomatic hypernatremia.

Appendix

F

Acid-Base Rules: General Guidelines

Disturbance	Change in pH	Compensatory Response	Results of Compensation
Respiratory acidosis			
Acute	pH ↓ 0.08 for every 10 mm Hg ↑ in $Paco_2$	Immediate release of tissue buffers (i.e., HCO_3^-)	1.0 mEq/L ↑ in HCO_3^- from patient's baseline for every 10 mm Hg ↑ in $Paco_2$
Chronic	Depends on renal compensation; often near normal	↑ renal reabsorption of HCO_3^-; clinically evident after 8 hr; maximal effect 3-5 days	3.5 mEq/L ↑ in HCO_3^- for every 10 mm Hg ↑ in $Paco_2$
Respiratory alkalosis			
Acute	pH ↑ 0.08 for every 10 mm Hg ↓ in $Paco_2$	Immediate release of tissue buffers	2.0 mEq/L ↓ in HCO_3^- from patient's baseline for every 10 mm Hg ↓ in $Paco_2$
Chronic	pH can be returned to normal if renal function is adequate	↓ renal reabsorption of HCO_3^-	Maximal renal compensation causes HCO_3^- to ↓ 5 mEq/L for every 10 mm Hg ↓ in $Paco_2$. Maximal effect can take 7-9 days and may *normalize* pH

Disturbance	Change in pH	Compensatory Response	Results of Compensation
Metabolic acidosis			
Acute	pH ↓ 0.15 for every 10 mEq/L ↓ in HCO_3^-	Hyperventilation occurs immediately	1.2 mm Hg ↓ in $Paco_2$ for every 1 mEq/L ↓ in HCO_3^-
Chronic	pH same as it would be if no respiratory compensation were present	Hyperventilation	The effects of hyperventilation last only a few days because the ↓ in $Paco_2$ causes a further ↓ in renal reabsorption of HCO_3^-
Metabolic alkalosis			
Acute	pH ↑ 0.15 for every 10 mEq/L ↑ in HCO_3^-	Hypoventilation occurs immediately	0.7 mm Hg ↑ in $Paco_2$ for every 1 mEq/L ↑ in HCO_3^-
Chronic	pH same as it would be if respiratory compensation were present	Hypoventilation	The effects of hypoventilation last for only a few days because the ↑ in $Paco_2$ causes ↑ renal excretion of H^+ and ↑ serum HCO_3^-

Appendix

G

Abbreviations Used in this Text

ABGs: Arterial blood gases
ACTH: Adrenocorticotropic hormone
ADH: Antidiuretic hormone
ANF: Atrial natriuretic factor
ARF: Acute renal failure
ATN: Acute tubular necrosis
ATP: Adenosine triphosphate
AV: Atrioventricular
BP: Blood pressure
BSA: Body surface area
BUN: Blood urea nitrogen
Ca^{2+}: Calcium ion
CAVH: Continuous arteriovenous hemofiltration
CBC: Complete blood cell count
CHF: Congestive heart failure
Cl^-: Chloride ion
cm H_2O: Centimeters of water
CNS: Central nervous system
CO: Cardiac output
CO_2: Carbon dioxide
COPD: Chronic obstructive pulmonary disease
CRF: Chronic renal failure
CVP: Central venous pressure
DI: Diabetes insipidus
DKA: Diabetic ketoacidosis
dl: Deciliter
ECF: Extracellular fluid
ECV: Effective circulating volume
EKG: Electrocardiogram (also abbreviated ECG)

Fio_2: Fraction of inspired oxygen
GI: Gastrointestinal
H^+: Hydrogen ion
H_2CO_3: Carbonic acid
HCl: Hydrochloric acid
HCO_3^-: Bicarbonate ion
Hct: Hematocrit
Hg, Hgb: Hemoglobin
HHNC: Hyperosmolar hyperglycemic nonketotic coma
HOB: Head of bed
HR: Heart rate
ICF: Intracellular fluid
IM: Intramuscular
I&O: Intake and output
ISF: Interstitial fluid
IU: International unit
IμU: International microunit
IV: Intravascular
K^+: Potassium ion
L: Liter
LOC: Level of consciousness
MAP: Mean arterial pressure
mEq: Milliequivalent
mg: Milligram
Mg^{2+}: Magnesium ion
ml: Milliliter
mm Hg: Millimeters of mercury
mOsm: Milliosmole
μg: Microgram
N: Nitrogen
Na^+: Sodium ion
ng: Nanogram
NG: Nasogastric
NH_3: Ammonia
NH_4^+: Ammonium
NS: Normal saline, i.e., isotonic solution of NaCl
O_2: Oxygen
OTC: Over-the-counter
P: Phosphorus
PA: Pulmonary artery
$Paco_2$: Carbon dioxide tension of arterial blood
Pao_2: Oxygen tension of arterial blood
PAP: Pulmonary artery pressure
PAWP: Pulmonary artery wedge pressure

pg: Picogram
PO_4^{2-}: Phosphate ion
PRBCs: Packed red blood cells
prn: As needed
PTH: Parathyroid hormone
PVC: Premature ventricular contractions
q: Every
RBC: Red blood cell
RDA: Recommended daily allowance
ROM: Range of motion
RR: Respiratory rate
SC: Subcutaneous
SIADH: Syndrome of inappropriate antidiuretic hormone
SOB: Shortness of breath
stat: Immediately
SVR: Systemic vascular resistance
TCF: Transcellular fluid
TKO: To keep open
TPN: Total parenteral nutrition
μg: Microgram
μIU: Microinternational unit
VS: Vital sign
WBC: White blood cell
WOB: Work of breathing

Appendix

H

Glossary

Acidemia: Change of pH in arterial blood to <7.40.

Acidosis: Abnormal accumulation of acid or loss of base from the body.

Acids: Substances that can give up a hydrogen ion (H^+).

Active transport: The movement of solutes across a cell membrane in the absence of a favorable electrochemical or concentration gradient; requires energy.

Acute renal failure (ARF): A sudden loss of renal function that is usually reversible.

Aldosterone: A mineralocorticoid hormone released by the adrenal cortex that increases the reabsorption (saving) of sodium and secretion and excretion of potassium and hydrogen by the kidneys.

Alkalemia: Increase in arterial pH to >7.40.

Alkalosis: Abnormal accumulation of bicarbonate or loss of acid in the body.

Anaerobic metabolism: Occurs when there is not enough oxygen available for metabolism and alternate pathways are used, resulting in an accumulation of organic acids (lactic acidosis).

Analog: A substance with structure and function similar to another substance.

Anascara: Severe generalized edema.

Angiotensin: A polypeptide found in the blood and formed by the action of renin on the α-2-globulin, angiotensinogen. *Angiotensin I* is converted to *angiotensin II*. Angiotensin II, a potent vasoconstrictor, acts on the adrenal cortex to stimulate the release of aldosterone.

Anions: Ions that develop a negative charge in solution. Examples of the body's most common anions include chloride ion (Cl^-), bicarbonate ion (HCO_3^-), and phosphate ion (PO_4^{3-}). Proteins are another important group of anions.

Anion gap: Reflection of the anions in plasma (e.g., phosphates, sulfates, and proteinates) that normally are unmeasured. Anion gap is helpful in the differential diagnosis of metabolic acidosis or mixed acid-base disorders.

Antidiuretic hormone (ADH): Produced by the hypothalamus and released by the posterior pituitary gland, it increases reabsorption (saving) of water by the kidneys, allowing excretion of a concentrated urine. In addition, ADH is an arterial vasoconstrictor that increases BP by increasing vascular resistance.

Anuria: The production of ≤ 100 ml of urine in 24 hours.

Arterial blood gases (ABGs): Measurement of pH, carbon dioxide tension, and oxygen tension of arterial blood to evaluate acid-base and pulmonary functions.

Asterixis: Hand-flapping tremor that occurs with extension of the arm and dorsiflextion of the wrist. It is often seen with metabolic disorders.

Atrial natriuretic factor (ANF): Hormone released by the cardiac atria in response to an increased vascular volume. Its actions include increased excretion of sodium and water by the kidneys secondary to increased filtration, decreased synthesis of renin, decreased release of aldosterone, and direct vasodilatation. ANF reduces vascular volume and BP.

Azotemia: Increased retention of metabolic wastes.

Baroreceptors: Pressure-sensitive nerve endings located in the carotid sinuses, aortic arch, cardiac atria, and renal vessels, which respond to changes in blood pressure via changes in stretch in the arterial wall, leading to changes in cardiac output, vascular resistance, thirst, and renal handling of sodium and water.

Bases: Substances that can take on a hydrogen ion (H^+).

Bicarbonate: The body's most important and abundant buffer. It is generated in the kidney and aids in excretion of H^+.

Buffers: Substances that combine with excess acid or base, resulting in a minimally altered pH.

Capillary membrane: Separates the intravascular fluid from the interstitial fluid.

Cardiac output (CO): Product of heart rate times stroke volume (i.e., the amount of blood moved with each contraction of the left ventricle/min). Normal value is 4 to 7 L/min for the adult.

Cations: Ions that develop a positive charge in solution and are attracted to negative electrons. Examples of the body's most common cations include sodium ions (Na^+), potassium ions (K^+), calcium ions (Ca^{2+}), magnesium ions (Mg^{2+}), and Hydrogen ions (H^+).

Cell membrane: Composed of lipids and protein, this membrane separates intracellular fluid (ICF) from the interstitial fluid.

Central venous pressure (CVP): Measurement of the right atrial pressure and right ventricular end-diastolic pressure via a catheter inserted in or near the right atrium.

Chronic renal failure (CRF): An irreversible loss of kidney function, also known as *end-stage renal disease (ESRD)*.

Chvostek's sign: A signal of tetany occurring with hypocalcemia or hypomagnesemia, it is considered positive when there is unilateral contraction of the facial and eyelid muscles in response to facial nerve percussion.

Colloid: In the medical vernacular, colloid is *an IV fluid that contains solutes that do not readily cross the capillary membrane.* Dextran, blood, albumin, mannitol, and plasma are all colloids. When combined with water, colloids do not form true solutions.

Concentration gradient: The concentration difference between an area of a high concentration and an area of low concentration of the same substance.

Crystalloid: In the medical vernacular, crystalloid is *an IV fluid that contains solutes that readily cross the capillary membrane.* Examples include dextrose or electrolyte solutions. When combined with water, crystalloids dissolve and form true solutions.

Diffusion: Random movement of particles through a solution or gas, in which the particles move from an area of high concentration to an area of low concentration. When diffusion of a particular solute is dependent on the availability of a carrier substance, it is termed *facilitated* diffusion. Diffusion not dependent on a carrier substance is termed *simple* diffusion.

Edema: Palpable swelling of the interstitial space that can be either localized or generalized.

Effective circulating volume (ECV): The portion of intravascular volume that actually perfuses the tissues. For example, in congestive heart failure intravascular volume increases because of sodium and water retention, yet ECV decreases because of the pooling of blood in the venous circuit.

Effective osmolality: Changes in osmolality that will cause water to move from one compartment to another. If a substance has an equal concentration on both sides of the membrane, there is no effective osmolality. *Tonicity* is another term for effective osmolality.

Electrolytes: Substances (solutes) that dissociate in solution and conduct an electrical current. Electrolytes dissociate into positive ions (cations) and negative ions (anions).

Epithelial membrane: Separates interstitial fluid (ISF) and intravascular fluid (IVF) from the transcellular fluid (TCF) and produces TCF.

Erythropoietin: A glycoprotein hormone released by the renal cells in response to low oxygen levels, which stimulates production of RBCs by the bone marrow.

Extracellular fluid (ECF): Fluid found outside the cells, comprising approximately one third of the body's fluid (in the adult).

Filtration: Movement of water and solutes from an area of high hydraulic pressure to an area of low hydraulic pressure.

Glomerular filtration rate (GFR): The volume of fluid crossing the glomerular membrane each minute.

Hemolysis: Breakdown of red blood cells (RBCs) that may occur if blood is exposed to a hypotonic solution.

Homeostasis: Physiologic balance in which there is relative constancy in the body's environment, maintained by adaptive responses.

Hydrostatic pressure: Pressure created by the weight of fluid.

Hydraulic pressure: One of the factors that affects the movement of fluid across the capillary membrane. It is a combination of hydrostatic pressure and the pressure created by the pump action of the heart. The terms *hydrostatic pressure* and *hydraulic pressure* are often used interchangeably.

Hypercapnia: Increased amounts of carbon dioxide in the blood caused by hypoventilation. It is also known as hypercarbia.

Hypertonicity: State in which a solution's effective osmolality is greater than that of the body's fluids.

Hyperventilation: Any process resulting in a decreased $Paco_2$.

Hypervolemia: Expansion of the ECF volume. Usually used to describe the expansion of the intravascular portion of the ECF.

Hypocapnia: Decreased amounts of carbon dioxide in the blood caused by hyperventilation. It is also known as hypocarbia.

Hypotonicity: State in which a solution's effective osmolality is greater than that of the body's fluids.

Hypoventilation: Any process resulting in an increased $Paco_2$.

Hypovolemia: A reduction in the ECF volume. Usually used to describe a reduction in the volume of the intravascular portion of the ECF.

Insensible fluid: Imperceptible loss of fluid through the skin or respiratory system via evaporation. Because it is nearly free of electrolytes, insensible fluid loss is considered pure water loss.

Interstitial fluid (ISF): The fluid surrounding the cells, including lymph fluid.

Intracellular fluid (ICF): Fluid contained within the cells, comprising approximately two thirds of the body's fluid (in the adult).

Intravascular fluid (IVF): Fluid contained within the blood vessels (i.e., plasma).

Isohydric principle: A change in the H^+ concentration will affect the ratio of acids to bases in all buffer systems.

Isotonic solutions: Fluids with the same effective osmolality as body fluids.

Kussmaul respirations: Rapid, deep, *sighing* breaths.

Mean arterial pressure (MAP): A reflection of the average pressure within the arterial tree throughout the cardiac cycle. The normal value is 70 to 105 mm Hg.

Metastatic calcifications: Precipitation and deposition of calcium phosphate in the soft tissue, joints, and arteries, also known as soft tissue calcifications.

Milk-alkali syndrome: Renal dysfunction and metabolic alkalosis resulting from chronic ingestion of excessive amounts of absorbable alkali (i.e., milk and calcium carbonate).

Minute ventilation: Respiratory rate × tidal volume.

Nonelectrolytes: Substances that do not dissociate (separate) in solution. Examples include glucose, urea, creatinine, and bilirubin.

Nonvolatile (fixed) acid: Any acid that cannot be vaporized and excreted by the lungs.

Oliguria: Urinary output of <400 ml/24 hr.

Oncotic pressure: Osmotic pressure exerted by protein.

Osmolality: The number of particles contained in body fluids, i.e., concentration. Measured in milliosmoles (mOsm)/kg of water.

Osmolarity: Like osmolality, it is a term used to describe the concentration of fluids. Measured in milliosmoles (mOsm)/L of solution.

Osmosis: Movement of water across a semipermeable membrane from an area of lower solute concentration to an area of higher solute concentration.

Osmotic diuresis: Increased urine output caused by such substances as mannitol, glucose, or contrast media, which are excreted in the urine and reduce water reabsorption.

Osmotic pressure: The pressure that *pulls* water across a semipermeable membrane when the membrane separates two solutions with different concentrations. See *osmosis*.

Oxygen saturation: The degree to which hemoglobin is combined with oxygen.

pH: Measurement of H^+ concentration in body fluids reflecting one of the following states: normal (7.40), acidic ($<$7.40), or alkalotic ($>$7.40).

Plasma: The fluid portion of the blood containing water, protein, and electrolytes.

Polyuria: Excessive urine output.

Pulmonary artery pressure (PAP): Pressure measured in the pulmonary artery. When pulmonary function is normal, it reflects the pressure within the left ventricle at the end of diastole. PAP is used to evaluate left ventricular function and fluid volume. Normal PAP is 20 to 30/8 to 15 mm Hg.

Pulmonary artery wedge pressure (PAWP): Measurement of the pulmonary capillary pressure by means of a balloon-tipped catheter passed into the distal pulmonary artery. It provides a more accurate reflection of left ventricular end-diastolic pressure than PAP. Normal PAWP is 6 to 12 mm Hg.

Renin: Proteolytic enzyme produced and released by specialized cells located in the arterioles of the kidney. Renin is released in response to decreased renal perfusion or stimulation of the sympathetic nervous system and is important in the formation of angiotensin.

Sensible fluid: Perceptible loss of body fluid (i.e., sweat) via the skin, which contains a significant amount of electrolytes.

Serum: Plasma minus the fibrinogen and other clotting factors; i.e., the fluid that remains after a blood specimen has been allowed to form a clot.

Sodium-potassium pump: A physiologic mechanism present in all body cell membranes that transports sodium from the inside of the cell to the outside and transports potassium from the outside of the cell to the inside. It requires energy and the presence of adequate magnesium.

Solutes: Dissolved particles found in body fluids. There are two types: electrolytes and nonelectrolytes.

Specific gravity: Measurement of the weight of a substance in relationship to water. Water = 1.000.

Substrate: A substance that is acted upon (and changed by) an enzyme during a chemical reaction.

Syndrome of inappropriate antidiuretic hormone (SIADH): A condition in which there is inappropriate hypothalamic production or enhanced action or ectopic production of ADH resulting in excess water retention.

Systemic vascular resistance (SVR): Clinical measurement of the resistance in vessels, which is used to determine workload of the left ventricle (afterload). Normal SVR is 900 to 1,200 $dynes/sec/cm^{-5}$.

Tachypnea: Increased respiratory rate. It is also called *hyperpnea.*

Third-space fluid shift: The loss of ECF into a normally nonequilibrating space. Although the fluid has not been lost from the body, it is temporarily unavailable to the ICF or ECF for its use.

Tidal volume: Normal resting volume of ventilation.

Tonicity: Another term for *effective osmolality.*

Transcellular fluid (TCF): Fluid secreted by epithelial cells. These fluids include cerebrospinal, pericardial, pleural, synovial, and intraocular fluids and digestive secretions.

Trousseau's sign: Ischemia-induced carpal spasm that occurs with hypocalcemia and hypomagnesemia. It may be elicited by applying a BP cuff to the upper arm and inflating it past systolic BP for 2 minutes.

Ventilation-perfusion mismatch: An inequality in the ratio between ventilation and perfusion that occurs with shunting of venous blood past unventilated alveoli.

Volatile acid: An acid that can be vaporized and eliminated by the lungs (i.e., CO_2).

Appendix

I

Laboratory Tests Discussed in this Textbook: Normal Values

Complete Blood Count (CBC)	Adult Normal Values
Hemoglobin	Male: 14-18 g/dl Female: 12-16 g/dl
Hematocrit	Male: 40%-54% Female: 37%-47%
Red blood cell (RBC) count	Male: 4.5-6.0 million/µl Female: 4.0-5.5 million/µl
White blood cell (WBC) count	4,500-11,000/µl
Neutrophils	54%-75% (3,000-7,500/µl)
Band neutrophils	3%-8% (150-700/µl)
Lymphocytes	25%-40% (1,500-4,500/µl)
Monocytes	2%-8% (100-500/µl)
Eosinophils	1%-4% (50-400/µl)
Basophils	0%-1% (25-100/µl)
Platelet count	150,000-400,000/µl

Normal values may vary significantly with different laboratory methods of testing.

Serum, Plasma, and Whole Blood Chemistry	Normal Values
ACTH	8 AM-10 AM <100 pg/ml
ADH	0-2 pg/ml/serum osmolality <285 mOsm/kg;
	2-12 pg/ml/serum osmolality >290 mOsm/kg
Albumin	3.5-5.5 g/dl
Aldosterone	Male: 6-22 ng/dl
	Female: 4-31 ng/dl
Ammonia	70-200 μg/dl
	56-150 μg/dl
Amylase	60-180 Somogyi U/dl
Base, total	145-160 mEq/L
Bicarbonate	22-26 mEq/L
Bilirubin	Total: 0.3-1.4 mg/dl
Blood gases, arterial	
pH	7.35-7.45
PaCO_2	35-45 mm Hg
PaO_2	80-95 mm Hg
O_2 saturation	95%-99%
Calcitonin	<100 pg/ml
Calcium	8.5-10.5 mg/dl;4.3-5.3 mEq/L
Chloride	95-108 mEq/L
Cortisol	8 AM-10 AM: 5-25 μg/dl
	4 PM-midnight: 2-18 μg/dl
CO_2 content (Total CO_2)	22-28 mEq/L
Creatine phosphokinase (CPK)	Male: 55-170 U/L
	Female: 30-135 U/L
Creatinine	0.6-1.5 mg/dl
Creatinine clearance	Male: 107-141 ml/min
	Female: 87-132 ml/min
Globulins, total	1.5-3.5 g/dl
Glucose, fasting	True glucose: 65-110 mg/dl
	All sugars: 80-120 mg/dl
Glucose, 2-hr postprandial	<145 mg/dl
Glucose tolerance, intravenous	Fasting: 65-110 mg/dl
	5 min: maximum 250 mg/dl
	60 min: decrease
	2 hr: <120 mg/dl
	3 hr: 65-110 mg/dl
oral	Fasting: 65-110 mg/dl
	30 min: <155 mg/dl
	1 hr: <165 mg/dl
	2 hr: <120 mg/dl
	3 hr: ≤65-110 mg/dl

Serum, Plasma, and Whole Blood Chemistry	Normal Values
17-OCHS	Male: 7-19 µg/dl
	Female: 9-21 µg/dl
Insulin	11-240 µlU/ml
	4-24 µU/ml
Iron	Total: 60-200 µg/dl
	Male, average: 125 µg/dl
	Female, average: 100 µg/dl
	Elderly: 60-80 µg/dl
Ketone bodies	2-4 µg/dl
Lactic acid	Arterial: 0.5-1.6 mEq/L
	Venous: 1.5-2.2 mEq/L
Magnesium	1.8-3.0 mg/dl
	1.5-2.5 mEq/L
Osmolality	280-300 mOsm/kg
Parathyroid hormone	<2000 pg/ml
Phosphatase, acid	0-1.1 U/ml (Bodansky)
	1-4 U/ml (King-Armstrong)
	0.13-0.63 U/ml (Bessey-Lowery)
Phosphatase, alkaline	1.5-4.5 U/dl (Bodansky)
	4-13 U/dl (King-Armstrong)
	0.8-2.3 U/ml (Bessey-Lowery)
Phosphorus	2.5-4.5 mg/dl; 1.7-2.6 mEq/L
Potassium	3.5-5.0 mEq/L
Renin	Normal sodium intake:
	Supine (4-6 hr): 0.5-1.6 ng/ml/hr
	Sitting (4 hr): 1.8-3.6 ng/ml/hr
	Low sodium intake:
	Supine (4-6 hr): 2.2-4.4 ng/ml/hr
	Sitting (4 hr): 4.0-8.1 ng/ml/hr
Sodium	137-147 mEq/L
Thyroid stimulating hormone (TSH)	4.6 µU/ml
Urea clearance Serum/24-hr urine	64-99 ml/min (maximum clearance)
	41-65 ml/min (standard clearance)
Urea nitrogen (BUN)	6-20 mg/dl
Uric acid	Male: 2.1-7.5 mg/dl
	Female: 2.0-6.6 mg/dl

Urine Chemistry		Normal Values
Albumin	Random	Negative
	24-hr	10-100 mg/24 hr
Amylase	2-hr	35-260 Somogyi U/hr
	24-hr	80-5,000 U/24 hr
Bilirubin	Random	Negative: 0.02 mg/dl
Calcium	Random	1+ turbidity; 10 mg/dl
	24-hr	50-300 mg/24 hr
Creatinine	24-hr	Male: 20-26 mg/kg/24 hr
		Female: 14-22 mg/kg/24 hr
Creatinine clearance		Male: 107-141 ml/min
		Female: 87-132 ml/min
Glucose	Random	Negative: 15 mg/dl
	24-hr	130 mg/24 hr
Ketone	24-hr	Negative: 0.3-2.0 mg/dl
Osmolality	Random	350-700 mOsm/kg
	24-hr	300-900 mOsm/kg
	Physiologic range	50-1,400 mOsm/kg
pH	Random	4.6-8.0
Phosphorus	24-hr	0.9-1.3 g; 0.2-0.6 mEq/L
Protein	Random	Negative: 2-8 mg/dl
	24-hr	40-150 mg
Sodium	Random	50-130 mEq/L
	24-hr	40-220 mEq/L
Specific gravity	Random	1.010-1.020
	After fluid restriction	1.025-1.035
Sugar	Random	Negative
Urea clearance	24-hr	64-99 ml/min (maximum)
		41-65 ml/min (standard)
Urea nitrogen	24-hr	6-17 g

Index

C

D

J

K

Q

R

S

T

W

X

Z